Inverse Problems in Atmospheric Constituent Transport

The critical role of trace gases in global atmospheric change makes an improved understanding of these gases imperative. Measurements of the distributions of these gases in space and time provide important information, but the interpretation of this information often involves ill-conditioned model inversions. Various techniques have therefore been developed in order to analyse these problems.

Inverse Problems in Atmospheric Constituent Transport is the first book to give comprehensive coverage of the work on this topic. The trace-gas-inversion problem is presented in general terms and the various approaches are unified by treating the inversion problem as one of statistical estimation. Later chapters demonstrate the application of these methods to studies of carbon dioxide, methane, halocarbons and other gases implicated in global climate change. Finally, the emerging field of down-scaling global inversion techniques to estimate fluxes on sub-continental scales is introduced.

This book is aimed at graduate students and researchers embarking upon studies of global atmospheric change and biogeochemical cycles in particular and, more generally, earth-systems science. Established researchers will also find it an invaluable resource due to its extensive referencing and the conceptual linking of the various techniques.

Cambridge Atmospheric and Space Science Series
Editors: A. J. Dessler, J. T. Houghton, and M. J. Rycroft

This series of upper-level texts and research monographs covers the physics and chemistry of the various regions of the earth's atmosphere, from the troposphere and stratosphere, up through the ionosphere and magnetosphere and out to the interplanetary medium.

IAN ENTING is a senior principal research scientist at CSIRO Atmospheric Research, Aspendale, Victoria, Australia. He is head of the greenhouse-gas-modelling team and specialises in diagnostic modelling in general and ill-conditioned inverse problems in particular. This work has resulted in numerous research papers and several chapters in books. He was also a lead author for the *Radiative Forcing Report* and the *Second Assessment Report* of the Intergovernmental Panel on Climate Change (IPCC).

Cambridge Atmospheric and Space Science Series

EDITORS

Alexander J. Dessler
John T. Houghton
Michael J. Rycroft

TITLES IN PRINT IN THE SERIES

Inverse Problems
in Atmospheric
Constituent Transport

I. G. Enting

CAMBRIDGE
UNIVERSITY PRESS

CAMBRIDGE UNIVERSITY PRESS
Cambridge, New York, Melbourne, Madrid, Cape Town, Singapore, São Paulo

Cambridge University Press
The Edinburgh Building, Cambridge CB2 2RU, UK

Published in the United States of America by Cambridge University Press, New York

www.cambridge.org
Information on this title: www.cambridge.org/9780521812108

First published 2002
This digitally printed first paperback version 2005

A catalogue record for this publication is available from the British Library

Library of Congress Cataloguing in Publication data

Enting, I. G.
Inverse problems in atmospheric constituent transport / I. G. Enting.
  p.   cm.
Includes bibliographical references and index.
ISBN 0 521 81210 0
1. Atmospheric diffusion – Mathematical models.   2. Dynamic meteorology – Mathematical models.
3. Inverse problems (Differential equations) I. Title.

QC880.4.A8 E68 2002
628.5′3′015118 – dc21   2001052977

ISBN-13 978-0-521-81210-8 hardback
ISBN-10 0-521-81210-0 hardback

ISBN-13 978-0-521-01808-1 paperback
ISBN-10 0-521-01808-0 paperback

Contents

Preface xi

Part A   **Principles** 1

Chapter 1   **Introduction** 3

1.1   Overview 3
1.2   Atmospheric inversion problems 5
1.3   Uncertainty analysis 8
1.4   Toy models 12

Chapter 2   **Atmospheric transport and transport models** 22

2.1   The structure and circulation of the atmosphere 22
2.2   Tracer transport 25
2.3   Mathematical modelling of transport 29
2.4   Transforming the mathematical model 31
2.5   Transport models and GCMs 33
2.6   Numerical analysis 37

Chapter 3   **Estimation** 41

3.1   General 41
3.2   Bayesian estimation 42
3.3   Properties of estimators 46

3.4   Bayesian least-squares as MPD 52
3.5   Communicating uncertainty 56

Chapter 4   **Time-series estimation** 61

4.1   Time series 61
4.2   Digital filtering 65
4.3   Time-series analysis 69
4.4   State-space estimation 73

Chapter 5   **Observations of atmospheric composition** 79

5.1   Measurement of trace constituents 79
5.2   Data processing 85
5.3   Data analysis 88
5.4   Error modelling 91

Chapter 6   **The sources and sinks** 98

6.1   Classification 98
6.2   What do we want to know? 103
6.3   Types of information 105
6.4   Process models 108
6.5   Statistical characteristics 110
6.6   Global constraints 112

Chapter 7   **Problem formulation** 116

7.1   Problems and tools 116
7.2   Studying sources 119
7.3   Data assimilation 122
7.4   Chemical-data assimilation 127
7.5   Related inverse problems 129

Chapter 8   **Ill-conditioning** 131

8.1   Characteristics of inverse problems 131
8.2   Classification of inverse problems 136
8.3   Resolution and discretisation 142

Chapter 9   **Analysis of model error** 150

9.1   Model error 150
9.2   Large-scale transport error 154

9.3   Selection of data in models 157
9.4   Errors in the statistical model 160

Chapter 10   **Green's functions and synthesis inversion** 163

10.1   Green's functions 163
10.2   A limiting Green function for atmospheric transport 168
10.3   Synthesis inversion 171
10.4   A history of development 175
10.5   Multi-species analysis 179
10.6   Adjoint methods 186

Chapter 11   **Time-stepping inversions** 192

11.1   Mass-balance inversions 192
11.2   State-space representations 196
11.3   Time-stepping synthesis 204

Chapter 12   **Non-linear inversion techniques** 207

12.1   Principles 207
12.2   Linear programming 211
12.3   Other special solutions 216
12.4   Process inversion 219

Chapter 13   **Experimental design** 223

13.1   Concepts 223
13.2   Applications 226
13.3   Network design 227

Part B   **Recent applications** 231

Chapter 14   **Global carbon dioxide** 233

14.1   Background 233
14.2   Multi-tracer analysis of global budgets 239
14.3   Time-dependent budgeting 246
14.4   Results of inversions 248

Chapter 15   **Global methane** 255

15.1   Issues 255
15.2   Inversion techniques 259

15.3   Chemical schemes 261
15.4   Results of inversions 263

Chapter 16   **Halocarbons and other global-scale studies** 267

16.1   Issues 267
16.2   Estimation of sources 272
16.3   Tropospheric and stratospheric sinks 274
16.4   Other surface fluxes 281

Chapter 17   **Regional inversions** 284

17.1   General issues 284
17.2   Types of inversion 286
17.3   Some applications 292
17.4   Carbon dioxide 295

Chapter 18   **Constraining atmospheric transport** 298

18.1   Principles 298
18.2   Stratospheric transport 301
18.3   Tropospheric transport 303

Chapter 19   **Conclusions** 310

            Appendices

A   **Notation** 314

B   **Numerical data** 321

C   **Abbreviations and acronyms** 324

D   **Glossary** 327

E   **Data-source acknowledgements** 334

            Solutions to exercises 336

            References 352

            Index 389

Preface

My first contact with ill-conditioned inverse problems was while flying to the 1977 Australian Applied Mathematics Conference at Terrigal, NSW. By chance, I was seated next to Bob Anderssen, who showed me his recent report [8] on regularisation expressed in terms of Fredholm integrals.

Ill-conditioned problems have now dominated the largest part of my scientific career. However, back in 1977, my (private) reaction to Bob's work was 'why would any reasonable person want to work on such perversely difficult problems?'. The answer, 'that this is the form in which we get most of our information about the real world', completely escaped me. In any case, my idealised modelling of phase transitions in lattice systems was seemingly only loosely related to the real world – even my percolation modelling of bubble trapping [125] lay many years in the future.

Two jobs later, I had left mathematical physics as a career and was working for CSIRO modelling the carbon cycle. In particular, I was trying to calibrate our model. My attempts at finding a best-fit parameter set were thrashing around in some poorly defined subspace, when a vague memory stirred. It took me about two days to find Bob's report again and start to realise what I was up against and what I should do about it.

One important point is that what Bob gave me was a technical report. This genre is much-maligned as 'grey-literature'. It can, however, have the advantage of telling 'the truth' and 'the whole truth', even if the lack of anonymous peer review means occasional failures of 'nothing but the truth'. As well as giving important details, technical reports can and do contain things such as motivation and false leads that for reasons of space, style and tradition are fiercely suppressed in the peer-reviewed scientific journals of today. Sir Peter Medawar has discussed this situation in his article 'Is the scientific paper a fraud?' [314].

This starting point meant that much of my initial influence came from seismology. In particular Jackson's paper [224] 'Interpretation of inaccurate, insufficient and inconsistent data' provided me with a concise introduction to the key mathematical concepts. This influence continued, with our network-design calculations [400] based directly on an example from seismology [194]. At times I have used the analogy of a seismologist, confined to the earth's surface and looking down, compared with an atmospheric chemist confined to the earth's surface (by budget constraints) and looking up. Other influences on my thinking on inverse problems have been Twomey [488], Wunsch [521] and, most importantly, Tarantola [471].

In studying biogeochemical cycles, much of the excitement and much of the frustration comes from the apparent need to know about everything: meteorology, oceanography, chemistry, biology, statistics, applied mathematics and many subfields of these major disciplines. In a small way, even my work on the lattice statistics of percolation became relevant when I came to study the trapping of air bubbles in polar ice [125].

This book aims to capture some of this diversity and pull together the techniques and applications of inversions of distributions of atmospheric trace gases. The emphasis is on the inversion calculations as techniques for producing statistical estimates. Breakout boxes cover some of the tangential matters that could not be easily included with any pretence of a linear ordering in the book.

The book is intended for graduate students and beginning researchers in global biogeochemical cycles. My hope is that researchers and students in the more general and growing area of earth-systems science will also find it useful. My expectation is that inverse calculations will play an important role in validating earth-system models by allowing the specification of interactions between earth-system components in terms of boundary conditions based on observations.

The use of this book for a course on the topic of tracer inversion depends on the time available and the background of the students.

- The minimal course would be the early parts of Chapter 3 on estimation, Chapter 10 on Green's functions (probably omitting adjoint techniques), Chapter 11 and then applications from one or more of Chapters 14, 15 and 16. Essentially, this was the content of my tutorial lecture for the Heraklion conference [133] from which this book has grown.

- For an expanded course, I would add Chapters 8 and 13. Chapter 7 could form an important part of integrating this book into a more general course on earth-system modelling.

- Chapter 12 is a potential mathematical framework for student projects that go beyond applying tested techniques to new problems. Chapter 9 would be part of the basis for such extensions and Chapter 17 (and to a lesser extent Chapter 18) shows some areas of application where such extensions may be possible. Depending on the problem, such work may draw on Chapters 4 and 6.

- Chapters 2 and 5 are background that is included to give some semblance of completeness. The coverage emphasises those aspects of most relevance to tracer inversions. I would expect that courses covering these areas would use other textbooks.

- Variations and extensions to a course could be based on the related topics that are introduced in some of the breakout boxes.

- The book includes a small number of exercises at the ends of the chapters. Mostly, these are mathematical derivations that illustrate or extend matters covered in the text. This is because these are the questions for which it is possible to provide answers and regarding which there is no need to make assumptions about what computer software is available. The exercises are confined to the chapters that cover the principles. Some of them are data-oriented and could be used, given appropriate data sets, in several different areas of application. Additional exercises can be constructed by applying some of the illustrative analyses used in the figures to new data sets.

Beyond the influence from seismology, my work has been shaped by many factors. The Applied Mathematics Conferences of the Australian Mathematics Society have influenced me far beyond my initial contact with Bob Anderssen. Because of the small number of scientists working in any one field in Australia, the Applied Mathematics Conferences emphasise communication across fields. This exposed me to a diverse range of mathematical techniques, which I found extremely beneficial when starting work in the relatively new field of atmospheric composition. In a very different way, the diversity fostered by the American Geophysical Union with its emphasis on breaking down barriers between aspects of earth and planetary science has, through its publications and meetings, been a continuing source of inspiration.

An important part of my work environment over 1993–2000 was the Cooperative Research Centre for Southern Hemisphere Meteorology (CRCSHM). The CRCSHM was funded through the Australian Government's Cooperative Research Centre programme as a seven-year initiative bringing together contributions from the CSIRO (divisions of Atmospheric Research and Applied Physics) (later Telecommunications and Industrial Physics), the Australian Bureau of Meteorology, the Mathematics Department of Monash University and Cray Research, Australia (later Silicon Graphics). Throughout its seven-year life, the CRCSHM provided an environment that fostered high-quality research and a vibrant graduate programme. In particular this reflects the efforts of the directors, David Karoly and Graeme Stephens. One of my aims in writing this book is to create a concrete reminder of what we achieved at the CRCSHM.

The cooperative nature of the CSIRO and CRCSHM working environments means that many of the concepts described here have involved shared or parallel development. There are two important cases where I have taken ideas about communication directly from unpublished ideas of my colleagues. First, there is Roger Francey's idea of presenting our understanding of the atmospheric carbon budget in terms of the last few decades' sequence of new concepts (see Section 14.1). Secondly, there is Peter Rayner's emphasis of the complementary nature of synthesis versus mass-balance inversions. Both involve spatial interpolation of data: synthesis does it in the source space and mass-balance does it in the concentration space.

Several colleagues have kindly provided either diagrams of their results or the data from which to produce examples and diagrams. Detailed acknowledgements are given in Appendix E, but the assistance from Rachel Law has been invaluable.

Examples that are based on our initial synthesis inversions [145] have mostly been recalculated for this book (based on the specifications in Box 10.1), but these calculations derive from the initial work by Cathy Trudinger.

Valuable collaboration has come from other present and former colleagues at the CSIRO and CRCSHM, particularly Jim Mansbridge, Cathy Trudinger, Reter Rayner and Rachel Law. Liz Davy and her team in the library have good-naturedly tolerated my fractal return rate and aided me in tracking down obscure references. At the CSIRO, the chief, Graeme Pearman, and my programme and project leaders, Ian Galbally and Paul Steele, have provided encouragement and support, as did Keith Ryan at the CRCSHM. From my first day at the CSIRO, Roger Francey has been a source of wisdom, support and scientific vision.

Beyond the CSIRO, Gary Newsam, George Kuczera and, most importantly, Tony Guttmann, have given me ongoing contact with broader areas of applied mathematics.

Because of the global nature of greenhouse-gas problems, I have had the pleasure of working with scientists from many parts of the world as collaborators; and as co-authors in the case of Martin Heimann, Pieter Tans, Inez Fung, Keith Lassey, Thomas Kaminski and Tom Wigley. My participation in the IPCC assessment process gave me the privilege of participating in a unique exercise in communicating science in response to urgent policy needs and I thank my co-authors for their part in what we achieved together and for all that I learnt from this. My fellow participants in the IGBP–GAIM TRANSCOM project have been a continuing source of ideas and inspiration.

While it often seems that biogeochemical modelling requires its practitioners to know about virtually everything, of course none of us can really do so. It is inevitable that many of the specialist fields have been dealt with in a manner that the relevant experts will find unsatisfactory. In a field lacking clear-cut boundaries some incompleteness is inevitable. I hope that the number of outright errors is small.

One area of possible criticism, for which I expect to remain totally unrepentant, is my adoption of a Bayesian approach. This is a contentious issue [79]. While critics scoff at the possibility of prior knowledge, proponents such as Jeffreys [231] argue that it is only the Bayesian approach that answers the questions that working scientists really ask. Tarantola [471] simply regards the Bayesian approach (somewhat generalised) as reflecting the normal process of natural science, continually building on past knowledge, albeit critically re-examined.

Bard [22] suggests that there are three cases in which the use of a prior distribution is not controversial:

(i) assigning $p_{\mathrm{prior}}(\mathbf{x}) = 0$ to physically impossible values of $\mathbf{x}$;

(ii) if $\mathbf{x}$ truly is a random variable then its distribution function should be used as $p_{\mathrm{prior}}(\mathbf{x})$; and

(iii) when $p_{\mathrm{prior}}(\mathbf{x})$ represents the results from earlier experiments.

The aim in this book is to try to bring the use of Bayesian estimation into the compass of case (iii). This may, of course, include (i) or (ii) as special cases identified on the basis of earlier experiments. In the context of this book, this is simply a recognition that the use of atmospheric-transport modelling to study trace gases does not occur in isolation from other areas of science.

References (in square brackets) are listed in chronological order of publication or, sometimes, in order of importance.

While the errors, omissions and obscurities in this book must remain my responsibility, I have been helped in reducing their number by several colleagues who have kindly read various chapters: Rachel Law, Cathy Trudinger, Paul Fraser, Ying-Ping Wang, Peter Rayner, Simon Bentley, Roger Francey, Ian Galbally, Mick Meyer and Colin Allison. John Garratt kindly gave me the benefit of his editorial experience by reading the full text.

The skill and support of the editorial and production team at Cambridge University Press has made the task of going from typescript to finished book refreshingly hassle-free.

The idea for writing a book like this has been bouncing around my head for several years. The real impetus came from writing my overview of synthesis inversion [133] for the *Workshop on Inverse Problems in Biogeochemistry*, held in Heraklion in 1998 – thanks to Martin Heimann and his fellow organisers, it was a great meeting.

My task in writing this book has been supported by the CSIRO and particularly by the CRCSHM. However, the effort has extended far beyond my normal work-time, which has meant a great reduction in time with my family. I thank them for their forbearance.

Part A

Principles

Chapter 1

Introduction

One cannot possibly study the disease unless one understands what it means to be
healthy. We probably only have a few more decades to study a 'healthy' earth.

R. F. Keeling: Ph. D. Thesis [251].

1.1   Overview

Human activity is changing the composition of the atmosphere. This goes beyond the
often obvious problems of local and regional pollution – even in remote locations
there are changes in concentrations of minor atmospheric constituents such as carbon
dioxide, methane and nitrous oxide. These and other long-lived gases affect the balance
of radiation of the earth – they are the so-called greenhouse gases. Other long-lived
gases are implicated in the decrease in concentration of ozone in the stratosphere.

The ability to understand the current atmospheric budgets of these trace gases is essential if we are to be able to project their future concentrations in the atmosphere. This
book concentrates on one group of techniques that are being used to improve our knowledge – the interpretation of spatial distributions of trace-gas concentrations. An important theme of this book is the use of a statistical approach as being essential to obtaining
realistic assessments of the uncertainties in the interpretation of trace-gas observations.

Modelling of the atmospheric transport of carbon dioxide (CO_2), methane (CH_4) and
other greenhouse gases is used to interpret the observed spatial distributions of these
gases. The spatial distribution of trace-gas concentrations represents a combination
of the effect of spatially varying sources and sinks and the effect of atmospheric
transport. Therefore a model of atmospheric transport is needed if the source/sink
distributions are to be deduced from observed concentrations. The main reason for

deducing the source/sink distributions is to help identify and quantify the processes responsible. We define 'tracer inversion' as the process of deducing sources and sinks from measurements of concentrations. We also consider a number of related inversion problems involving trace atmospheric constituents. This use of modelling is termed 'diagnostic' – the model is being used to interpret observations. The alternative use of models is in 'prognostic' operation, in which the model is used to make projections of future conditions.

Modelling of global atmospheric change has progressively widened its scope from the physical properties of the atmosphere to include atmospheric chemistry and bio-geochemistry and is progressing to the currently emerging area of 'earth-system sci-ence'. This increase in scope has been motivated by the recognition of causal links between the components of the earth system. Realistic projections have to consider these connections and model their evolution in time. In contrast, diagnostic modelling is able to analyse components of the earth system, defining the linkages in terms of observations. Therefore, we can expect that inverse modelling in general (and inverse modelling of the atmosphere in particular) will become an increasingly important part of the development of earth-system science and the validation of earth-system models.

Recognition of the information present in global-scale spatial differences in con-centration of CO_2 came soon after the establishment of high-precision measurement programmes at Mauna Loa (Hawaii) and the South Pole in 1958. The CO_2 records revealed a persistent difference and this mean spatial gradient has increased over the subsequent decades. Much of this difference is due to fossil-fuel use, which occurs mainly in the northern hemisphere. It is a measure of the difficulty of interpretation that there remain competing interpretations for the residual.

In order to achieve local air-quality objectives, many jurisdictions have established regulations controlling emissions. On a larger scale, cross-border transport of sulfur compounds has led to international agreements in some cases. On a global scale, the two objectives have been the control of ozone-depleting chemicals through the Montreal Protocol (see Box 16.3) and the restrictions on emission of greenhouse gases through the still-to-be-ratified Kyoto Protocol (see Box 6.1). The existence of these agreements creates a need to ensure that they are based on sound science, in order to ensure that the prescribed actions achieve the objectives of the agreements.

This book aims to capture my own experience and that of my colleagues in us-ing atmospheric-transport modelling to help understand the global carbon cycle and other similar biogeochemical cycles. Since this activity is composed of so many inter-linked parts, this introduction is designed to serve as a road-map to what lies ahead. The main division of this book is into Part A, which surveys general principles, and Part B, which reviews recent applications.

The components of tracer inversions are (i) a set of observations, (ii) an atmospheric-transport model and (iii) a set of mathematical/statistical techniques for matching observations to model results. This book is mainly about the matching process. It takes its context from the specific issues raised by the nature of atmospheric transport, the types of observations that are available and what we would like to learn about trace-gas fluxes.

Two important issues that we identify in developing practical inversion calculations
are

- ill-conditioning, as introduced in Section 1.3, whereby the inversions are
  highly sensitive to errors and uncertainties in the inputs and assumptions; and
- the use of a statistical approach to the assessment of uncertainty.

1.2 Atmospheric inversion problems

As noted above, this book is divided into two parts, covering principles and applications, respectively. Nevertheless, principles need illustrative examples and most of the
developments of techniques of trace-gas inversion have been in response to specific
problems. The main classes of trace-gas inverse problem are the following.

Estimation of atmospheric transport. Inversion calculations to determine
   atmospheric transport have played a relatively small role in trace-gas studies.
   An exception is early studies of ozone as a tracer of atmospheric motion. A few
   tracer studies have concentrated on estimating key indices of transport, such as
   interhemispheric exchange times. Some of these are reviewed in Chapter 18.

Estimation of sources and sinks of halocarbons. Studies of the various
   halocarbons have mainly been motivated by their role in ozone depletion.
   Initially, studies of chlorofluorocarbons (CFCs) concentrated on estimating the
   loss rates, expressed in terms of atmospheric lifetimes. Studies of methyl
   chloroform (CH_3CCl_3), for which there are good concentration data and quite
   good estimates of emissions, also aim to estimate the loss rate. CH_3CCl_3 is
   removed from the troposphere by reaction with the hydroxyl radical (OH) and
   so the CH_3CCl_3 loss rate can characterise the loss by reaction with OH of other
   trace gases, particularly methane [377]. More recently, studies of halocarbons
   have attempted to estimate the strengths and locations of unreported emissions.
   Inversions of halocarbon distributions are discussed in Chapter 16.

Estimation of sources and sinks of CO_2. The key issue in studies of CO_2 is the
   atmospheric carbon budget and, in particular, the partitioning of CO_2
   exchanges between oceanic and biospheric processes. Atmospheric CO_2
   inversions aim to use the spatial distribution of CO_2 to infer the spatial
   distribution of surface fluxes, the objective being to obtain sufficient detail to
   distinguish terrestrial from ocean fluxes. (Note that this book uses the term flux
   to mean both (i) exchange of mass per unit area, generally in the context of
   partial differential equations, and (ii) area-integrated exchange of mass, in
   contexts involving finite areas.) CO_2 inversions are discussed in Chapter 14.

Estimation of sources and sinks of CH_4. As with CO_2, the important questions
   for CH_4 are those concerning the atmospheric budget. Consequently, the main
   atmospheric inverse problem is that of estimating the spatial distribution of
   methane fluxes, mainly from surface-concentration data. The sink in the free
   atmosphere is an additional complication. Methane inversions are discussed in
   Chapter 15.

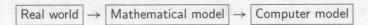

Figure 1.1 A schematic diagram of the relation among the real world, the mathematical model and the computer model. We adopt the terminology [423] of using 'validation' for testing the mathematical model against the real world and 'verification' for testing the computer model against the mathematical model.

Global-scale inversions of other trace gases are noted in Section 16.4 and regional-scale inversions are discussed in Chapter 17.

We consider the most common tracer inversion problem, that of deducing sources and sinks from concentration data. As noted above, this requires the use of a model of atmospheric transport. Figure 1.1 represents the relation among (i) the real world, (ii) a mathematical model of (some aspect of) the world and (iii) a computer implementation of the mathematical model. Identifying the mathematical model as an explicit intermediate step in model building allows us to use a wide range of mathematical techniques to analyse the modelling process. Much of this book is written in terms of such mathematical models.

The general mathematical form of the transport equation for a trace constituent describes the calculated rate of change with time of $m(\mathbf{r}, t)$, the (modelled) atmospheric concentration:

$$\frac{\partial}{\partial t} m(\mathbf{r}, t) = s(\mathbf{r}, t) + \mathcal{T}[m(\mathbf{r}, t), t] \tag{1.2.1}$$

where $s(\mathbf{r}, t)$ is the local source and $\mathcal{T}[., .]$ is a transport operator. Equation (1.2.1) expresses the rate of change of a trace-gas concentration at a point, $\mathbf{r}$, and time, t, as the sum of the net local source-minus-sink strength at that point, plus a contribution due to trace-gas transport from other locations. The transport is usually modelled with an advective component, $\nabla \cdot (\mathbf{v}m)$, often with a diffusive component to represent sub-grid-scale processes.

We can identify two main classes of inversion, which we denote 'differential' and 'integral'. The former works with equation (1.2.1). The latter uses Green's functions obtained by (numerical) integration of the transport equations. There are also various 'hybrid' techniques.

  (a) **Differential inversions.** These are based on rewriting the transport equation (1.2.1) as

$$\hat{s}(\mathbf{r}, t) = \frac{\partial}{\partial t} \hat{m}(\mathbf{r}, t) - \mathcal{T}[\hat{m}(\mathbf{r}, t), t] \tag{1.2.2}$$

where $\hat{s}$ and $\hat{m}$ denote statistical estimates. The most common application is deducing surface sources from surface observations, so (1.2.2) is used at surface grid points, while (1.2.1) is numerically integrated throughout the free atmosphere. Equation (1.2.2) is applied with $\hat{m}(\mathbf{r}, t)$ as a statistically smoothed version of the observed concentration field, $c(\mathbf{r}, t)$ (hence the notation $\hat{m}$). This technique is described as a 'differential' form because of the $(\partial / \partial t)\hat{m}$ term – it is often referred to as the 'mass-balance' technique since the transport equations both in the original and in transformed forms are expressing local conservation of mass. Mass-balance inversion techniques are reviewed in Section 11.1.

(b) **Green-function methods.** These are expressed formally through the Green function, $G(\mathbf{r}, t, \mathbf{r}', t')$, relating modelled concentrations, $m(\mathbf{r}, t)$, to source strengths, $s(\mathbf{r}, t)$,

$$m(\mathbf{r}, t) = m_0(\mathbf{r}, t) + \int_{t_0}^{t} G(\mathbf{r}, t, \mathbf{r}', t')s(\mathbf{r}', t')\, d^3r'\, dt' \qquad (1.2.3)$$

where $m_0(\mathbf{r}, t)$ describes the way in which the initial state, $m(\mathbf{r}, t_0)$, evolves in the absence of sources. Of necessity, actual calculations are performed using some discretisation of (1.2.3). This is expressed as the generic relation

$$c_j = \sum_{\mu} G_{j\mu}s_\mu + \epsilon_j = m_j + \epsilon_j \qquad (1.2.4)$$

where c_j is an item of observational data, m_j is the model prediction for this item of data, ϵ_j is the error in c_j, s_μ is a source strength and $G_{j\mu}$ is a discretisation of $G(\mathbf{r}, t, \mathbf{r}', t')$.

The discretisation is based on decomposing the sources as

$$s(\mathbf{r}, t) = \sum_{\mu} s_\mu \sigma_\mu(\mathbf{r}, t) \qquad (1.2.5)$$

so that the $G_{j\mu}$ are the responses (for observation j) to a source defined by the distribution $\sigma_\mu(\mathbf{r}, t)$. (Often, for convenience, $G_{j\mu}$ includes pseudo-sources defining the m_0 of (1.2.3), which is assumed to be constant for each species.) The sources are estimated by using (1.2.4) to fit the coefficients, s_μ. For this reason, these Green-function methods that work in terms of pre-defined components, $\sigma_\mu(\mathbf{r}, t)$, have been termed 'synthesis' calculations [165], since the source estimate is synthesised from these pre-defined components.

The most important point for the development of Green-function methods is that (1.2.1) defines a linear relation between the concentrations, $m(\mathbf{r}, t)$, and the sources, $s(\mathbf{r}, t)$, so the full machinery of linear algebra can be applied to solving (1.2.4). Green-function techniques are discussed in Chapter 10.

(c) **Hybrid techniques.** These techniques lie between the differential (mass-balance) and the integral (synthesis) inversions. Generally, they take the form of synthesis inversions over a sequence of relatively short time-steps. Examples of this are the techniques of Brown [56], Hartley and Prinn [197] and Ramonet *et al.* [394]. These and other similar techniques are reviewed in Section 11.2. In addition, there is an exploratory discussion in Chapter 12 of techniques involving non-linear estimation.

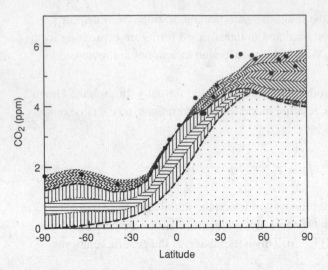

Figure 1.2 A schematic representation of synthesis inversion as analogous to a jigsaw, fitting an unknown number of differently shaped components (fossil, ocean and terrestrial) to find the best fit to observations (solid points).

The emphasis given to Green-function, or synthesis, techniques in this book primarily reflects the scope for error analysis. This also underlies the second reason, which is the extensive experience of synthesis inversions in our research group. In its simplest form, the synthesis approach corresponds to multiple regression: a function $c(x)$ is expressed as a linear combination of specified functions, $G_\mu(x)$, in terms of unknown coefficients, s_μ, by fitting a set of observations at points x_j as $c(x_j) \approx \sum_\mu s_\mu G_\mu(x_j)$.

A visual illustration of the technique can be obtained by regarding the fitting process as a 'jigsaw'. Figure 1.2 gives a schematic representation of the regression (in terms of latitudinal variation only) for CO_2 distributions. Rather than fit pieces of unknown size but known shape, the 'jigsaw' analogy approximates this by fitting an unknown number of pieces of fixed shape and size, shown by alternating hatching. Figure 1.2 demonstrates fitting an unknown number of land pieces (five in this example, above the upper dashed line and shown with diagonal hatching) and an unknown number of ocean pieces (three in this case, between the dashed lines) plus a fairly well-known fossil piece (below the lower dashed line with dot fill) to fit the observed spatial distribution (solid points). (Note that the higher concentrations in the 'land' pieces at high southern latitudes reflect the transport of northern air southwards through the upper troposphere.)

1.3   Uncertainty analysis

A key focus of this book is the estimation of uncertainties. It is particularly important in ill-conditioned problems that are subject to large error-amplification. Uncertainty analysis is required on general grounds of scientific integrity and the needs of policy-related science, as well as for input into further levels of scientific (or policy-related) analysis. In addition, as described in Chapter 13, we have used a systematic uncertainty analysis as the basis of experimental design.

The underlying principle is that *any statistical analysis requires a statistical model.* Generally we need to *assume* such a model. We can test (but never prove) the validity of the assumptions. In discussing the various types of error that can affect trace-gas inversions, Enting and Pearman [141] noted that "any variability that cannot be modelled [· · ·] must be treated as part of the 'noise' [· · ·]". In other words, the noise in the statistical model is whatever is not being modelled deterministically. Statistical estimation is described in Chapter 3; the special case of time series is described in Chapter 4.

This book follows Tarantola [471] in being firmly based on the use of prior information. This is both an optimal use of available information and an essential part of stabilising ill-conditioned problems. The standard Bayesian formalism has proved adequate for the problems that we have encountered in practice, without the need to adopt the extensions proposed by Tarantola (see Box 12.1).

The emphasis on the statistical modelling of uncertainty leads us to express the results from inversion calculations in the terminology of statistical estimation. The results of inverting (1.2.4) are *estimates*, denoted $\hat{s}_\mu$, of the source components, s_μ, or more generally estimates, $\hat{x}_\mu$, of parameters, x_μ. As noted above, the essential requirement for any statistical analysis is that one must have a statistical model of the problem.

The word 'model' has been used both for the transport model and for the statistical model. This multiple usage needs to be recognised since both types of model are needed for tracer inversions. The transport model represents a deterministic relation between the sources and the concentrations. However, our overall knowledge of the sources and concentrations is incomplete. Most obviously, our knowledge is not infinitely precise. This incompleteness in our knowledge is expressed in statistical terms. With this terminology, the quote from Enting and Pearman [141] needs to be reworded as *any variability that cannot be modelled deterministically* [· · ·] *must be treated as part of the 'noise' and modelled statistically.*

In order to address these issues of uncertainty, inversion studies need to use an overall model with both deterministic and statistical aspects. A common form of the combined model is one in which the statistical aspects appear as noise added to the outputs of a deterministic model and (when one is using a Bayesian approach) finite uncertainties on the inputs. With such a structure, there can be an apparent distinction between the deterministic (transport) model and the statistical model. The deterministic model never 'sees' the statistical model – all the statistical analysis occurs somewhere outside. Conversely, the statistical model sees the deterministic model as a functional relation and need take no account of the immense complexity that may lie inside a functional representation of atmospheric transport. This convenient separation between deterministic and statistical models becomes rather less tenable when we wish to apply a statistical approach to considering uncertainties in the deterministic model itself. State-space modelling (see Section 4.4) provides one framework in which statistical and deterministic aspects of modelling can be integrated.

The simplest statistical model of the observations is to assume independent normally distributed errors. For the case of linear relations between observations and parameters, this leads to a weighted least-squares fit giving the optimal (minimum-variance) estimates. Conversely, adopting a least-squares fitting procedure and associated error analysis is equivalent to assuming, whether implicitly or explicitly, that the errors are normally distributed. Errors with a multivariate normal distribution lead to a weighted least-squares fit with 'off-diagonal' weights (see Section 3.3) as the optimal estimates. Other error distributions (and associated non-linear estimation) have been considered in a highly simplified zonally averaged inversion [130]. Additional discussion of non-linear estimation is given in Chapter 12.

Many of the inversions use a Bayesian approach, i.e. independent prior estimates of the sources are included in the estimation procedure. Detailed discussions of applications of Bayesian estimation in carbon-cycle studies have been given for global model calibration [140], for synthesis inversion [143, 145], for methane [242, 206] and in subsequent work.

Some of the key aspects of error analysis are the following.

Measurement error. Most error analyses in trace-gas studies have assumed that the errors in the various observations, c_j, are independent. To the extent that 'error' includes the effect of small-scale sources that are omitted from the model, the assumption of independence of distinct sites can easily fail. Probably, a more important omission is the time correlation in the errors for a single site. Most inversion studies have ignored this problem. Two early exceptions are the three-dimensional model study reported by Bloomfield [34], in which an autoregressive error model was used, and the synthesis inversion by Enting *et al.* [145], in which the issue of autocorrelation in the data was avoided because time-dependence was expressed in the frequency domain.

Model error. The problem of determining the effect of model error remains largely unsolved. The difficulty is particularly great in ill-conditioned inverse problems with their large sensitivity to errors. Tarantola [471] describes a formalism in which model error becomes an extra component added to observational error (see equation (9.1.2)). Enting and Pearman [141] considered such a formalism with particular reference to the 'truncation error' when the small-scale degrees of freedom are excluded from the process of synthesis inversion (see also Section 8.3). One difficulty with this approach is that these errors are unlikely to be independent and there is little basis for defining an appropriate error covariance.

Numerical modelling involves an initial discretisation of the spatial and temporal variations both of sources and of concentrations. Further discretisation of source distributions may be needed because of the limited information contained in a sparse observational network and the loss of information associated with the ill-conditioning of the inversion. The 'synthesis-inversion' approach defined by equation (1.2.4) is usually based on a coarse discretisation of source/sink processes.

In recent years there has been a collaborative project of the International Geosphere–Biosphere Program (IGBP) known as TRANSCOM, which compares some of the atmospheric-transport models used to study CO_2 [398, 277; see also Section 9.2 and Box 9.1]. These studies have confirmed the importance of the problem but have not yielded a 'magic-bullet' solution. A further aspect of 'model error' that must be considered is that of errors in the statistical model.

Source statistics. The Bayesian approach requires, as one of its inputs, prior statistics of the source strengths. One of the most critical issues is the time correlation of these prior source estimates in time-dependent Bayesian inversions. Initially, Rayner *et al.* [402] used time pulses whose prior distributions were assumed to be independent. Later calculations (e.g. those in Figure 14.9) used 'mean-plus-anomaly' representation of data error and prior estimates. Mulquiney *et al.* [332] used a random-walk model for the prior statistical characterisation of sources.

In addition, consideration of the spatial statistics of the sources is required in order to assess the discretisation error inherent in the synthesis approach. The particular importance of spatial statistics is quite explicit for inversions using adjoint calculations [240], in which very large numbers of source components are involved. Assumptions of independent uncertainties for a large number of small regions could imply an unrealistically small uncertainty in the total. Similar issues of spatial statistics are implicit in synthesis inversions based on a small number of highly aggregated components.

The process of deducing trace-gas fluxes from concentration data is severely hampered by a mathematical characteristic known as ill-conditioning. Figure 1.3 gives a schematic representation of the source of the difficulty: dissipative processes in the real world lead to a loss of detail in the information available for analysis. In cases in which the model calculations (or, more precisely, the model-based inferences) are in the opposite direction to the real-world chain of causality, the attenuation of detailed

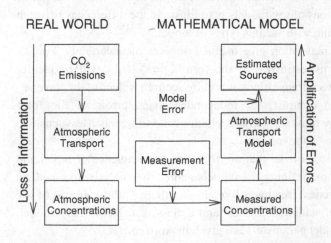

REAL WORLD　　　MATHEMATICAL MODEL

Figure 1.3 The origin of ill-conditioning in the tracer inversion problem. Figure from [132].

information requires a corresponding amplification to try to reconstruct the source distribution. This will also amplify errors introduced by the observational process and errors in the transport model.

The loss of information in ill-conditioned inverse problems can be quantified in terms of how rapidly the attenuation in the forward problem changes as the length-scale decreases. In terms of an inverse length-scale specified by a wave-number, k, the low-frequency response of the surface concentration to a surface flux behaves as k^{-1}. The asymptotic behaviour of global-scale transport can be (and was) derived from an analytical purely diffusive model, because the limiting behaviour is determined by the most dissipative process. Calculations with numerical models show that the k^{-1} behaviour applies quite accurately for $k > 1$ [132, 136]. The ill-conditioned nature of the atmospheric-tracer inversion problem has been known since the work of Bolin and Keeling [37], although the k^{-2} response that they found applies to vertically averaged concentrations rather than surface concentrations.

Similar ill-conditioned inverse problems are common in many areas of the earth sciences. Much of our work has drawn on analogies from seismology, particularly the network design study [400] noted in Chapter 13. Chapter 8 revisits these issues of ill-conditioning. For the present, the important point is that the limited ability to recover details from indirect information is an important reason for using Bayesian estimation.

1.4   Toy models

This book makes frequent use of 'toy models' to illustrate aspects of tracer inversion. Toy models are highly simplified models that capture only a few important attributes of a system. The value of these models comes from the insights that can be obtained from analytical or semi-analytical solutions and/or the ability to explore solutions for a large number of different conditions quickly. For example, identifying the sensitivities to key model parameters is an important application of toy models. In this book, one of the most important applications of toy models is to illustrate the differences between forward and inverse modelling with various types of model.

In addition, toy models may often give useful estimates of uncertainty. For ill-conditioned inverse problems, estimates of the form $\hat{x} = \sum a_j c_j$ will frequently involve cancellations and so accurate estimates require accurate values of the 'inverse-model' coefficients, a_j. In contrast, for independent data, errors of the form $\mathrm{var}\,\hat{x} = \sum |a_j|^2 \mathrm{var}\, c_j$ involve sums without cancellations and so are much less sensitive to errors and approximations in the model.

There are five classes of toy model that are followed in a sequence of boxes and exercises throughout the book, refining the models and/or the analysis, to elucidate aspects of the general theoretical treatment. We classify them as three groups of toy transport model, a group of statistical models and a class of toy chemistry model. Suggested values for the model parameters are given in Appendix B.

Box 1.1. Toy transport model A.1. The one-reservoir atmosphere

In representing the atmosphere as a small number, N, of reservoirs, we start with $N = 1$. Transport is irrelevant and the only changes in concentration are due to the net effect of sources and sinks.

We write the rate of change of concentration as

$$\frac{d}{dt}m = \lambda(t)m(t) + s(t) \tag{1}$$

where $s(t)$ is an 'external' source and $-\lambda$ represents an atmospheric decay rate (or inverse 'lifetime') for a decay process, such as chemical reaction or radioactive decay.

For a specified initial condition, $m(t_0)$ at time t_0, equation (1) has the formal solution

$$m(t) = m(t_0)\exp\left(\int_{t_0}^{t}\lambda(t')\,dt'\right)$$

$$+ \exp\left(\int_{t_0}^{t}\lambda(t')\,dt'\right)\int_{t_0}^{t}\exp\left(\int_{t_0}^{t''}-\lambda(t')\,dt'\right)s(t'')\,dt'' \tag{2}$$

$$= m(t_0)\xi(t) + \int_{t_0}^{t}\xi(t)/\xi(t')s(t')\,dt'$$

where

$$\xi(t) = \exp\left(\int_{t_0}^{t}\lambda(t')\,dt'\right) \tag{3}$$

is termed an integrating factor. Relation (2) is a special case of the Green-function formalism

$$m(t) = m_0(t) + \int_{t_0}^{t}G(t, t')s(t')\,dt' \tag{4}$$

(see Section 10.1) whereby an inhomogeneous differential equation is 'solved' in terms of an integral operator that is the inverse of the differential operator.
For constant λ, (2) reduces to

$$m(t) = m(t_0)e^{(t-t_0)\lambda} + \int_{t_0}^{t}e^{(t-t'')\lambda}s(t'')\,dt'' \tag{5}$$

A. **Few-reservoir representations of atmospheric transport.** These models allow us to look at broad-scale features at a low level of discretisation. From a mathematical perspective, the models are defined by sets of ordinary differential equations (ODEs). This simplifies some of the analyses.

B. **Diffusive transport models.** At the opposite extreme to the highly discretised models, the purely diffusive model uses a continuum view of the atmosphere. Tracer distributions are modelled with partial differential equations (PDEs).

Box 1.2. Toy transport model B.1. The diffusive atmosphere

The first representation of an atmosphere with purely diffusive transport distorts the geometry in order to simplify the mathematics. We consider anisotropic diffusion in a rectangular region. Schematically it represents a zonal average of the atmosphere. The flux across the lower boundary is specified in terms of a flux-gradient relation and zero flux is prescribed on the other three boundaries. The lower boundary is at $y = 1$ and the upper boundary is at $y = 0$ so that y is acting like a pressure coordinate and x is analogous to $0.5[1 + \sin(\text{latitude})]$.

The diffusion equation is

$$\frac{\partial m}{\partial t} = \kappa_x \frac{\partial^2 m}{\partial x^2} + \kappa_y \frac{\partial^2 m}{\partial y^2} \tag{1}$$

in the domain $x \in [0, 1]$ and $y \in [0, 1]$ and subject to

$$\frac{\partial m}{\partial x} = 0 \quad \text{at } x = 0 \text{ and } x = 1$$

$$\frac{\partial m}{\partial y} = 0 \quad \text{at } y = 0 \qquad \text{and} \qquad \frac{\partial m}{\partial y} = s(x) \quad \text{at } y = 1$$

For the steady-state case, the boundary conditions imply solutions of the form

$$m(x, y) = \sum_{n=0}^{\infty} m_n \, \Xi_n(x, y) \tag{2a}$$

with

$$\Xi_n(x, y) = \cos(n\pi x) \cosh(\gamma_n y) \tag{2b}$$

and the differential equation implies

$$\kappa_x n^2 \pi^2 - \kappa_y \gamma_n^2 = 0 \quad \text{whence} \quad \gamma_n = n\pi \sqrt{\kappa_x / \kappa_y} \tag{2c}$$

From the boundary conditions, the sources are

$$s(x) = \sum_n s_n \, \Xi_n(x, 1) \quad \text{with} \quad s_n = \gamma_n \tanh(\gamma_n) \, m_n$$

The $m_n \sim s_n / n$ attenuation illustrated here remains a common feature of the problem of deducing surface sources from surface data as the diffusive model is refined (Box 2.2) and also characterises realistic advective–diffusive models [136, 132].

This model is of historical interest because of its early use (1963), averaged over height and longitude, by Bolin and Keeling [37]. However, the greatest importance of this model is that it gives a realistic estimate of the rate at which information about small-scale details of the fluxes is attenuated by atmospheric mixing.

Box 1.3. Toy transport model C.1. An advective 'atmosphere'

The model domain is from $x = 0$ to $x = 1$ and $y = 0$ to $y = 1$. The transport is purely advective and is expressed as a time-varying stream function:

$$\chi = \sin(\pi y)\left[\sin(2\pi x) + \alpha \sin(\pi x)\cos(\omega t)\right] \tag{1}$$

This represents a system of two counter-rotating cells with a time-varying modulation of the relative sizes of the cells.

The transport in this system is not amenable to analytical solution – indeed the purpose of this toy model is to illustrate how simple time-varying flow fields can lead to chaotic transport of matter.

The stream function, $\chi(x, y)$, defines the velocity components as

$$v_x = \frac{\partial \chi}{\partial y} = \pi \cos(\pi y)\left[\sin(2\pi x) + \alpha \sin(\pi x)\cos(\omega t)\right] \tag{2a}$$

$$v_y = -\frac{\partial \chi}{\partial x} = -\pi \sin(\pi y)\left[2\cos(2\pi x) + \alpha \cos(\pi x)\cos(\omega t)\right] \tag{2b}$$

This transport formalism can be used in two ways, either in a Lagrangian mode to advect individual particles with velocity $[v_x, v_y]$ or in an Eulerian mode to describe the evolution of a concentration field, $m(x, y, t)$, as

$$\frac{\partial}{\partial t}m(x, y, t) = -\frac{\partial}{\partial x}(v_x m) - \frac{\partial}{\partial y}(v_y m) = \frac{\partial \chi}{\partial x}\frac{\partial m}{\partial y} - \frac{\partial \chi}{\partial y}\frac{\partial m}{\partial x} \tag{3}$$

C. **Atmospheric transport modelled by advective-stirring.** This is also presented as a continuum view of the atmosphere. The transport representation is the opposite of that in model B: model C is purely advective, whereas model B is purely diffusive. The role of model C is to explore the statistics of chaotic processes and the conditions under which chaotic transport can appear as diffusive.

D. **Statistics of 'signal plus noise'.** These models are used to illustrate the underlying statistical principles. The common framework is one of a deterministic signal plus random noise. The standard problem is that of estimating the signal in the presence of the noise. Initially the 'signal' is simply a mathematical function. Later refinements of the model incorporate deterministic components specified by the 'toy' transport models.

E. **Toy chemistry.** This is a 'whole-atmosphere' representation of the CH_4–CO–OH balance, with lumped production/loss rates characterising all other reactions.

Box 1.4. **Toy statistical model D.1. Linear trend plus white noise**

The sequence of statistical models is of the general form signal plus noise. We start with N data values, c_n for $n = 1$ to N modelled with a simple signal: a linear trend and simple noise: independent normally distributed white noise, ϵ_n, with zero mean and known variance, Q. Thus the statistical model is

$$c_n = \alpha + n\beta + \epsilon_n \tag{1}$$

with $E[\epsilon_n] = 0$ and $E[(\epsilon_n)^2] = Q$.

Ordinary least-squares (OLS) estimates of α and β are obtained by minimising

$$J_{\text{OLS}} = \sum (c_n - \alpha - n\beta)^2 \tag{2}$$

giving the estimates

$$\hat{\alpha} = \frac{\sum (n^2) \sum c_n - \sum n \sum (nc_n)}{N \sum (n^2) - (\sum n)^2} \tag{3a}$$

$$\hat{\beta} = \frac{N \sum (nc_n) - \sum n \sum c_n}{N \sum (n^2) - (\sum n)^2} \tag{3b}$$

Using (1), these can be rewritten as

$$\hat{\alpha} = \alpha + \frac{\sum (n^2) \sum \epsilon_n - \sum n \sum (n\epsilon_n)}{N \sum (n^2) - (\sum n)^2} \tag{4a}$$

$$\hat{\beta} = \beta - \frac{\sum n \sum \epsilon_n - N \sum (n\epsilon_n)}{N \sum (n^2) - (\sum n)^2} \tag{4b}$$

which emphasises that, once the data are modelled as random variables, estimates derived from these data will also be random variables. Since the ϵ_n have zero mean and appear linearly in (4a) and (4b), the estimates $\hat{\alpha}$ and $\hat{\beta}$ are unbiased, i.e. $E[\hat{\alpha}] = \alpha$ and $E[\hat{\beta}] = \beta$, where $E[.]$ denotes the mean over the error distribution, i.e. the expected value averaging over an arbitrarily large ensemble of realisations of the random noise process, ϵ_n.

Equations (4a) and (4b) also give the starting point for calculating the variance of these estimates as $E[(\hat{\alpha} - \alpha)^2]$ and $E[(\hat{\beta} - \beta)^2]$ (problem 2 of this chapter) as well as the covariance, $E[(\hat{\alpha} - \alpha)(\hat{\beta} - \beta)]$.

This regression can also be solved by a recursive technique that goes back to C. F. Gauss [525]. This is a special case of the Kalman filter (see equations (11.2.3a)–(11.2.3c) and Box 4.4).

Box 1.5. **Toy chemistry model E.1**

Several workers have used highly simplified models of the major balance of tropospheric chemistry among methane (CH_4), carbon monoxide (CO) and the hydroxyl radical (OH). These were expressed in terms of the global totals, m_{CH_4}, m_{CO} and m_{OH} of CH_4, CO and the OH radical. Several of the models are special cases of the general form

$$\frac{d}{dt} m_{CH_4} = -k_1 m_{CH_4} m_{OH} - \lambda_{CH_4} m_{CH_4} + S_{CH_4} \tag{1a}$$

$$\frac{d}{dt} m_{CO} = -k_2 m_{CO} m_{OH} + \zeta k_1 m_{CH_4} m_{OH} - \lambda_{CO} m_{CO} + S_{CO} \tag{1b}$$

$$\frac{d}{dt} m_{OH} = -k_1' m_{CH_4} m_{OH} - k_2' m_{CO} m_{OH} - \lambda_{OH} m_{OH} + S_{OH} \tag{1c}$$

where k_1 and k_2 are the rate constants for the reactions for oxidation (by OH) of CH_4 and CO, respectively, and using $k_1' = k_1/\xi$ and $k_2' = k_2/\xi$ allows m_{OH} to be expressed in mm^{-3} and equation (1c) to have time units of seconds. The factor ζ represents the proportion of oxidation of CH_4 that generates CO. For CH_4, CO and OH, S_{CH_4}, S_{CO} and S_{OH} are the respective rates of production and λ_{CH_4}, λ_{CO} and λ_{OH} are the respective loss rates, in each case excluding the reactions described by k_1 and k_2. Because of its very short lifetime, the concentration of OH can be treated as being in a steady state defined by

$$m_{OH} = \frac{S_{OH}}{\lambda_{OH} + k_1' m_{CH_4} + k_2' m_{CO}} \tag{2}$$

Guthrie [186] considered the case $\zeta = 1$, $\lambda_{OH} = \lambda_{CO} = 0$ and found that the system was unstable unless $2 S_{CH_4} > (\xi S_{OH} - S_{CO}) > 0$.

   Prather [366] considered the case $\lambda_{CH_4} = \lambda_{CO} = 0$, $\zeta = 1$ in his analysis of adjustment times for methane perturbations.

   Krol and van der Woerd [265] considered the case $\lambda_{OH} = 0$ and $\zeta = 0.8$, with S_{CO} including indirect sources from non-methane hydrocarbons.

   The units that we use for numerical examples generally follow those used in the IMAGE-2 description [265], time is in years, so that the λ_η are in yr^{-1}. The concentrations m_{CH_4} and m_{CO} are in ppm while, unlike the IMAGE units, we use m_{OH} in mm^{-3}. Thus k_1 and k_2 are in $mm^3 \, yr^{-1}$. The sources are expressed as in IMAGE in terms of emissions in mass units per year multiplied by conversion factors of 0.202×10^{-3} ppm/(Tg CO) and 0.353×10^{-3} ppm/ (Tg CH_4).

Further reading

As I noted in the preface, one of the joys and difficulties of studying biogeochemical cycles is the need to know about so many fields of science. The following list of suggested reading reflects my own experience of what has been useful. In most cases the references are chosen for depth rather than breadth.

Atmospheric chemistry. A comprehensive account of atmospheric chemistry has recently been produced by Brasseur *et al.* [51]. Although inverse problems are given only a brief discussion, many of the chapters parallel some of the 'background' sections of the present book. In particular, the chapter *Atmospheric dynamics and transport* [168] gives an account that goes well beyond the overview in Chapter 2, *Observational methods* [301] expands on Section 5.1 and *Modeling* [50] goes beyond Chapter 2. Additional overview accounts of atmospheric chemistry are given in the IPCC assessments [369, 370] and WMO ozone assessments [516].

Atmospheric circulation and dynamics. There are many books. For the purposes of understanding tracer distributions, that of James [230] gives a good balance between the description of the circulation and the dynamical processes that cause it. The books by Holton [212], Gill [177] and Green [184] are notable examples of books on dynamical meteorology.

Bayesian statistics. The classic textbook on Bayesian estimation is Box and Tiao [47]. As an older book, Jeffreys [231] has the interest of being written from a more defensive position, taking note of the foundations of mathematics and theories of knowledge. It justifies the Bayesian formalism as being that which answers the questions that scientists actually ask. Tarantola [471] takes the Bayesian approach as a necessary 'given' and extends the formalism (see Box 12.1).

Biogeochemical cycles. Schlesinger [424] reviews biogeochemistry firstly in terms of processes and then by considering the respective cycles of water, carbon, nitrogen, phosphorus and sulfur. Broecker and Peng [54] consider the carbon cycle with an emphasis on the role of the oceans. Lovelock [292] introduced an entirely different perspective with his 'Gaia hypothesis', proposing that the biogeochemical system is self-regulating, with living organisms acting to provide negative feedbacks that maintain the earth in conditions suitable for maintaining life.

Carbon cycle. The global carbon cycle is an active area of research, since CO_2 is the most important of the anthropogenic greenhouse gases. The assessments by the Intergovernmental Panel on Climate Change (IPCC) include assessments of the state of science for the greenhouse gases. CO_2 and the carbon cycle have been reviewed in the Radiative Forcing Report [422] and, most recently, in the Third Assessment Report [371]. Several volumes from 'summer schools', e.g. [200, 507], provide valuable introductions to the field.

Computer programming. Computer-programming techniques have evolved considerably since the days when I initially trained in Fortran and Cobol. I have found Michael Jackson's book [225] to give a lot of valuable concepts about program structure. In addition, *Elements of Programming Style* [255] gives a number of rules (with informed justification), of which my favourite is 'Make it right before you make it faster'.

Earth-system modelling. The 'earth-system' view goes back at least to the NASA report [334]. Harvey [198] has written a textbook that uses simple modelling to quantify influences contributing to global change. Integrated assessment models tend to include at least the faster-acting components of the earth system, although often in a highly parameterised manner. The IMAGE model volume [4] describes one such model. The volume by Martens and Rotmans [306] covers both model formulations and the implications of model-based decision making.

Inverse problems. For a comprehensive presentation of the mathematical formulation of inverse problems, Tarantola's book [471] stands out. In the same way as that in which all western philosophy has been described as footnotes to Plato, the whole of the present book can be regarded as footnotes to Tarantola – identifying the specific quantities to insert into one or other of Tarantola's estimation techniques to solve a biogeochemical inverse problem.

Numerical techniques. The book *Numerical Recipes* by Press *et al.* [373] has become extremely popular and editions are now available for Fortran-77, C, Pascal and Fortran-90. In this case, the coverage reflects breadth rather than depth. For depth, the 'references therein' provide a starting point. There has been a series of discussions on the internet, criticising the quality of some of the algorithms. My overall assessment is that, generally, using a 'numerical recipe' is better for a non-specialist than writing your own routine from an abstract mathematical formulation. Certainly they should be adequate for the exercises in this book. For problems that are large, or otherwise difficult, routines from a specialised mathematical subroutine library may be needed. Acton's book [1] *Numerical Methods That Work* gives advice on pitfalls in some common calculations. For the specific area of numerical optimisation applied to inverse problems, Tarantola [471] describes the main algorithms.

Time-series analysis. There are many books on time-series analysis. Priestley's book [374] stands out for comprehensiveness, although achieving such coverage in 'only' 890 pages makes for a generally compact presentation. Even within the field of atmospheric science, there are many books on time-series analysis; that by von Storch and Zwiers [455] is both comprehensive and recent. In the specialized area of Kalman filtering Gelb's book [173] is a classic. Young's book [525] is also a good introduction to recursive estimation.

Spatial statistics. Cressie's book [84] is the best general reference that I know of for spatial statistics. Techniques relevant to climate research are covered by von Storch and Zwiers [455]. Kagan [238] describes the statistical aspects of averaging observational data. Mandelbrot [300] provides something entirely different. First, there is the style (which he describes as a scientific essay: dealing with a subject from a personal point of view). Secondly, there was the highly innovative topic of self-similar probability distributions (fractals), with arguments for the ubiquity of such distributions in nature.

Atmospheric tracer inversion. Writing this book was motivated by the lack of books covering this field. There are, however, some review papers: Mulquiney *et al.* [331] give a summary of the main inversion techniques, with a little background on applications; Prinn and Hartley [376] concentrate on state-space modelling; and Heimann and Kaminski [203] give a more recent overview. The AGU monograph [244] from the Heraklion workshop contains a number of articles that review particular inversion techniques [375, 133, 176, 429, 97, 58, 20, 449]. In addition there are several other descriptions that focus on applications of tracer inversion [430, 460, 356]. The present book grew from my own contribution to the Heraklion conference [133]. My aims in writing this book have been the following: to achieve a more unified presentation; to emphasise the inversions as problems in statistical estimation; and to explore the boundaries of current practice as a guide to possible future developments.

Others. A number of other books are important references for topics less closely related to tracer inversions. Morgan *et al.* [324] discuss uncertainty, including aspects of uncertainty that have little relevance to atmospheric science but very great relevance to policy-related analyses. Daley's book gives a comprehensive coverage of the assimilation of meteorological data into operational weather-forecasting models [94]. Trenberth's account of climate modelling [485] is a comprehensive introduction. The *Concise Encyclopedia of Environmental Systems* [526] is a useful reference to a range of statistical techniques.

Applied mathematics. Finally, in the case of 'applied mathematics', rather than selecting particular books, I will try to identify the level of mathematics required. The key requirements are a knowledge of vectors and matrices, the calculus of vector fields and an understanding of probability distributions.

In most cases, additional suggestions for further reading are given at the ends of the chapters.

Notes on exercises

Exercises illustrating various aspects of the presentation are included at the end of each of Chapters 1–13, i.e. those chapters dealing with techniques. Many of them are mathematical in nature. For those that require numerical values, Appendix B gives values of real-world constants and toy-model parameters. It also gives information about electronic access to observed concentration data and emission inventories. Other exercises involve writing simple computer programs, particularly to implement the Kalman filter, at least for the special case of recursive regression (see Box 4.4). Other useful generic components are a Monte Carlo inversion unit and a singular-value-decomposition (SVD) routine. Routines for SVD (and its use in regression analysis) are included in *Numerical Recipes* [373] and these versions are included in IDL (Interactive Data Language, from Research Systems Inc.). In a class situation, instructors may be able to provide such components.

Exercises for Chapter 1

1. By how much does the annual burning of 6 Gt (1 Gigatonne $= 10^{15}$ g) of fossil carbon change the mass of the atmosphere if 50% of the carbon remains in the atmosphere and the rest is taken up by the oceans?

2. Use the expressions in Box 1.4 to obtain an expression (valid for large N) for the variance of the estimated trend β, in terms of R, the variance of the ϵ_n. How much does this change if (as is suggested for monthly mean CO_2 data – see Section 5.4.3) the covariance of consecutive errors is $R/2$?

3. (Numerical problem, no solution given) Write a computer program (preferably with graphical display) to implement two or more particles advected according to toy model C.1 of Box 1.3. Note the pattern of residence times in each hemisphere. If multiple colours can be displayed, simulate the evolution of an initial colour gradient.

Chapter 2

Atmospheric transport and transport models

[...] they replace the atmosphere by a different atmosphere, which Napier Shaw described in [...] *Manual of Meteorology*, as a fairy tale, but which today we would call a model.

E. N. Lorenz: *The Essence of Chaos* [290].

2.1 The structure and circulation of the atmosphere

The atmosphere exists as a thin, almost spherical, shell around the earth. Within the atmosphere differences in the thermal structure characterise a sequence of layers from the troposphere, adjacent to the earth's surface, through to the thermosphere and beyond. Figure 2.1 shows an average profile of the temperature of the atmosphere, indicating the layers and the boundaries between them.

In the study of greenhouse gases, the most important distinction is that between the troposphere and the stratosphere. The troposphere comprises about the lowest 85% of the atmosphere (in terms of mass) and is characterised by a negative lapse rate, i.e. temperature generally decreases with height. In contrast, in the stratosphere, temperature generally increases with height. While Figure 2.1 shows a mean distribution, the height of the tropopause varies significantly with latitude, being higher (typically 16 km) in the tropics than it is at high latitudes (typically 8 km).

Table 2.1 lists some of the components of the dry atmosphere. The main constituents, those in the upper group in the table, are distributed virtually uniformly in space and time, although it has recently become possible to measure the very small changes in oxygen content and use them to interpret aspects of the carbon cycle. This low variability is one end of the general inverse relation between variability and atmospheric

Table 2.1. *The composition of the dry atmosphere in terms of molar proportions. Values in the upper group are from the US Standard Atmosphere [340], apart from the rate of change in concentration of oxygen. This rate of change and the lower group of values are from the IPCC Third Assessment Report [3]. Rates of change (which are notionally for 1996) are as proportions of the total atmosphere. Negative values indicate decreases in concentration.*

Constituent	Formula	Molar fraction (1996)	Change yr^{-1}
Nitrogen	N_2	0.780 84	0
Oxygen	O_2	0.209 476	-3×10^{-6}
Argon	Ar	0.009 34	0
Neon	Ne	0.000 018 18	0
Helium	He	0.000 005 24	0
Krypton	Kr	0.000 001 14	0
Carbon dioxide	CO_2	365×10^{-6}	1.5×10^{-6}
Methane	CH_4	1.745×10^{-6}	8.4×10^{-9}
Nitrous oxide	N_2O	314×10^{-9}	0.8×10^{-9}
CFC-11	CCl_3F	268×10^{-12}	-1.4×10^{-12}

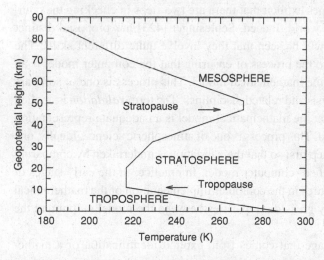

Figure 2.1 The mean temperature profile of the atmosphere (for mid-latitudes) based on the US Standard Atmosphere [340]. The geopotential height levels are levels of equal changes in potential energy, i.e. geometric heights, adjusted for the decrease of gravity with height.

lifetime: the major gases, namely O_2, N_2 and the inert gases, have very small turnover rates and their concentrations vary little in space and time. Much greater spatial and temporal variability is displayed by gases with either a shorter natural lifetime, or shorter time-scales of anthropogenic forcing, such as those in the lower part of Table 2.1. The interpretation of this variability is the main topic of this book.

For some gases, this variability is characterised by the Junge relation [236]:

$$\sqrt{\operatorname{var} c_\eta}/\bar{c}_\eta \approx 0.14 \lambda_\eta$$

where λ_η is an inverse lifetime, in years, of a gas η, with concentration c_η. However, this relation has to be regarded as illustrative: quantitative analysis of trace gases needs to be based on a more mechanistic approach, rather than on statistical relations whose domains of validity are unknown.

In the introduction, the transport of atmospheric tracers was expressed in the form

$$\frac{\partial}{\partial t}m(\mathbf{r}, t) = T[m(\mathbf{r}, t), t] + s(\mathbf{r}, t) \tag{2.1.1}$$

In discussing mathematical modelling of atmospheric transport, it is helpful to follow Figure 1.1 and identify three distinct entities:

(i) **the real-world system,** in this case actual atmospheric transport of trace species;

(ii) **a mathematical model of the system,** in this case a set of equations describing atmospheric transport; and

(iii) **a computer implementation of the mathematical model,** generally with a number of approximations being needed.

This separation needs to be done both for the modelling of transport and for the modelling of the dynamical relations that determine the transport.

The identification of a mathematical model as an intermediate step has a number of advantages. First, it becomes explicit that there are two steps in checking the computer model against the real world. Indeed, Schlesinger [423] has proposed distinct terms for these steps and it will be seen that they involve quite different skills. The term *verification* is applied to the process of ensuring that the computer model is an adequate implementation of the mathematical model. This process is one of software engineering, numerical analysis and related disciplines. The term *validation* is applied to the process of ensuring that the mathematical model is an adequate representation of the system in the real world. This process is one of atmospheric science. Ideally, one undertakes the verification step first so that the validation is undertaken by comparing the real world with the 'verified' computer model. In practice, in the early stages of development of a model, failures in the computer implementation of the mathematical model are often identified by gross discrepancies between the model output and the real world.

A second type of advantage that comes from explicit identification of a mathematical model as an intermediate step is that one can manipulate or transform the mathematical model into alternative forms appropriate to different problems and then produce computer implementations of these transformed models. A simple example is the transformation of $(\partial/\partial t)m(\mathbf{r}, t) = s(\mathbf{r}, t) + T[m(\mathbf{r}, t), t]$, as used for integration, into $s(\mathbf{r}, t) = (\partial/\partial t)m(\mathbf{r}, t) - T[m(\mathbf{r}, t), t]$, as used for calculating sources. This transformation is obvious when it is expressed in terms of the mathematical model, while developing the corresponding transformation of Fortran or other code, without the example of the above equations (or their conceptual equivalent) is definitely not obvious.

The distinction among the real world, the mathematical model and the computer implementation forms the basis for this chapter. The following section reviews some of the most relevant aspects of atmospheric transport and dynamics. Section 2.3 describes mathematical models and Section 2.4 describes some of the ways in which they can be transformed. Section 2.5 describes the computer models: tracer transport models, chemical transport models and general circulation models (GCMs). Section 2.6 discusses some of the important aspects of numerical analysis involved in developing such models.

2.2   **Tracer transport**

For the interpretation of trace-gas distributions, what matters is how the atmosphere advects passive tracers. In this context, 'passive' refers to the tracer not affecting the dynamics. This excludes treating heat as a passive tracer. Additionally, H_2O cannot be regarded as a passive tracer due to the thermal effects of phase changes. In principle, the radiatively active gases also have thermal effects, but, for gases with little spatial dependence of concentration, there is little direct connection. Ozone is one of the few cases in which there is a significant chemistry-to-dynamics coupling on a less-than-global scale.

The energy driving the global circulation of the atmosphere is radiation from the sun. The latitudinal and seasonal differences determine its structure. Large-scale convective motions in the atmosphere and oceans transport heat from low latitudes, where incoming energy exceeds outgoing energy, to high latitudes, where outgoing energy exceeds incoming energy. This transport is modified by the rotation of the earth so that north–south motion induces east–west accelerations, relative to the rotating earth. This can be readily understood in terms of conservation of angular momentum: as an airmass with zero east–west velocity moves away from the equator, it has to develop an eastward velocity relative to the earth (i.e. westerly wind) in order to preserve angular momentum.

The differences in thermal structure mean that the troposphere and the stratosphere exhibit major differences in the way this heat transport occurs. The zonal mean transport in the stratosphere is dominated by a single circulation cell with air rising in the summer hemisphere and descending in the winter hemisphere. The other main features of the stratospheric circulation [213] are a high tropical tropopause, a lower high-latitude tropopause, a surf zone of wave breaking separating these regions and polar vortices, particularly in winter and particularly in the south. In the extra-tropics, two-way exchange between the troposphere and the lower stratosphere is possible through the processes of horizontal mixing across the 'tropopause gap' and 'tropopause folding', associated with synoptic weather systems. Elsewhere, the stability of the stratosphere

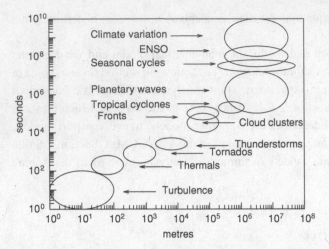

Figure 2.2 Space- and time-scales of dynamical processes in the atmosphere.

acts as a barrier to mixing and leads to large gradients in concentration for many trace constituents [296].

In contrast to the stratosphere, tropospheric processes occur on a wide range of space- and time-scales. Figure 2.2 gives a schematic representation of the relation between space- and time-scales for particular classes of process. This figure is based on Orlanski [347] and variants by other workers, e.g. [455]. On the largest scales, the convective circulation still exhibits a seasonal oscillation, but the dominance of the cell in the 'winter hemisphere' is less complete than it is in the stratosphere. More strikingly, at mid-latitudes, much of the transport of energy (and tracer mass) occurs through eddy motions rather than the zonal mean circulation. Taking an Eulerian mean over longitude produces a circulation with three cells per hemisphere: a direct Hadley cell, an intermediate indirect Ferrel cell and a polar cell. However, as noted in the following section, a Lagrangian mean shows a single cell in each hemisphere.

Originally statistics of the general circulation were assembled by averaging and interpolating observations. More commonly, contemporary practice is to determine the statistics from numerical models into which the observations have been incorporated, the model performing the roles of interpolation and ensuring dynamical consistency. This process, termed 'data assimilation', is the basis for incorporating observations into weather-forecasting models – it is described in Section 7.3. Recent accounts of the general circulation of the atmosphere are the companion papers by Hurrell *et al.* [220] emphasising the 'static' aspects, such as pressure, temperature and water content, and that of Karoly *et al.* [243], which emphasises the transports of mass, energy, momentum and moisture.

Given the role of radiation as the driver of the atmospheric circulation, it is not surprising that the circulation exhibits a strong seasonal variation in response to seasonality in the heating. As well as the seasonal forcing of the convective circulation, there are seasonally oscillating monsoon circulations driven by the land–sea temperature gradients varying seasonally.

The other case of periodicity in the forcing is the diurnal variation. This affects the atmosphere mainly on relatively small spatial scales. However, the dynamics of the atmosphere evolves chaotically (see Box 7.2), generating variability on a wide range of time-scales. Some of these are amplified because they correspond to preferred times-scales of variation. The synoptic time-scales of the familiar mid-latitude weather systems are one of the most familiar to those of us living in mid-latitudes. Additional characteristic time-scales exist, such as the Madden–Julian oscillation (MJO) in the tropical troposphere and the quasi-biennial oscillation (QBO) in the stratosphere. Many of the relations between space- and time-scales in Figure 2.2 represent such wave-like phenomena. An important part of the study of atmospheric dynamics addresses the issues of conditions under which various types of wave can grow. Longer time-scale effects such as El Niño (or the southern oscillation) [6] and longer-term interdecadal variability reflect chaotic behaviour of the combined atmosphere–ocean system. Still-longer time-scales involve coupling of the atmosphere–ocean system to the cryosphere and the lithosphere.

An additional characteristic of the troposphere is the boundary layer at the earth's surface [169]. The role of the boundary layer is particularly important for measurement of concentrations of trace gases at the earth's surface. The extent to which such measurements are representative of large-scale air-masses depends to a large degree on the extent to which the boundary layer acts to isolate air near the surface. This can cause serious problems in relating observations of surface concentrations to large-scale model concentrations (see Section 9.2). It can also be the basis of alternative techniques for estimating trace-gas fluxes (see Box 17.2).

For the purposes of interpreting trace-gas distributions, it is useful to consider atmospheric transport in its integrated (Green-function) form (see Chapter 10):

$$m(\mathbf{r}, t) = m_0(\mathbf{r}, t) + \int G(\mathbf{r}, t, \mathbf{r}', t') s(\mathbf{r}', t') \, dt' \, d^3 r' \tag{2.2.1a}$$

This Green-function representation is a mapping from the space–time distribution of the sources $s(\mathbf{r}, t)$ onto the space–time distribution of the concentrations, $m(\mathbf{r}, t)$. One of the important issues in tracer studies is which aspects of the sources are mapped onto which aspects of the concentrations. Equation (2.2.1b) gives a schematic representation of this mapping. Notionally, source variations that reflect small scales in space and time of the sources produce variations in concentration with small space- and time-scales. If we represent (2.2.1a) in terms of frequencies and wave-vectors,

$$m(\mathbf{k}, \omega) = m_0(\mathbf{k}, \omega) + \int G(\mathbf{k}, \omega, \mathbf{k}', \omega') s(\mathbf{k}', \omega') \, d\omega' \, d^3 k' \tag{2.2.1b}$$

then we would expect that $G(\mathbf{k}, \omega, \mathbf{k}', \omega')$ might be dominated by 'diagonal' contributions. This can only be an approximation since such 'diagonal' relations in $\mathbf{k}$ or ω would occur only if the transport were invariant under translations in space or time, respectively.

While the formulation (2.2.1b) is introduced only for illustrative purposes, it does demonstrate some of the key issues:

the degree of variability in the source–sink relation is modulated by the scales of
variability in atmospheric transport; and

concentration signals predominantly associated with a particular scale of
variability will nevertheless include contributions from other, often unwanted,
scales of source variability.

The earliest tracer inversions typically used observations representative of monthly
time-scales. This was achieved by selecting sites at which such representative 'baseline' concentrations occurred and then selecting such baseline data by either targeted
sampling or selection from continuous records. In terms of Figure 2.2, truncating
the smaller time-scales corresponds to truncating the smaller spatial scales. For tracers, this is only partly true because, as shown by toy model C.1, time variation in
large-scale transport fields can generate small-scale structure in tracer fields. However, Bartello [23] shows that, for a turbulent flow field, the small-scale structure of a
tracer field derived from only large-scale advection will have an incorrect distribution
of variability. Corresponding to small-scale temporal structure there will be small-
scale spatial structure. Nevertheless, the truncations involved in using baseline data
mean that much of the spatial variation is being lost. It is this characteristic that makes
the atmospheric transport *as seen through baseline data* appear diffusive on small
scales.

Although the integral form is introduced in (2.2.1a) and (2.2.1b) as a schematic representation, there have been several proposals for quantifying integrated
aspects of atmospheric transport as a basis for modelling or for inter-comparison of
models.

Exchange times. These are intended to characterise the time-scales of large-scale
exchanges between large domains, especially between the hemispheres or
between stratosphere and troposphere (see Chapter 18 for examples).

Transport times. In order to refine the concept of 'exchange time', Plumb and
MacConalogue [359] developed the concept of the transport-time as an
approximation to the Green function for long-lived tracers. This is reviewed in
Section 10.2.

Age-distribution. Hall and Waugh [191] generalised a single transport-time into
an 'age-distribution'. This is of greatest utility in the stratosphere, where the
tropical tropopause acts as a single 'source region' and so the age-distribution
is a function of a single position coordinate and (ignoring seasonality) a single
time difference (see Chapter 18).

Other ways of characterising transport are the 'transport modes' noted by Prather
[367] and the multi-exponential representation of Green's functions used by Mulquiney
et al. [330, 332]. There are also special transport characteristics relevant for particular
tracers such as the strength of the rectifier effect [105] affecting CO_2.

2.3 Mathematical modelling of transport

We now need to address the question of the appropriate description for the transport operator, $T[.]$. All of the atmospheric processes in Figure 2.2 (and others not given there) can affect the distributions of trace constituents. However, apart from molecular diffusion, all these processes represent some form of advective transport. Using the explicit expression for advective transport, (2.1.1) becomes

$$\frac{\partial}{\partial t} m(\mathbf{r}, t) = -\nabla \cdot [\mathbf{v}(\mathbf{r}, t) m(\mathbf{r}, t)] + \text{molecular diffusion} + s(\mathbf{r}, t) \qquad (2.3.1)$$

However, when we come to develop a numerical model of atmospheric transport, considering atmospheric transport as advection plus molecular diffusion is not a practical way to proceed. For global or even regional models, the feasible scale of discretisation is very much larger than that on which molecular diffusion becomes dominant. Many small-scale advective processes are inevitably excluded from numerical models of transport. Since we can neither observe nor model the advection on all scales, we need to replace (2.3.1) and represent transport as an 'effective' transport operator, T_{eff}, defined as the combination of advection with a smoothed velocity field $\bar{\mathbf{v}}$ and a residual transport that is a parameterised representation of the small-scale effects:

$$\frac{\partial}{\partial t} m(\mathbf{r}, t) = T_{\mathrm{eff}}[m(\mathbf{r}, t), t] + s(\mathbf{r}, t) \qquad (2.3.2)$$

This raises the two questions '*what do we require?*' and '*how do we achieve this?*'. The answer to the first question is that we require the operator T_{eff} to have a Green function $\mathcal{G}_{\mathrm{eff}}(\mathbf{r}, t, \mathbf{r}', t')$ that is similar to that of the transport operator T, at least when it is smoothed over $\mathbf{r}$ and/or t. As noted at the end of the previous section, the question of whether the averaged representation is sufficiently similar to the true average will depend on the type of observations that are to be described.

While the mathematical modelling of atmospheric dynamics is largely outside the scope of this book, there are a few issues that need to be considered.

The dynamics of the atmosphere can be described in terms of three-dimensional fields for density, ρ; velocity, $\mathbf{v}$; pressure, p; temperature, T; and humidity. The velocity is often expressed in locally Euclidean coordinates $\mathbf{v} = [u, v, w]$ for the east–west, north–south and vertical components. The equations describing the evolution of the atmosphere are Newton's laws subject to mass continuity and the thermodynamic equation of state. The forcings are radiative absorption/emission, surface friction and water flux. It is frequently convenient to use the thermodynamic equation of state to define a 'potential temperature', Θ, as

$$\Theta = T(p/p_{\mathrm{surface}})^{\kappa} \qquad (2.3.3)$$

where $\kappa + 1$ is the ratio of specific heats of air at constant volume and constant pressure.

The zonal mean velocity $\bar{u}$ is strong westerlies at most latitudes, particularly at mid-to-high latitudes in the upper troposphere. This rapid east–west transport can limit the scope of tracer inversions for resolving longitudinal differences. The slower north–south transport is accomplished by the $[v, w]$ components. However, a zonal mean $[\bar{v}, \bar{w}]$ has a multi-cell structure, which seems inconsistent with information from tracer distributions that implies the existence of a single cell in each hemisphere. This discrepancy reflects the dominant role of eddy transport at mid-latitudes. A correct averaging of transport requires Lagrangian averaging, following transport, rather than Eulerian averaging in a fixed spatial-coordinate system.

An approximation to Lagrangian averaging is to average along lines of constant potential temperature, Θ, since these tend to follow the eddies. This produces what is known as the diabatic circulation, which is approximately equal to the Lagrangian mean so long as transient and dissipative eddies are small.

A further approximation known as the 'residual mean circulation' is a circulation designed to ensure that energy transport is correct, under the assumption that using the energy transport to determine eddy transport will lead to transport applicable to other tracers.

The residual mean circulation is defined [230] by

$$\bar{v}^* = \bar{v} - \frac{1}{\rho_0} \frac{\partial}{\partial z} \left(\frac{\rho_0 \overline{v'\Theta'}}{\Theta_{Rz'}} \right) \qquad (2.3.4a)$$

$$\bar{w}^* = \bar{w} + \frac{\partial}{\partial y} \left(\frac{\rho_0 \overline{v'\Theta'} \cos \phi}{\Theta_{Rz'}} \right) \qquad (2.3.4b)$$

where the overbar denotes zonal means, the $'$ denotes deviations from zonal means and the denominator is a static stability parameter.

Equations known as the transformed Eulerian mean (TEM) equations [13, 230] describe the dynamics of the residual mean circulation in terms of a combined heat-plus-momentum balance. The residual mean circulation is non-divergent and so it can be defined by a stream function (see, for example, Figure 2.7 of Karoly *et al.* [243]).

The tracer transport equation becomes

$$\frac{\partial \bar{m}}{\partial t} + \bar{v}^* \frac{\partial \bar{m}}{\partial y} + \bar{w}^* \frac{\partial \bar{m}}{\partial z} = s + \text{correlation terms} \qquad (2.3.4c)$$

The correlation terms involve the concentrations, velocities and potential temperature, but are generally small and vanish for steady eddies in the absence of sources and sinks of the tracer. If there are tracer sources and sinks, then the correlation terms include a contribution from 'chemical eddy transport'.

This type of effective circulation has been used in inversion calculations [136, 137] using transport fields deduced by Plumb and Mahlman [360] with additional transport characterised by diffusion. This was particularly important in the troposphere, since the assumptions underlying the TEM equations and the correlations in (2.3.4c) being small are primarily applicable in the stratosphere. With the increasingly widespread use of three-dimensional models in tracer inversions, the residual mean circulation is of interest mainly for conceptual understanding of tracer distributions rather than for quantitative calculations.

2.4   Transforming the mathematical model

An important reason for identifying the mathematical model as an intermediate stage between the real world and the computer implementation is the ability to define transformations of the model. When the question that arises is 'what type of mathematical model?', one answer is to define a sequence of related mathematical models. The following sequence (starting with those closest to the real world and progressing to those closer to the computer model) captures some of the important features of the modelling process. In particular, it emphasises the issues involved in representing the system on a relatively coarse grid in space and time.

A set of partial differential equations in space and time. A set of PDEs is the starting point adopted in the previous section. These differential equations will generally involve a combination of advection and diffusion with the diffusive processes arising from the parameterisation of the sub-grid-scale processes. Conceptually what is being achieved is a continuum model but with sub-grid parameterisation representing a spatial smoothing that converts the model into a form suitable for spatial discretisation. This is represented schematically by the transformation from (2.4.1a) below to (2.4.1b). An advantage of working in terms of PDEs is the body of specialised knowledge of numerical analysis for these problems. Note, however, that some models represent sub-grid convective transport as a 'non-local' mass transfer that is not in the PDE form.

A set of differential equations in time. This can be a useful type of mathematical model to consider since there is an extensive body of knowledge about such sets of differential equations. For atmospheric transport, a combination of numerical analysis and numerical experiment indicates the advantages and disadvantages of various integration schemes converting the differential equations into difference equations for computer implementation. In addition, there is the scope for transformations such as the use of Green's functions to develop alternative forms of the mathematical model.

A set of difference equations in time. This is very close to what is implemented on a computer. The primary verification issue is that of numerical precision, but this form of model is of little interest for further transformation. The choices of the form of difference equation are discussed in the following section.

One issue that arises in these different forms is that of boundary conditions. If the spatial discretisation is taken as a 'given' then the set of N first-order differential equations in time will require N boundary conditions in order to specify the solution uniquely. In contrast, a set of PDEs that is of first order in time and second order in space will require one boundary condition in the temporal coordinate and two boundary conditions in each spatial coordinate. These are usually specified in terms of concentrations and/or fluxes. In terms of polar coordinates (r, θ, ϕ), the boundary condition is for concentrations and gradients to be continuous between $\phi = 0$ and 2π and non-singular at $\theta = 0$ and π. The boundary conditions in the vertical generally specify zero gradient at the upper boundary and specify either concentrations or gradients at the lower boundary.

As noted above, all of the forms of transport associated with the processes shown in Figure 2.2 come down to only two: advective mass-transport and molecular diffusion. However, as shown by Figure 2.2, advective mass-transport occurs on a wide range of scales, including space- and time-scales much smaller than it is practical to include in a global or even regional model.

The first transformation is from PDEs

$$\frac{\partial}{\partial t}m(\mathbf{r}, t) = -\nabla \cdot [\mathbf{v}(\mathbf{r}, t)m(\mathbf{r}, t)] + s(\mathbf{r}, t) + \text{molecular diffusion} \qquad (2.4.1a)$$

to a set of PDEs defined for an effective transport:

$$\begin{aligned}\frac{\partial}{\partial t}m(\mathbf{r}, t) &= \mathcal{T}_{\text{eff}}[m(\mathbf{r}, t), t] + s(\mathbf{r}, t) \\ &= -\nabla \cdot [\bar{\mathbf{v}}(\mathbf{r}, t)m(\mathbf{r}, t)] + \mathcal{T}_{\text{param}}[m(\mathbf{r}, t), t] + s(\mathbf{r}, t)\end{aligned} \qquad (2.4.1b)$$

with a specific form of parameterised sub-grid-scale transport, $\mathcal{T}_{\text{param}}[m(\mathbf{r}, t), t]$ such as

$$\frac{\partial}{\partial t}m(\mathbf{r}, t) = -\nabla \cdot [\bar{\mathbf{v}}(\mathbf{r}, t)m(\mathbf{r}, t)] + \nabla \cdot (\mathbf{K}_{\text{diff}}\nabla m) + s(\mathbf{r}, t) \qquad (2.4.1c)$$

where $\bar{\mathbf{v}}$ is a smoothing of the small-scale variability in the advective field and $\mathbf{K}_{\text{diff}}$ is a diffusion tensor chosen to approximate the effect of the sub-grid-scale advection given by $\mathbf{v} - \bar{\mathbf{v}}$. Since diffusion is not equivalent to advection, we cannot expect equations (2.4.1a) and (2.4.1b) to be equivalent representations of how a continuum field, $m(\mathbf{r}, t)$, evolves in time. The best that we can hope for is that, if $m(\mathbf{r}, t)$ is averaged in space (and possibly in time), then the solutions of the two PDEs will be close. Again, it falls to a combination of numerical analysis and numerical experiment to determine when this sort of approximation is adequate.

As noted above, for some processes, such as small-scale convection, a better approximation may be to work in terms of a set of ODEs with exchange of mass between grid cells:

$$\frac{\mathrm{d}}{\mathrm{d}t}m_j(t) = \sum_k T_{jk}m_k(t) \qquad (2.4.1d)$$

where the coefficients T_{jk} are more general then those arising from discretisation of the advective–diffusive equation. This discretisation also represents a step on the path between the continuum (PDE) representation and the discrete representation in a numerical model.

The further level of discretisation is when the processes happen on time-scales that are comparable to, or shorter than, the model time-step and so the evolution cannot be fully described by continuously evolving DEs:

$$m_j(t + \Delta t) = \sum_k T_{jk}m_k(t)\Delta t + \sum_k S_{jk}m_k(t) \qquad (2.4.1e)$$

where the S_{jk} characterise processes such as convection with episodic mass transfer. In addition to specific parameterisations that require the form (2.4.1d), it is the form in which the other mathematical representations are implemented in numerical models.

2.5 Transport models and GCMs

Tracer transport is generally modelled in Eulerian terms – the tracer is represented in terms of coordinates fixed in space. The alternative, Lagrangian modelling, is used in chemical box modelling, following a single parcel of air as described below. The approach can be extended to global distributions through the use of an ensemble of advected particles as in the GRANTOUR model of Walton et al. [496]. Taylor [472] has produced a simplified Lagrangian model by using a stochastic representation of eddy transport.

A range of models with zero, one, two and three spatial dimensions has been developed for studies of atmospheric composition in general, including global-scale transport of trace atmospheric constituents.

Zero-dimensional models can be used to describe the global mean atmosphere (see toy model A.1, Box 1.1). Another form of zero-dimensional model is the 'advected-box' form of model which follows the chemical evolution of an air parcel as it is advected through various conditions of radiation and temperature, with mixing between air parcels being treated as negligible (or highly parameterised) [49].

The main one-dimensional models resolve only the vertical. This is generally done in models for which photochemistry is the dominant process and the vertical resolution is required because a detailed representation of radiative transfer is more important than a detailed representation of mass transport [50]. The analysis by Bolin and Keeling [37] is an example of a one-dimensional model resolving only latitude.

Two-dimensional models resolving only latitudinal and vertical dimensions were used in early CO_2 inversions [136, 465]. They have also played a significant role in studies such as assessments of ozone, in which the complexity of chemical reactions imposes computational demands that preclude the use of three-dimensional models for long runs for the assessment of global change.

Following the pioneering study by Mahlman and Moxim [297], virtually all of the global-scale three-dimensional models of atmospheric transport have been based on GCMs. These are the numerical models that integrate the equations of atmospheric dynamics. The motion of the atmosphere is calculated from these basic equations of motion for air, supplemented by the transport and thermodynamics of water and by radiative transfer. GCMs include a representation of tracer transport because they need to account for the distribution of water vapour.

Box 2.1. Toy model A.2. The two-box atmosphere

Extending the N-box atmospheric model from $N = 1$ (in Box 1.1) to $N = 2$ allows the inclusion of a minimal characterisation of transport. The variables m_N and m_S represent the masses of tracer in the two hemispheres. Often these are taken as tropospheric contents only, i.e. the stratosphere is ignored. The time-evolution is modelled as

$$\frac{d}{dt} \begin{bmatrix} m_N \\ m_S \end{bmatrix} = \begin{bmatrix} -\kappa - \lambda_N & \kappa \\ \kappa & -\kappa - \lambda_S \end{bmatrix} \begin{bmatrix} m_N \\ m_S \end{bmatrix} + \begin{bmatrix} s_N \\ s_S \end{bmatrix} \tag{1}$$

where transport is parameterised by an 'exchange coefficient', κ, and λ_N and λ_S are loss rates for boxes N and S with sources s_S and s_N.

The two-box model is particularly convenient because analytical expressions can be obtained for the eigenvalues, Λ, of the model evolution matrix, by solving

$$(\kappa + \lambda_N + \Lambda)(\kappa + \lambda_S + \Lambda) - \kappa^2 = 0 \tag{2}$$

For $\lambda_N = \lambda_S = \lambda$ we have $\Lambda = -\lambda$ or $\Lambda = -\lambda - 2\kappa$ and we can transform the equations to

$$\frac{d}{dt} \sum m = -\lambda \sum m + \sum s \tag{3}$$

and

$$\frac{d}{dt} \Delta m = -\lambda \, \Delta m - 2\kappa \, \Delta m + \Delta s \tag{4}$$

where $\sum m = m_N + m_S$ and $\Delta m = m_N - m_S$.

If the transport coefficients are constant, equations (3) and (4) have simple analytical solutions, otherwise they can each be integrated formally as described in Box 1.1.

For $\lambda = 0$, there are solutions of the form

$$\Delta m \propto e^{-2\kappa t} \tag{5}$$

so that $(2\kappa)^{-1}$ is the time-scale of decay of an interhemispheric gradient.

In some studies, the time-scales are characterised by the hemispheric 'turnover time' expressed as mass/flux $= m_j/(\kappa m_j) = \kappa^{-1}$ (e.g. Table 18.1).
This two-box model is used in a number of the exercises [403] in the volume from the Heraklion conference. It has also been applied to carbon budgeting by Fan *et al.* [150].

There are two main ways in which general circulation models are used; namely in weather-forecasting models and climate models. Although forecasting models and climate models have very many common features, the different applications lead to differences in a number of important characteristics.

Box 2.2. Toy transport model B.2. Diffusive transport in a spherical shell

In Box 1.2 (model B.1) boundary concentrations give estimates of sources on the same boundary in a two-dimensional rectangular geometry. The analysis can be extended to the more realistic geometry of a thin spherical shell, as is used to study the ill-conditioning of global-scale tracer inversions [339, 138, 130], cf. Section 8.2.

In spherical polar coordinates a tracer mixing ratio, c, evolves as

$$\frac{\partial c}{\partial t} = \frac{\partial}{\partial p}\left(\kappa_p \frac{\partial c}{\partial p}\right) + \frac{\kappa_\theta}{R_e^2 \sin^2\theta} \frac{\partial}{\partial \theta}\left(\sin^2\theta \frac{\partial c}{\partial \theta}\right) + \frac{\kappa_\theta(1 + \alpha \sin^2\theta)}{R_e^2 \sin^2\theta} \frac{\partial^2 c}{\partial \phi^2}$$

(1)

where the coordinates are colatitude $\theta \in [0, \pi]$, longitude $\phi \in [0, 2\pi]$ and pressure $p \in [p_1, p_0]$. The transport is characterised by diffusion coefficients κ_p for the vertical direction and κ_θ in the north–south direction, enhanced by a factor $1 + \alpha \sin^2\theta$ in the east–west direction.

The boundary conditions are $\partial c/\partial p = 0$ at p_1 and $K\kappa_p\, \partial c/\partial p = s_0(\theta, \phi, t)$ at p_0, with $K = N_{\text{atmos}}/(A_e p_0)$ and $A_e = 4\pi R_e^2$.

The concentrations and fluxes are expanded as

$$c(p, \theta, \phi, t) = \sum_{kmn} c_{kmn} \Xi_{kmn}(p, \theta, \phi, t)$$

and

$$s_0(\theta, \phi, t) = \sum_{kmn} s_{kmn} \Xi_{kmn}(p_0, \theta, \phi, t)$$

with

$$\Xi(p, \theta, \phi, t) \propto \cosh\left(\frac{\gamma_{kmn}(p - p_1)}{p_0 - p_1}\right) P_n^m(\cos\theta)e^{im\phi}e^{2\pi ikt/T}$$

and

$$\gamma_{kmn}^2 = \{\kappa_\theta[n(n+1) + \alpha m^2]/(\kappa_p R_e^2) + 2\pi ik/(\kappa T)\}(p_0 - p_1)^2$$

The lower boundary condition gives the relation

$$s_{kmn} = \frac{K\kappa_p}{p_0 - p_1}\gamma_{kmn} \tanh(\gamma_{kmn})c_{kmn}$$

For all but the largest-scale variations, the $\tanh \gamma$ term is close to unity and so the relation between sources and concentrations exhibits a $1/n$ attenuation.

Weather-forecasting models. These are used for short-term integrations starting from initial conditions based on observations. Forecasting systems require the capability for assimilation of data (see Section 7.3) whereby meteorological observations are used to update fields in the model in a way that is consistent with atmospheric dynamics.

Climate models. These are used for long integrations of decades or centuries or more. Apart from the need for computational efficiency, there is also a need to ensure that numerical errors do not accumulate in a way that leads to an artificial 'climatic drift'. Unlike forecasting models, climate models need to incorporate the evolution of the oceans rather than being able to use sea-surface temperatures based on observation. Calculations of equilibrium climate conditions can make do with simplified representations such as thin 'slab oceans', since the equilibrium temperature is determined primarily by the thermal/radiative structure of the atmosphere. However, calculations of the time-evolution of climate change need to include an explicit representation of the oceans since the large heat capacity of the oceans will have a very large effect on the rate at which equilibrium is approached. Modern climate models are based on coupled atmosphere–ocean GCMs (AOGCMs).

GCMs contribute to atmospheric transport modelling in two main ways. First, the computer routines used for atmospheric tracer-transport are often derived from a GCM. In general, GCMs will include explicit tracer-transport calculations for water, which are often used as the basis for computer routines to model the transport of other constituents. Secondly, GCMs are generally the source of the transport fields used in tracer-transport models.

Most of the global-scale three-dimensional transport models currently in use have been derived from climate models rather than forecasting models. This most probably reflects the research orientation of climate modelling, compared with the operational orientation of weather forecasting. However, transport fields from a climate model can expect, at best, to be correct only in an average sense. Specific synoptic events will not be represented at the correct times, if at all. In contrast, transport fields from a forecasting model will represent actual meteorological conditions for each particular place and time.

The relation between three-dimensional transport models and GCMs suggests two alternative models of operation:

on-line, whereby the transport calculations are undertaken in step with the dynamical calculations in a single run of the program; and

off-line, whereby the results of dynamical calculations are archived and used as input to a separate run of the transport model.

Some of the relative advantages and disadvantages are the following.

Reasons for running off-line
■ Off-line models are generally faster, because there is no need to integrate the dynamical equations. In addition, the transport component may be able to work with a longer time-step than would be required for stable integration of the dynamics. The relative advantage of off-line models decreases as the number

of tracers increases. This will happen in multi-species studies or when one is calculating responses to a number of source components for synthesis inversion.

■ Off-line models can use analysed winds. In contrast, repeating the cycle of analysis/assimilation in a GCM is impractical as a technique for routine modelling of tracers.

Reasons for running on-line

■ In a GCM, the meteorological data for sub-grid-scale transport are available and will be consistent with the larger-scale transport. In order to reduce the amount of data required, off-line models often use simplified average characteristics of sub-grid-scale transport.

■ GCMs generally provide better time-resolution and, perhaps more importantly, uniformity in time. Off-line models almost invariably use transport fields that are updated at intervals much larger than the integration time-step.

■ There are several ways of overcoming the apparent disadvantage of GCMs not using analysed winds. One way of having tracer studies with GCMs represent some aspects of real-world variability is to use observed sea-surface temperatures, i.e. an AMIP-type run [170] whereby the AGCM runs with a prescribed boundary condition based on observations. Another technique is to have the GCM 'nudged' towards analysed fields. This can be regarded as a partial data assimilation.

Some of the important features with which to characterise global scale transport models are

■ the spatial resolution, both in the horizontal and in the vertical direction;
■ the source of the transport fields;
■ the representation, if any, of the boundary layer; and
■ the advection scheme, including any mass-fixer (see the following section).

Summary details of transport models are contained in various reports on inter-comparisons of models: the small-time-scale inter-comparison (mainly using radon) [227]; TRANSCOM for transport of CO_2 [398]; and TRANSCOM-2 using SF_6 [106].

2.6   Numerical analysis

This section notes some of the issues involved in going from the DEs or PDEs to a set of difference equations to be coded in the computer model. Rood [413] has given an overview of these issues of numerical analysis, including a review of a number of earlier comparative studies. The present section aims merely to flag some of the important issues. These are accuracy, including numerical diffusion; stability; conservation of mass; and negative masses.

For Eulerian modelling, the main choice of representation is between grid-point and spectral representations. Spectral representations describe the horizontal distribution in terms of spherical harmonics $P_n^m(\theta)e^{im\phi}$. This simplifies some aspects of GCM (and transport) calculations, although some processes require conversion to a spatial grid. The need to transform back and forth from spectral to grid representation means that the grids are invariably chosen to optimise the transformation: 2^n longitudes are used so that the fast Fourier transform can be used. Latitudinal grids use Gaussian coordinates (defined by the zeros of the Legendre polynomials) so that Gaussian quadrature can be used to perform the integrations required by the transformation.

In the vertical direction, there are several different choices of coordinate, such as height, z, and pressure, p. One common choice is the sigma-coordinate, defined by $\sigma = p/p_{surface}$. Hybrid sigma-coordinates use σ at lower levels and pressure at upper levels, with a smooth transition in between.

The model grid represents a network of cells, generally with the topology of a cuboid (apart from at the poles). If the concentration values define the vertices of these cells, then there is the potential to place the other model variables, velocity, diffusion coefficients, temperatures, humidity and atmospheric mass, on the vertices, edges, faces or centres of the cells. These choices are associated with different finite-difference representations of the PDEs. The semi-Lagrangian technique has become widely used in transport models. This updates the concentration fields at specified grid points on the basis of transport from a source point that is determined by the advective flow field. The concentrations at the source points are determined by interpolation from the concentrations at grid points.

Modelling tracer transport is complicated by the fact that, even in one dimension, numerical integration of the advection equation is particularly difficult. Numerical schemes are subject to what is known as 'numerical diffusion', which leads to unphysical spreading of advected pulses. More generally, the accuracy of numerical schemes is often assessed in terms of how they handle spatially periodic variations. Numerical diffusion appears as unphysical attenuation of higher-frequency variations. Unphysical phase shifts are the other limit to accuracy that needs to be considered.

Stability of the numerical scheme becomes an important issue through its close association with the time-step. This is characterised by the Courant number:

$$\alpha_c = v\Delta t/\Delta x \tag{2.6.1}$$

which gives the fraction of a grid cell over which an air parcel is advected in a single time-step. In many numerical schemes, the range of stability is determined by the Courant–Friedrich–Lewy (CFL) condition of $\alpha_c < 1$.

An additional characteristic of most numerical advection schemes is the possibility of producing negative values for the tracer concentrations. In modelling chemical reactions, or the role of water, negative concentrations must be avoided. Computational techniques, called 'mass-fixers', are introduced in order to do this, generally by introducing an extra transfer of mass into regions with negative concentrations from nearby regions with positive concentrations. Mass-fixers can also be needed in order to enforce conservation of mass in cases in which this is not an automatic consequence of the numerical scheme. This process of mass-fixing is inherently non-linear. A more sophisticated approach is to construct advection schemes with an explicit non-linearity that is designed to reduce the occurrence of negative concentrations and so reduce the need for explicit mass-fixing.

In contrast, when one is computing responses for a synthesis calculation, it is desirable to prevent any 'mass-fixing' in order to preserve the linearity and, more specifically, to calculate a correct response to source components that have negative fluxes at some places and/or times. Similarly, explicitly non-linear advection schemes violate the assumptions underlying synthesis inversions.

A final issue is that of the units in which the trace constituents are modelled. The main choices are actual masses, mixing ratios and concentrations. It is the concentration in the strict sense (meaning 'mass per unit volume', rather than a short-hand for 'mixing ratio') that is required in chemical rate expressions. In numerical models, the choice is likely to reflect the way in which air-mass is represented and is usually taken from the way in which moisture is represented in the parent GCM.

Further reading

■ Chapter 1 noted a number of references on dynamical meteorology [212, 177, 184]. The book by James [230] has a useful balance between dynamics and description of the atmospheric circulation. Descriptions of the atmospheric circulation are given by Hurrell *et al.* [220] and Karoly *et al.* [243], covering the 'static' and 'dynamical' aspects, respectively.

■ The volume from the NATO Advanced Study Institute on *Numerical Modeling of the Global Atmosphere in the Climate System* gives an excellent tutorial coverage of important issues in climate modelling. Individual chapters address specific processes, covering boundary-layer processes, convection, cloud processes, radiation, gravity-wave-drag, land-surface processes and hydrology. Chapters of particular relevance to tracer inversions are those on inter-comparisons of models [35], modelling the whole climate system [45], numerical schemes [508] and chemistry and aerosols [153].

Exercises for Chapter 2

No exercises based on the two-box model of Box 2.1 are included here. Many exercises based on this model are given in the AGU volume [403] from the Heraklion conference.

1. Prove that, in the absence of stirring, i.e. $\alpha = 0$, a solution with $m(x, y) = f(\chi(x, y))$ is in a steady state for any arbitrary function $f(.)$.

2. Derive the 'transport modes' defined by $(\partial/\partial t)c = -\Lambda c$ for the zonally averaged case of toy model A.2.

3. (Numerical exercise). Write a program for numerical integration of one-dimensional advective transport and test for the case when v is a constant such that the Courant number $v\,\Delta t/\Delta x$ is comparable to unity.

Chapter 3

Estimation

> It is sometimes considered a paradox that the answer depends not only on the observations but on the question; – it should be a platitude.
>
> H. Jeffries: *Theory of Probability* [231].

3.1 General

In dealing with the real world, we very quickly find that it fails to exhibit the exact relations of applied mathematics. All of the quantities involved are subject to uncertainty and, in the case of ill-conditioned inverse problems, some quantities are subject to very great uncertainty. We need to represent our uncertainty, i.e. our lack of exact information, in statistical terms. Our analysis of trace-gas distributions will be in terms of estimates and these estimates will be regarded as being subject to statistical uncertainty. This chapter discusses the process of obtaining such estimates, first by setting up a general framework and then by considering several important special cases.

Our terminology for defining the general problem of interpreting indirect information is that there is a set of observations, denoted by a vector $\mathbf{c}$, suggesting concentration data, from which we wish to infer some model parameters, denoted by a vector $\mathbf{x}$. (These model parameters will often be more general than a set of source strengths, $\mathbf{s}$.)

The simplest form of analysis is that we wish to use the data, $\mathbf{c}$, to give us an estimate of the parameters, $\mathbf{x}$, making use of a theoretical relation between the observations and the parameters. Since we do not expect the theoretical relations to predict the real-world values exactly, we distinguish these theoretical predictions by denoting them $\mathbf{m}$, or more specifically $\mathbf{m}(\mathbf{x})$ when we wish to denote their dependence on the parameters.

The notation $\hat{\mathbf{x}}$ is used to denote estimates of $\mathbf{x}$. If these estimates are based on the observations, $\mathbf{c}$ (such estimation being the subject of this chapter), then we can denote this dependence with the notation $\hat{\mathbf{x}}(\mathbf{c})$. A further piece of terminology is that of an *estimator*. This is a procedure for calculating an estimate. We can extend our notation to distinguish among different types of estimate with the notation $\hat{\mathbf{x}}_X(\mathbf{c})$, where a subscript such as X denotes the particular estimator that is used to obtain the estimate. A special class of estimates consists of linear estimates, i.e. those for which the estimate $\hat{\mathbf{x}}$ is specified as a linear function of the observations. Because of their simplicity, these are the most commonly used. They are discussed in Section 3.3.

The need for statistical modelling arises from our inability to represent the observations exactly using a deterministic model. This failure is almost universal in practice, regardless of what may theoretically be possible. We require a statistical model of the observations, i.e. one that treats the observations as random variables. As an example, one of the commonest assumptions is to model the observations as being statistically independent and normally distributed about the predictions $\mathbf{m}(\mathbf{x})$ obtained using the true values, $\mathbf{x}$, of the model parameters. From the treatment of $\mathbf{c}$ as random variables, it follows that the estimates $\hat{\mathbf{x}}(\mathbf{c})$ will also be random variables (see, for example, Box 1.4). Since they are random variables, we can ask questions about the mean and variance of the estimates or more generally about the probability distributions of the estimates. The probability distribution for an estimate is known as the *sampling distribution*. The sampling distribution of an estimate derives from the form of $p(\mathbf{c})$ and the functional form of the particular estimator, $\hat{\mathbf{x}}_X(\mathbf{c})$.

A rather different approach is to consider the estimation problem in terms of the probability distribution of the parameters. This makes sense only if one regards probability distributions as representing information. If one has a real-world parameter, with an unknown 'true' value, then, in an objective sense, the probability of any particular value is either 0 or 1. However, in terms of information, it is meaningful to define a probability distribution for parameters.

This 'subjective' view, in terms of information, is the basis of the Bayesian formalism described in the following section. This is expressed in terms of probability distributions of the parameters. In these terms, the statistical representation of the inverse problem is that of determining $p(\mathbf{x}|\mathbf{c})$, the 'probability' (actually the probability density) of the parameter values being $\mathbf{x}$, given the measured values, $\mathbf{c}$.

3.2   Bayesian estimation

Bayesian analysis derives from the statistical representation of the inverse problem as that of determining $p(\mathbf{x}|\mathbf{c})$, the probability density for the parameter values $\mathbf{x}$, conditional on measured values, $\mathbf{c}$.

Formally, the solution comes from the expression for the joint probability density:

$$p(\mathbf{x}, \mathbf{c}) = p(\mathbf{c}|\mathbf{x})p(\mathbf{x}) \tag{3.2.1a}$$

or

$$p(\mathbf{x}, \mathbf{c}) = p(\mathbf{x}|\mathbf{c})p(\mathbf{c}) \tag{3.2.1b}$$

with

$$p(\mathbf{x}) = \int p(\mathbf{x}, \mathbf{c}) \, d\mathbf{c} \tag{3.2.1c}$$

and

$$p(\mathbf{c}) = \int p(\mathbf{x}, \mathbf{c}) \, d\mathbf{x} \tag{3.2.1d}$$

whence we obtain Bayes' result:

$$p(\mathbf{x}|\mathbf{c}) = \frac{p(\mathbf{c}|\mathbf{x})p(\mathbf{x})}{\int p(\mathbf{c}|\mathbf{x})p(\mathbf{x}) \, d\mathbf{x}} \tag{3.2.1e}$$

However, as emphasised by Tarantola [471], in terms of probability theory the equations above are simply definitions of the conditional probability densities $p(\mathbf{x}|\mathbf{c})$ and $p(\mathbf{c}|\mathbf{x})$ and the 'marginal' densities $p(\mathbf{c})$ and $p(\mathbf{x})$. The applications come from the context in which they are used.

We consider the information available prior to a set of measurements. Prior knowledge of the parameters, expressed in terms of $p_{\text{prior}}(\mathbf{x})$ (which may incorporate information from earlier measurements), is combined with our theoretical knowledge of the relation between the parameters and the observable quantities, expressed as $p_{\text{theory}}(\mathbf{c}|\mathbf{x})$. Therefore, before the measurements are made, our knowledge of the joint distribution of $\mathbf{x}$ and $\mathbf{c}$ is expressed as

$$p_{\text{prior}}(\mathbf{x}, \mathbf{c}) = p_{\text{theory}}(\mathbf{c}|\mathbf{x}) \, p_{\text{prior}}(\mathbf{x}) \tag{3.2.2a}$$

If we have a specific set of measured data values, $\mathbf{c}'$, the joint probability distribution collapses to a distribution over $\mathbf{x}$ proportional to $p_{\text{prior}}(\mathbf{x}, \mathbf{c}')$. This tells us what we know about $\mathbf{x}$ given the measured data vector, $\mathbf{c}'$. The alternative expression for $p_{\text{prior}}(\mathbf{x}, \mathbf{c})$ in terms of conditional probabilities allows us to re-express this in terms of a distribution over $\mathbf{x}$, which we denote $p_{\text{inferred}}(\mathbf{x}|\mathbf{c}')$ (and to define the proportionality factor):

$$p_{\text{prior}}(\mathbf{x}, \mathbf{c}') = p_{\text{inferred}}(\mathbf{x}|\mathbf{c}')p_{\text{prior}}(\mathbf{c}') \quad \text{for fixed } \mathbf{c}' \tag{3.2.2b}$$

With these definitions, the definition of the marginal distribution (3.2.1c) is essentially a tautology:

$$p_{\text{prior}}(\mathbf{x}) = \int p_{\text{prior}}(\mathbf{x}, \mathbf{c}) \, d\mathbf{c} = p_{\text{prior}}(\mathbf{x}) \int p_{\text{theory}}(\mathbf{c}|\mathbf{x}) \, d\mathbf{c} \tag{3.2.2c}$$

reflecting the normalisation of $p_{\text{theory}}(\mathbf{c}|\mathbf{x})$. The other marginal distribution (3.2.1d) defines the normalising factor for (3.2.2b):

$$p_{\text{prior}}(\mathbf{c}) = \int p_{\text{prior}}(\mathbf{x}, \mathbf{c}) \, d\mathbf{x} \tag{3.2.2d}$$

With these interpretations, the Bayes result takes the form

$$p_{\text{inferred}}(\mathbf{x}|\mathbf{c}') = \frac{p_{\text{theory}}(\mathbf{c}'|\mathbf{x})\,p_{\text{prior}}(\mathbf{x})}{\int p_{\text{theory}}(\mathbf{c}'|\mathbf{x})\,p_{\text{prior}}(\mathbf{x})\,d\mathbf{x}} \tag{3.2.2e}$$

Conventionally, the distribution $p_{\text{inferred}}(\mathbf{x}|\mathbf{c}')$ is termed the 'posterior' distribution. This book retains this terminology, but uses the notation $p_{\text{inferred}}(.)$ to emphasise that the posterior distribution is the tool for Bayesian inference.

In terms of a 'causal' view of the world (the left-hand side of Figure 1.3), relation (3.2.2a) is primary: we have a set of real-world quantities, $\mathbf{x}$, about which we are initially uncertain and this uncertainty is expressed in terms of a probability density $p_{\text{prior}}(\mathbf{x})$. For any particular parameter set, $\mathbf{x}$, we have a 'theoretical' prediction (generally subject to further uncertainty) of the results that would be obtained for a particular set of measurements. We express the (generally uncertain) theoretical prediction as $p_{\text{theory}}(\mathbf{c}|\mathbf{x})$, the probability density for obtaining results, $\mathbf{c}$, given parameters $\mathbf{x}$. Parameter inference represents the right-hand side of Figure 1.3, expressed as (3.2.2e). In practice, virtually all of the discussion in this book shall involve interpreting (3.2.1a)–(3.2.1e) in the manner described in connection with (3.2.2a)–(3.2.2e). Therefore we will sometimes dispense with the descriptive subscripts.

This theoretical framework is extended in Chapter 9, in which we consider $p_{\text{theory}}(\mathbf{c}|\mathbf{x})$ in more detail. The important aspect for tracer inversions is recognising that $p_{\text{theory}}(\mathbf{c}|\mathbf{x})$, which describes the distribution of observations, must include both measurement error and errors (and any statistical uncertainty) in the 'model' relations that are used to predict $\mathbf{c}$ from $\mathbf{x}$. Box 12.1 summarises Tarantola's extension of the Bayesian formalism to cover the case when observations take the form of a distribution, rather than a value with associated uncertainty.

We follow Tarantola [471] and regard equation (3.2.2e) as *the solution* to the inverse problem, but what does one actually do with it? Specifically, (i) how does one construct the various components of the right-hand side and then, (ii) having done so and determined the left-hand side, how does one analyse it?

Before exploring the possibilities in general mathematical terms, we consider how the questions relate to our main problems of estimating trace-gas fluxes. We can represent the relations in terms of a 'map', shown in Figure 3.1, of how the components of tracer inversions fit together.

Question (i): 'what are the appropriate probability densities to use in (3.2.2e)?' defines much of the frontier of research in atmospheric-tracer-inversion studies and various aspects of this question form the basis of a number of later chapters. Some key steps are shown schematically in Figure 3.1. The factor $p_{\text{prior}}(\mathbf{x})$ represents the *prior* information about the parameters, $\mathbf{x}$. In the tracer-inversion problem, we would expect that such information would come from outside the field of atmospheric science. To the extent that this is based on actual measurements, the use of such information in the Bayesian formalism falls into the main class that Bard [22] regards as uncontroversial. As noted in the preface, the aim is to have our estimation fall, as far as possible, into this uncontroversial class. Chapter 6 discusses specification and characterisation of sources, i.e. the type of information from which we can construct $p_{\text{prior}}(\mathbf{x})$. Issues of

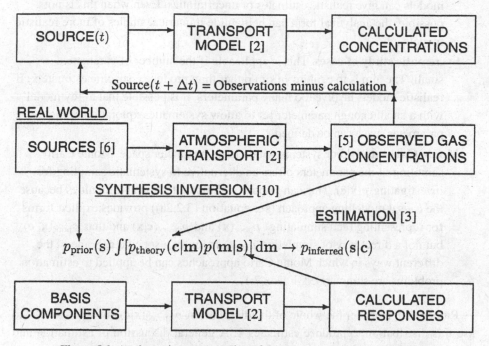

Figure 3.1 A schematic representation of the components of the tracer-inversion problem and the discussion in this book. Numbers in square brackets refer to chapters of this book. Chapter 7 includes other ways of formulating the inversion calculations.

$p(\mathbf{c}|\mathbf{x})$ are discussed in the following subsection. This partitions the associated uncertainties into measurement error, discussed in Chapter 5, and model error, discussed in Chapter 9. The issue of 'model/data representativeness' is an ambiguous case since it could be regarded as a model error (too little space–time resolution) or as a data error (failure of measurements to be representative). It is discussed separately in Section 9.3.

The mathematical procedures for answering question (ii) (i.e. how should $p_{\text{inferred}}(\mathbf{x}|\mathbf{c})$ be analysed?) are reviewed by Tarantola (often with more generality than we need here) and by many other authors. There are two main choices.

- Characterise the probability distribution in terms of a representative value, an *estimate* denoted $\hat{\mathbf{x}}$, and consider the extent to which $p_{\text{inferred}}(\mathbf{x}|\mathbf{c})$ is concentrated around $\hat{\mathbf{x}}$. These issues of estimation are represented on the right-hand side of Figure 3.1.

- Consider the posterior probability distribution $p_{\text{inferred}}(\mathbf{x}|\mathbf{c})$ as a whole. An intermediate case is to consider estimating confidence intervals.

For analysis of the posterior distribution, $p_{\text{inferred}}(\mathbf{x}|\mathbf{c})$, Tarantola proposes several choices.

Analytical solution. When analytical solutions are possible they can give considerable insight. Section 1.4 introduced various 'toy models' because of the insight that can be gained from analytical solutions. In some cases, toy models can give realistic estimates of uncertainties. Even when this is not possible, the analytical form can help guide numerical studies of more realistic models.

Systematic study of cases. This is applicable if the number of parameters is small. The aim is to explore $p(\mathbf{x}|\mathbf{c})$ rather than produce a specific estimate $\hat{\mathbf{x}}$. If realistic models involve too many parameters, it is possible that a 'toy model' with a small enough parameter set to allow systematic exploration of the parameter space can be defined.

Monte Carlo study. For systems with a large parameter space, Monte Carlo sampling of the parameters offers an alternative to systematic sampling for investigating $p(\mathbf{x}|\mathbf{c})$. This can impose significant technical difficulties because the causal modelling approach (see equation (3.2.2a)) provides explicit forms for representing (and simulating) $p_{\text{prior}}(\mathbf{x})$ and $p_{\text{theory}}(\mathbf{c}|\mathbf{x})$ and thus $p_{\text{prior}}(\mathbf{x}, \mathbf{c})$ but not a direct way of simulating $p_{\text{inferred}}(\mathbf{x}|\mathbf{c})$ for a specific $\mathbf{c}$. Some of the different ways in which Monte Carlo approaches can be applied to estimation problems are summarised in Box 7.1.

Rather than consider the whole of the distribution $p_{\text{inferred}}(\mathbf{x}|\mathbf{c})$, an alternative is to use this distribution to produce estimates $\hat{\mathbf{x}}$. A general discussion of estimation and linear estimates in particular is given in the following section. Some aspects that are particular to linear Bayesian estimates are discussed in Section 3.4.

3.3   Properties of estimators

3.3.1   Generic properties

As in the case of studies of $p_{\text{inferred}}(\mathbf{x}|\mathbf{c})$, there are several different approaches for obtaining estimates, $\hat{\mathbf{x}}(\mathbf{c})$, and for studying their statistical properties.

Analytical solution. As with analysis in terms of the whole distribution, analytical solutions can give considerable insight and simplicity.

Numerical optimisation. In cases in which the estimate is defined as the minimum of some penalty function (e.g. least-squares), numerical techniques can be used to locate this minimum. Different types of information lead to a range of different approaches, many of which are described by Tarantola [471] with appropriate computational algorithms.

Stochastic techniques. Various stochastic search algorithms can be used to obtain estimates.

If we are modelling the observations as random variables, with distribution $p(\mathbf{c}|\mathbf{x})$, then $\hat{\mathbf{x}}(\mathbf{c})$, being a function of random variables, will itself be a random variable. The distribution of $\hat{\mathbf{x}}$ is termed the sampling distribution and satisfies

$$p_{\text{sample}:X}(\hat{\mathbf{x}}_X(\mathbf{c})|\mathbf{x})\,d\hat{\mathbf{x}} = p_{\text{theory}}(\mathbf{c}|\mathbf{x})\,d\mathbf{c} \tag{3.3.1}$$

Thus, in conventional (i.e. non-Bayesian) estimation, the statistical properties of estimators are defined in terms of the probability distribution $p_{\text{theory}}(\mathbf{c}|\mathbf{x})$. Since we are dealing with several distinct probability distributions, we need to identify the distribution with respect to which the expectation is defined. For conventional estimation we denote

$$E_{\text{sample}:X}[\hat{\mathbf{x}}] = \int \hat{\mathbf{x}}_X(\mathbf{c})\,p_{\text{theory}}(\mathbf{c}|\mathbf{x})\,d\mathbf{c} \tag{3.3.2}$$

Fisher [154] introduced a number of important characteristics of such estimators.

(i) **Bias.** The (mean) amount of bias is given by

$$E_{\text{sample}:X}[\hat{\mathbf{x}} - \mathbf{x}] = \int [\hat{\mathbf{x}}(\mathbf{c}) - \mathbf{x}]\,p_{\text{theory}}(\mathbf{c}|\mathbf{x})\,d\mathbf{c} \tag{3.3.3a}$$

An estimator is termed unbiased if the mean bias is zero.

(ii) **Consistency.** An estimator is termed 'consistent' if (with probability unity) it converges to the true value as the amount of data, N, tends to infinity. Thus a consistent estimator becomes unbiased for large N. (The main exception that appears in this book is spectral estimation. The sample periodogram is not a consistent estimator of the spectral density – see Chapter 4.)

(iii) **Efficiency.** The relative efficiency, η, of two estimators, $\hat{x}_A$ and $\hat{x}_B$ is the ratio of their mean-square errors:

$$\eta = E[(\hat{x}_A - x)^2]/E[(\hat{x}_B - x)^2] \tag{3.3.3b}$$

An 'efficient' estimator is an unbiased estimator whose variance is smaller than the variance of any other unbiased estimator. For estimates, $\hat{\mathbf{x}}$, of vectors, $\mathbf{x}$, efficient estimators are those whose covariance matrix is given by the Rao–Cramer *minimum-variance bound* [22].

An additional important property of estimators is robustness; which is variously described as the degree of insensitivity to the assumptions of the distribution or, more particularly, as insensitivity to large outliers.

For vector estimates, the mean-square error is given by the trace of the matrix

$$E_{\text{sample}:X}[(\hat{\mathbf{x}}_X - \mathbf{x}) \otimes (\hat{\mathbf{x}}_X - \mathbf{x})] = \int [\hat{\mathbf{x}}(\mathbf{c}) - \mathbf{x}] \otimes [\hat{\mathbf{x}}(\mathbf{c}) - \mathbf{x}]\,p_{\text{theory}}(\mathbf{c}|\mathbf{x})\,d\mathbf{c} \tag{3.3.4a}$$

where $\otimes$ denotes the 'outer-product' operator. The mean-square error can be decomposed into

$$\begin{aligned}
E_{\text{sample}:X}[(\hat{\mathbf{x}} - \mathbf{x}) \otimes (\hat{\mathbf{x}} - \mathbf{x})] \\
= E_{\text{sample}:X}[(\hat{\mathbf{x}} - \mathbf{x})] \otimes E_{\text{sample}:X}[(\hat{\mathbf{x}} - \mathbf{x})] \\
+ E_{\text{sample}:X}[(\hat{\mathbf{x}} - E_{\text{sample}:X}[\hat{\mathbf{x}}]) \otimes (\hat{\mathbf{x}} - E_{\text{sample}:X}[\hat{\mathbf{x}}])]
\end{aligned} \tag{3.3.4b}$$

The diagonals of the first term give the mean-square bias and the second term gives the covariance matrix for the estimates.

Some of the most important types of estimate are the following.

Means. The general expression for the mean is $\hat{\mathbf{x}}_{\text{MEAN}} = \int \mathbf{x} \, p_{\text{inferred}}(\mathbf{x}|\mathbf{c}) \, d\mathbf{c}$; it is the solution that has the minimum variance.

Modes. The maximum-likelihood estimate $\hat{\mathbf{x}}_{\text{ML}}$ is the solution that maximises $p(\mathbf{c}|\mathbf{x})$. The Bayesian equivalent is the MPD, i.e. 'mode of posterior distribution', the $\mathbf{x}$ that maximises $p_{\text{inferred}}(\mathbf{x}|\mathbf{c})$. In the simplest cases these modal solutions appear as one aspect of an overall analytical solution. Otherwise, a numerical solution is required. Tarantola describes a number of computational techniques, in a number of cases making use of special characteristics of $p(\mathbf{x}|\mathbf{c})$ in particular problems.

Least-squares. Least-squares estimates can be useful even when they are not the maximum-likelihood estimates, especially if they approximate the maximum-likelihood solution. We use $\hat{\mathbf{x}}_{\text{OLS}}$ to denote estimates obtained from 'ordinary least-squares', $\hat{\mathbf{x}}_{\text{WLS}}$ to denote estimates obtained from 'weighted least-squares' and $\hat{\mathbf{x}}_{\text{BLS}}$ for estimates obtained by using Bayesian least-squares. In the last case, squares of deviations from prior estimates of the parameters are included in the penalty function. For Bayesian estimation with multivariate normal distributions, $\hat{\mathbf{x}}_{\text{MEAN}} = \hat{\mathbf{x}}_{\text{MPD}} = \hat{\mathbf{x}}_{\text{BLS}}$.

3.3.2   Linear estimators

A special class of estimators consists of linear estimators: those in which the estimate is constructed as a linear combination of the observed values. Specific linear estimators arise from least-squares criteria, but for the moment we consider general linear estimators. We use the generic notation

$$\hat{\mathbf{x}} = \mathbf{Hc} \tag{3.3.5}$$

for such an estimate.

Following Jackson [224], we consider the additional specialisation that occurs when the theoretical model on which this estimate is based is also linear:

$$\mathbf{c} \approx \mathbf{m} = \mathbf{Gx} \tag{3.3.6}$$

For a 'good' estimator, we should have $\mathbf{H}$ as an approximate inverse of $\mathbf{G}$, i.e. $\mathbf{GH} \approx \mathbf{I}$ and $\mathbf{HG} \approx \mathbf{I}$. As discussed below, exact equality may be undesirable (because of the approximate nature of the relation (3.3.6)) and in any case it is impossible unless $\mathbf{G}$ is a square matrix.

For a general linear estimate, we can write the error in the estimate as

$$\hat{\mathbf{x}} - \mathbf{x} = [\mathbf{HG} - \mathbf{I}]\mathbf{x} - \mathbf{H}[\mathbf{c} - \mathbf{Gx}] \tag{3.3.7}$$

Box 3.1. The null space

In the linear-estimation problem defined by solving $\mathbf{m} = \mathbf{Gx}$ for $\mathbf{x}$, there is the possibility that there may be null vectors $\mathbf{x}^*$ such that $\mathbf{Gx}^* = 0$. Since any linear combination of such vectors will have the same 'null' property, such null vectors form a vector space known as the null subspace of operator $\mathbf{G}$. These are the sets of parameters about which the observations give absolutely no information. The null space does not depend on the normalisation of the operator $\mathbf{G}$. If we define transformed vectors $\mathbf{x}' = \mathbf{Ax}$ and $\mathbf{m}' = \mathbf{Bm}$ then we will have $\mathbf{m}' = \mathbf{G}'\mathbf{x}'$ with $\mathbf{G}' = \mathbf{BGA}^{-1}$. The null space is invariant under such transformations in the sense that $\mathbf{Gx} = 0$ if and only if $\mathbf{G}'\mathbf{x} = 0$.

However, when we wish to characterise the subspace of parameter combinations that can be estimated, a specified normalisation is essential. The obvious choice of taking the parameter space orthogonal to the null space requires the definition of an inner product so that orthogonality can be defined. Without additional criteria, an infinite number of inner products can be defined on a real vector space.

To illustrate the issue, consider functions $f(x)$ defined on the interval $[-1, 1]$ as polynomials of degree ≤ 8. These are for a vector space that can be expressed as combinations of basis vectors in various ways. Possible basis sets are in terms of powers $[1, x, x^2, x^3, x^4, x^5, x^6, x^7, x^8]$, Legendre polynomials $[1, P_1(x), \ldots, P_8(x)]$ or other orthogonal polynomials. As an example of an operator $\mathcal{G}$ we take the third derivatives. The null space is clearly the set of polynomials of degree ≤ 2, i.e. those with zero third derivatives. However, the set of polynomials (of degree ≤ 8) with non-zero third derivatives is not a subspace. Looking at the basis sets above and excluding basis vectors in the null space, it is clear that sets of basis vectors $[x^3, x^4, x^5, x^6, x^7, x^8]$ and $[P_3(x), \ldots, P_8(x)]$ span different subspaces.

The first term on the right-hand side is a fixed bias (which vanishes only if $\mathbf{HG} - \mathbf{I}$ is zero, or at least has $\mathbf{x}$ entirely in its null space). It is fixed, in that it is not a random variable characterised by the sampling distribution. The second term on the right-hand side of (3.3.7) captures the randomness from the sampling distribution.

An important class of linear estimators derives from the singular-value decomposition of $\mathbf{G}$

$$G_{j\mu} = \sum_p V_{jp} \lambda_p U_{\mu p} \tag{3.3.8a}$$

with $\lambda \geq 0$ and $\mathbf{U}$ and $\mathbf{V}$ as orthonormal matrices (see Box 3.2). The class of approximate inverses is expressed as

$$H_{\mu j} = \sum_p V_{jp} f_{\text{reg}}(\lambda_p) U_{\mu p} \tag{3.3.8b}$$

where $f_{\text{reg}}(x) \approx 1/x$.

Box 3.2. The singular-value decomposition

The singular-value decomposition (SVD) states that any real $N \times M$ matrix $\mathbf{G}$ can be decomposed in the form

$$G_{i\mu} = \sum_q U_{iq} V_{\mu q} \lambda_q \tag{1}$$

with the λ_q all non-negative and P of the λ_q non-zero, with $P \leq N$ and $P \leq M$, where V and U are square matrices with the orthogonality properties

$$\sum_{i=1}^{N} U_{ip} U_{iq} = \delta_{pq} \tag{2a}$$

and

$$\sum_{\mu=1}^{M} V_{\mu p} V_{\mu q} = \delta_{pq} \tag{2b}$$

The SVD will depend on the choice of normalisation of G since the orthogonality relations (2a) and (2b) depend on the basis chosen to define the indices i and μ in their respective vector spaces. The space spanned by vectors from one or more of the λ_q is not invariant under the type of linear transformation described in Box 3.1. The exception is the null spaces of $\mathbf{G}$ and $\mathbf{G}^{\mathsf{T}}$: the space spanned by vectors with $\lambda_q = 0$.

An important use of the SVD in atmospheric science is the construction of what are known as empirical orthogonal functions. A space–time dependence of some quantity is decomposed as

$$X_{\mathbf{r}t} = \sum_q u_q(\mathbf{r}) \lambda_q v_q(t) \tag{3}$$

Here the various $u_q(\mathbf{r})$ represent spatial patterns and the $\lambda_q v_q(t)$ represent the variation with time of the contribution of the qth pattern to the total variation.

This class of estimate is further considered in Section 8.1. Particular cases are the solution defined using the pseudo-inverse, which corresponds to $f_{\text{reg}}(0) = 0$ and $f_{\text{reg}}(\lambda) = \lambda^{-1}$ if $\lambda > 0$. Other cases use a higher cut-off and put $f_{\text{reg}}(\lambda) = 0$ for $\lambda \leq \lambda_c$. With appropriate normalisations, Bayesian estimation corresponds to an approximate inverse defined by $f_{\text{reg}}(\lambda) = \lambda/(1 + \lambda^2)$ as indicated by equation (3.3.15b).

As noted above, the least-squares formalism leads to linear estimators. The weighted-least-squares estimates, $\hat{\mathbf{x}}_{\text{WLS}}$, are defined as the vectors that minimise the objective function

$$J_{\text{WLS}} = (\mathbf{c} - \mathbf{Gx})^{\mathsf{T}} \mathbf{X} (\mathbf{c} - \mathbf{Gx}) \tag{3.3.9a}$$

where $\mathbf{X}$ is the covariance matrix of $p(\mathbf{c}|\mathbf{m})$. Ordinary least-squares (OLS) corresponds to the special case $\mathbf{X} = \mathbf{I}$.

The WLS estimate is given by

$$\hat{\mathbf{x}}_{\text{WLS}} = [\mathbf{G}^{\mathsf{T}}\mathbf{X}\mathbf{G}]^{-1}[\mathbf{G}^{\mathsf{T}}\mathbf{X}\mathbf{c}] \tag{3.3.9b}$$

In terms of (3.3.7), $\mathbf{HG} = \mathbf{I}$ and so, if $E_{\text{sample}:X}[\mathbf{c}] = \mathbf{Gx}$, then weighted-least-squares estimates are unbiased with respect to the distribution of the observations.

3.3.3 Bayesian least-squares

The Bayesian least-squares estimate $\hat{\mathbf{x}}_{\text{BLS}}$ is obtained as the vector that minimises the objective function

$$J_{\text{BLS}} = (\mathbf{c} - \mathbf{Gx})^{\mathsf{T}}\mathbf{X}(\mathbf{c} - \mathbf{Gx}) + (\mathbf{x} - \mathbf{z})^{\mathsf{T}}\mathbf{W}(\mathbf{x} - \mathbf{z}) \tag{3.3.10}$$

As expected, from our justification of Bayesian estimation as equivalent to fitting additional data, we can write the problem in a form that parallels the WLS estimation. Bayesian estimation is represented as fitting an extended data set

$$\mathbf{y} = \mathbf{Ax} \tag{3.3.11a}$$

with

$$\mathbf{y} = \begin{bmatrix} \mathbf{c} \\ \mathbf{z} \end{bmatrix} \qquad \mathbf{A} = \begin{bmatrix} \mathbf{G} \\ \mathbf{I} \end{bmatrix} \qquad \mathbf{Y} = \begin{bmatrix} \mathbf{X} & \mathbf{0} \\ \mathbf{0} & \mathbf{W} \end{bmatrix} \tag{3.3.11b}$$

The objective function, J_{BLS}, becomes

$$J_{\text{BLS}}(\mathbf{x}) = (\mathbf{y} - \mathbf{Ax})^{\mathsf{T}}\mathbf{Y}(\mathbf{y} - \mathbf{Ax}) \tag{3.3.12a}$$

and leads to the estimate

$$\hat{\mathbf{x}}_{\text{BLS}} = (\mathbf{A}^{\mathsf{T}}\mathbf{YA})^{-1}\mathbf{A}^{\mathsf{T}}\mathbf{Yy} \tag{3.3.12b}$$

This form is convenient for actual calculations and is readily shown to be equivalent to the form

$$\hat{\mathbf{x}}_{\text{BLS}} = [\mathbf{G}^{\mathsf{T}}\mathbf{X}\mathbf{G} + \mathbf{W}]^{-1}[\mathbf{G}^{\mathsf{T}}\mathbf{X}\mathbf{c} + \mathbf{W}\mathbf{z}] \tag{3.3.12c}$$

which follows from minimising (3.3.10) directly.

The process of Bayesian estimation can be addressed in three ways:

(i) directly from minimising (3.3.10), which in turn can be derived as the MPD of $p_{\text{inferred}}(\mathbf{x}|\mathbf{c})$ in the case of multivariate normal distributions, as described in the following section;

(ii) as an extended-data least-squares fit, as indicated by (3.3.12a); and

(iii) as a biased estimation of corrections to prior estimates as described below.

Rewriting the Bayesian least-squares estimation problem in a 'correction' form helps clarify the role of the prior information. The analysis is expressed in terms of

$$q_j = c_j - \sum_{\mu} G_{j\mu}z_{\mu} \tag{3.3.13a}$$

and

$$f_\mu = x_\mu - z_\mu \tag{3.3.13b}$$

so that (3.3.10) is expressed

$$J(\{f_\mu\}) = \sum_{jk} \left[\left(q_j - \sum_\mu G_{j\mu} f_\mu \right) X_{jk} \left(q_k - \sum_\nu G_{k\nu} f_\nu \right) \right] + \sum_{\mu\nu} f_\mu X_{\mu\nu} f_\nu$$

$$= (\mathbf{q} - \mathbf{Gf})^\mathsf{T} \mathbf{X} (\mathbf{q} - \mathbf{Gf}) + \mathbf{f}^\mathsf{T} \mathbf{Wf} \tag{3.3.13c}$$

Proving the equivalence to (3.3.10) is left as an exercise (problem 2 below).

The role of the prior information in Bayesian estimation becomes apparent when we represent $\mathbf{G}$ in a basis in which $\mathbf{W}$ and $\mathbf{X}$ are identity matrices – this is simply a re-scaling if $\mathbf{W}$ and $\mathbf{X}$ are initially diagonal. In this basis $\mathbf{G}$ is denoted $\tilde{\mathbf{G}}$ and expanded using the singular-value decomposition as

$$\tilde{G}_{j\mu} = \sum_p V_{jp} \lambda_p U_{\mu p} \tag{3.3.14}$$

with $\lambda \geq 0$ and $\mathbf{U}$ and $\mathbf{V}$ as orthonormal matrices.

The solution that minimises (3.3.13c) is

$$f_\mu = \sum_{pj} U_{\mu p} f_{\text{reg}}(\lambda_p) V_{jp} q_j \tag{3.3.15a}$$

with

$$f_{\text{reg}}(\lambda) = \frac{\lambda}{\lambda^2 + 1} \tag{3.3.15b}$$

This is a special case of the general form based on (3.3.9b) with a smooth transition from cases in which the estimates are based mainly on $\mathbf{c}$ (large λ_q) and components such that the estimates mainly reflect the priors (small λ_q). In this Bayesian form, the source estimates are significantly influenced only by those components of $\mathbf{q}$ that correspond to singular values (of $\tilde{\mathbf{G}}$) that are of order unity or greater. Biased estimation is often used as a regularisation technique for ill-conditioned inverse problems (see Box 8.1). However, Draper and Smith [114] suggest that such biased estimation is most readily justified if it derives from a Bayesian approach, as is done explicitly in the following section.

3.4   Bayesian least–squares as MPD

The linear-estimation problem was introduced in the previous section in order to illustrate some key properties of estimators *per se*, as well as the derivation of linear estimators from least-squares objective functions, without explicit justification from statistical models. In this section, we revisit linear estimation for the case of Bayesian estimation in which the distributions $p_{\text{prior}}(\mathbf{x})$ and $p_{\text{theory}}(\mathbf{c}|\mathbf{x})$ are multivariate normal.

This statistical modelling allows us to investigate the sampling distribution of the estimates.

As in the previous section, the model relation is taken as linear, $\mathbf{m}(\mathbf{x}) = \mathbf{Gx}$, and the distributions $p(\mathbf{c}|\mathbf{x})$ and $p(\mathbf{x})$ are multivariate normal with inverse covariance matrices $\mathbf{X}$ and $\mathbf{W}$, respectively, and respective means $\mathbf{m} = \mathbf{Gx}$ and $\mathbf{z}$.

Specifically:

$$p_{\text{prior}}(\mathbf{x}) = \sqrt{\frac{|\mathbf{W}|}{(2\pi)^M}} \exp\left[-\tfrac{1}{2}(\mathbf{x} - \mathbf{z})^{\mathsf{T}}\mathbf{W}(\mathbf{x} - \mathbf{z})\right] \tag{3.4.1a}$$

$$p_{\text{theory}}(\mathbf{c}|\mathbf{x}) = \sqrt{\frac{|\mathbf{X}|}{(2\pi)^N}} \exp\left[-\tfrac{1}{2}(\mathbf{c} - \mathbf{Gx})^{\mathsf{T}}\mathbf{X}(\mathbf{c} - \mathbf{Gx})\right] \tag{3.4.1b}$$

$$p_{\text{prior}}(\mathbf{c}, \mathbf{x})$$

$$= \sqrt{\frac{|\mathbf{X}||\mathbf{W}|}{(2\pi)^{M+N}}} \exp\left[-\tfrac{1}{2}(\mathbf{c} - \mathbf{Gx})^{\mathsf{T}}\mathbf{X}(\mathbf{c} - \mathbf{Gx}) - \tfrac{1}{2}(\mathbf{x} - \mathbf{z})^{\mathsf{T}}\mathbf{W}(\mathbf{x} - \mathbf{z})\right]$$

$$\tag{3.4.1c}$$

Therefore we have

$$p_{\text{inferred}}(\mathbf{x}|\mathbf{c}) \propto \exp\left[-\tfrac{1}{2}(\mathbf{c} - \mathbf{Gx})^{\mathsf{T}}\mathbf{X}(\mathbf{c} - \mathbf{Gx}) - \tfrac{1}{2}(\mathbf{x} - \mathbf{z})^{\mathsf{T}}\mathbf{W}(\mathbf{x} - \mathbf{z})\right]$$

$$= \exp(-J_{\text{BLS}}/2) \tag{3.4.2a}$$

with

$$J_{\text{BLS}}(\{x_\mu\}) = \sum_{jk}\left(c_j - \sum_\mu G_{j\mu}x_\mu\right)X_{jk}\left(c_k - \sum_\nu G_{k\nu}x_\nu\right)$$

$$+ \sum_{\mu\nu}(x_\mu - z_\mu)W_{\mu\nu}(x_\nu - z_\nu) \tag{3.4.2b}$$

As noted above, Tarantola suggests that the distribution $p_{\text{inferred}}(.)$ can be regarded as *the solution* to the inverse problem. More commonly, $p_{\text{inferred}}(\mathbf{x}|\mathbf{c})$ is used to obtain an *estimate*, denoted $\hat{\mathbf{x}}$, of the parameters, $\mathbf{x}$. Such estimates are derived from the observations, $\mathbf{c}$, and this is expressed as a functional relation $\hat{\mathbf{x}} = \hat{\mathbf{x}}(\mathbf{c})$. The MPD estimate is defined as the maximum of $p_{\text{inferred}}(\mathbf{x}|\mathbf{c})$, which corresponds to the minimum of J_{BLS}.

In the present analysis of the MPD, we can ignore the value of the normalising factor in (3.4.2a) because it is a fixed value that does not affect the MPD estimate. The normalising factor is given explicitly in (3.4.10) below – it follows from the normalisation condition on p_{inferred}. The full expression for p_{inferred} is used in Section 12.1 as a starting point for analysing non-linear-estimation problems with unknown variances.

Box 3.3. **Toy statistical model D.2. A constant source in model A.1 (Green's functions)**

The example is a single time series of concentrations due to a constant source, s. The model may be time-varying. The observed time series is c_1 to c_N; these values have measurement error with variance $1/X$. The source, s, is the sole parameter to be estimated. In a Bayesian formalism, we give it an *a priori* value of 0 with an *a priori* variance of $1/W$.

  The deterministic model is used to calculate the responses, G_j, at time j.

  The statistical model is

$$c_n = G_n s + \epsilon_n$$

Following the formalism of Section 3.2, for the batch-mode regression

$$\mathbf{A}^\mathsf{T} = [G_1, G_2, \ldots, G_N, 1]$$

and

$$\mathbf{Y} = \begin{bmatrix} X & \cdots & 0 & 0 \\ \vdots & \ddots & \vdots & 0 \\ 0 & \cdots & X & 0 \\ 0 & \cdots & 0 & W \end{bmatrix}$$

and

$$\mathbf{y}^\mathsf{T} = [c_1, c_2, \ldots, c_N, 0]$$

so that the estimate of the source strength, s, is

$$\hat{s} = (\mathbf{A}^\mathsf{T}\mathbf{Y}\mathbf{A})^{-1}\mathbf{A}^\mathsf{T}\mathbf{Y}\mathbf{y} = \sum_{j=1}^{N} G_j X c_j / f_N$$

and the variance of this estimate is

$$\mathrm{Var}(\hat{s}) = 1 \left/ \left(\sum_{j=1}^{N} G_j X G_j + W \right) \right. = 1/f_N$$

where

$$f_n = \sum_{j=1}^{n} G_j X G_j + W$$

The dimensionality of the matrix that needs to be inverted is given by the number of source components: one in this case.

With specific distributions for $p_{\text{theory}}(\mathbf{c}|\mathbf{x})$ and $p_{\text{prior}}(\mathbf{x})$ we are able to characterise the statistics of the estimates. The degree of uncertainty in the estimates is expressed in terms of how much the estimates $\hat{\mathbf{x}}$ are expected to differ from the unknown 'true' values, $\mathbf{x}$. For unbiased estimates, the average difference is zero, the uncertainty is measured by taking the mean-square difference (for each component) and covariances between differences.

The statistical model giving distributions (3.4.1a) and (3.4.1b) is

$$\mathbf{c} = \mathbf{Gx} + \boldsymbol{\epsilon} \tag{3.4.3a}$$

and

$$\mathbf{x} = \mathbf{z} + \varepsilon \tag{3.4.3b}$$

or

$$\mathbf{y} = \mathbf{Ax} + \mathbf{d} \tag{3.4.4a}$$

with

$$\mathbf{d} = \begin{bmatrix} \boldsymbol{\epsilon} \\ -\varepsilon \end{bmatrix} \tag{3.4.4b}$$

Thus, from (3.3.12b),

$$\hat{\mathbf{x}} - \mathbf{x} = (\mathbf{A}^\mathsf{T}\mathbf{YA})^{-1}\mathbf{A}^\mathsf{T}\mathbf{YAx} + (\mathbf{A}^\mathsf{T}\mathbf{YA})^{-1}\mathbf{A}^\mathsf{T}\mathbf{Yd} - \mathbf{x} = (\mathbf{A}^\mathsf{T}\mathbf{YA})^{-1}\mathbf{A}^\mathsf{T}\mathbf{Yd} \tag{3.4.5}$$

The outer product of the error vector has expectation

$$E_{\text{prior}}[(\mathbf{d} \otimes \mathbf{d}^\mathsf{T})] = \mathbf{Y}^{-1} \tag{3.4.6}$$

from the definition of $\mathbf{Y}$. Therefore

$$E_{\text{prior}}[(\hat{\mathbf{x}} - \mathbf{x}) \otimes (\hat{\mathbf{x}} - \mathbf{x})^\mathsf{T}] = ((\mathbf{A}^\mathsf{T}\mathbf{YA})^{-1}\mathbf{A}^\mathsf{T}\mathbf{Y})\mathbf{Y}^{-1}((\mathbf{A}^\mathsf{T}\mathbf{Y})^\mathsf{T}(\mathbf{A}^\mathsf{T}\mathbf{YA})^{-1})$$
$$= (\mathbf{A}^\mathsf{T}\mathbf{YA})^{-1} \tag{3.4.7}$$

Clearly, since the random variables $\boldsymbol{\epsilon}$ and ε that form the components of $\mathbf{d}$ come from different distributions, the expectation has to be taken over the full distribution.

In these terms we have

$$E_{\text{prior}}[\mathbf{c}] = \mathbf{Gz} \tag{3.4.8a}$$

so that

$$E_{\text{prior}}[\hat{\mathbf{x}}_{\text{BLS}}] = \mathbf{z} = E_{\text{prior}}[\mathbf{x}] \tag{3.4.8b}$$

Thus, in terms of the respective distributions $p_{\text{theory}}(\mathbf{c}|\mathbf{x})$ and $p_{\text{prior}}(\mathbf{x}, \mathbf{c})$, both $\hat{\mathbf{x}}_{\text{OLS}}$ and $\hat{\mathbf{x}}_{\text{BLS}}$ are unbiased, but, relative to $p_{\text{theory}}(\mathbf{c}|\mathbf{x})$, $\hat{\mathbf{x}}_{\text{BLS}}$ is biased since

$$E_{\text{sample}:X}[\hat{\mathbf{x}}_{\text{BLS}}] = \mathbf{x} - [\mathbf{G}^\mathsf{T}\mathbf{XG} + \mathbf{W}]^{-1}\mathbf{W}[\mathbf{z} - \mathbf{x}] \tag{3.4.9}$$

These results allow us to write the explicit form

$$p_{\text{inferred}}(\mathbf{x}|\mathbf{c})$$

$$= \sqrt{\frac{|\mathbf{G}^\mathsf{T}\mathbf{XG} + \mathbf{W}|}{(2\pi)^M}} \exp\left[-\tfrac{1}{2}(\mathbf{x} - \hat{\mathbf{x}}(\mathbf{c}, \mathbf{z}))^\mathsf{T}(\mathbf{G}^\mathsf{T}\mathbf{XG} + \mathbf{W})(\mathbf{x} - \hat{\mathbf{x}}(\mathbf{c}, \mathbf{z}))\right] \tag{3.4.10}$$

which is derived from (3.4.2a) by ensuring that the quadratic and linear dependences are equivalent. Writing the argument of the exponential as a quadratic form (absorbing any other constant terms into the normalisation) ensures that the overall normalising factor is as given in (3.4.10).

These linear estimates have the special property that the range of uncertainty of the estimates does not depend on the data, $\mathbf{c}$. This property can be exploited in experimental design studies as described in Chapter 13. In such investigations we can consider hypothetical new data with a specified precision (or hypothetical improvements in the precision of existing data) and calculate the range of uncertainty that would be achieved if such data were available.

3.5   Communicating uncertainty

As noted above, in Bayesian terms *the solution* to the inverse problem is the posterior probability distribution function, $p_{\text{inferred}}(\mathbf{x}|\mathbf{c})$. This is a full specification of the uncertainty (as we perceive it) in our knowledge of the parameters, $\mathbf{x}$. In complex atmospheric inversion problems the challenge is to communicate this in a meaningful way, since $p_{\text{inferred}}(\mathbf{x}|\mathbf{c})$ will be a function defined on the multi-dimensional parameter space. Even if the information in $p_{\text{inferred}}(\mathbf{x}|\mathbf{c})$ is expressed as an estimator $\hat{\mathbf{x}}(\mathbf{c})$, we still need techniques for communicating the degree of uncertainty in $\hat{\mathbf{x}}$.

The communication of uncertainty has been discussed by Morgan *et al.* [324]. In particular they noted studies that analysed the effectiveness of various ways of communicating uncertainty. The present section reviews a number of numerical and visual ways of communicating uncertainties in the results of atmospheric inversion problems. These are summarised in Table 3.1, listing numerical and visual characterisations, including the special case of (multivariate) normal distributions.

If the quantity $\mathbf{x}$ is actually a scalar, x, communicating the result is relatively straightforward. The 'solution' can be communicated by plotting $p_{\text{inferred}}(x|\mathbf{c})$ or the cumulative distribution $P(x \leq y) = \int_{-\infty}^{y} p(x|\mathbf{c}) \, dx$. In particular, the cumulative distribution can be used in cases in which $p(x|\mathbf{c})$ has δ-function components. (Morgan *et al.* also suggest that the probability density and the cumulative distribution should often both be used since, although the information in each is formally equivalent, they emphasise different properties.) Numerical specification can be in terms of selected percentiles; normal distributions can be characterised by the mean and variance.

The difficult communication problems arise when we go beyond estimates of scalars. If $\mathbf{x}$ has M components, then, even in the case of multivariate normal distributions, specification of the distribution requires M numbers to specify the mean of $\mathbf{x}$ and $(M + 1)M/2$ independent covariance components that specify the extent of the uncertainty. In inverse problems in which the estimates are likely to be highly correlated, simply quoting the M means and their M variances gives a seriously incomplete picture of the statistical characteristics of the estimation.

Table 3.1. *A summary of some numerical and visual techniques for the characterisation and communication of uncertainty.*

Quantity	Numerical	Numerical (normal)	Visual
Single scalar	Mean and percentiles	Mean and standard deviation	$p(x)$ $\int^x p(y)\,dy$
Two scalars		Means and covariances	Contours
More than two scalars		Means and covariances	Contours for subsets
Time series		Mean and spectrum	Plot of spectrum
(stationary)			Plot of autocovariance realisations
Fields in two dimensions			Contour plots of means, variances and realisations
Fields in more than two dimensions			Contour plots of sections: means, variances and realisations

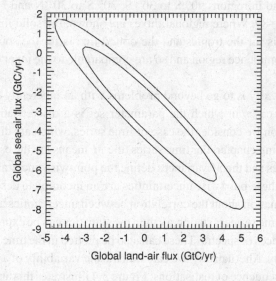

Figure 3.2 Confidence regions for net fluxes of CO_2 at 68% and 90%, based on prior estimates and observed rates of increase in concentration of CO_2.

Morgan *et al.* [324] suggest a number of alternatives. For the case of $M = 2$ the joint (posterior) distribution of x_1 and x_2 can be represented as a two-dimensional contour plot. Figure 3.2 shows the joint confidence regions for CO_2 fluxes to the oceans and terrestrial biota, using only CO_2 data. These reflect relatively precise knowledge of the sum (as the difference between fossil input and atmospheric increase) but rather less ability to distinguish between the two fluxes.

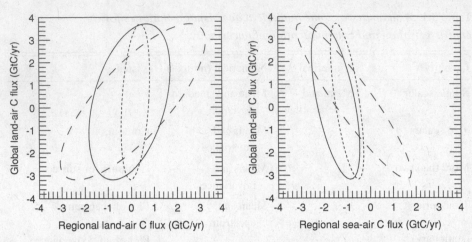

Figure 3.3 An example showing multiple confidence regions for selected pairs of regional and global oceanic and terrestrial fluxes of CO_2. For cases involving regional estimates, the solid line is for north of 30° N, the dashed line is for 30° S to 30° N and the dotted line is for south of 30° S. Further details and additional plots of confidence intervals are given in Chapter 14.

Cases with more parameters can be illustrated by showing multiple combinations. Figure 3.3 shows estimates for oceanic and terrestrial carbon fluxes obtained using the type of Green-function synthesis inversion described in Chapter 10. The fluxes are shown both globally and in regions 90° S to 30° S, 30° S to 30° N and 30° N to 90° N, averaged over 1980–89. Where regional fluxes are shown, the solid line is for the north, the dashed line is for the tropics and the dotted line is for the south. The contours are for the 68% confidence region and so are comparable to the inner contour from Figure 3.2.

The next level of complexity is to go beyond problems with an inherently discrete parameter set, to consider cases in which the parameter set is a discretisation of a continuum. As an illustration we consider the case of time series, which are discussed in more detail in the following chapter. For time series, the M means are the estimated values at a sequence of times and the M variances define the pointwise uncertainties in these estimates. However, these pointwise uncertainties are an incomplete description because they give no information about the correlation between uncertainties at different times. If, however, the time series is stationary then the statistics can be expressed in terms of either autocovariances or spectra. These can also be plotted in the time and frequency domain, respectively. Alternatively, we can indicate the variability of a random function by constructing a sequence of realisations. Figure 3.4 illustrates this approach.

The example in Figure 3.4 shows a set of estimates of a time series for which the stationary error distribution is taken as an AR(1) process with correlation 0.85 and standard deviation 0.2. This uncertainty can be expressed as a ± 0.2 pointwise confidence interval, but this gives an incomplete indication of the types of variability that can occur. The autocorrelation and spectrum are both known as analytical expressions for AR(1) processes (see Section 4.1) but Figure 3.4 gives a visual representation of the extent of variability.

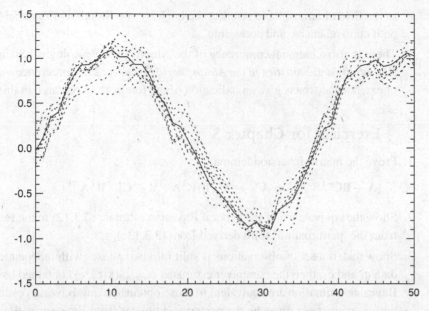

Figure 3.4 An example of the use of random realisations to illustrate the range of functions allowed by a particular statistical characterisation. The covariance matrix has $E[e_n e_m] = 0.2 \times 0.85^{|n-m|}$.

These techniques for time series can also be applied to problems in the spatial domain. However, in the spatial domain the problems of illustrating uncertainties are more complex. There is often the need to consider several dimensions, but, even if only one dimension is involved, the description in terms of a stationary process will usually be inapplicable except possibly on small scales. In more complicated cases, presenting synthetic realisations from the posterior distribution may be one of the few options for conveying the form of the uncertainty in a function.

In Chapter 8 various inversions are characterised as having characteristics of numerical differentiation. Commonly this will lead to large errors, which, beyond a local range of effective smoothing, have strong negative correlations. In these circumstances, plotting the integral of a function (for which the pointwise uncertainties are likely to be smaller and less correlated) may give a better visual representation of what is actually known about a function.

Further reading

- ■ There are, of course, many books on statistics, statistical modelling and statistical estimation. I have made particular use of Deutsch [109] and Bard [22].

- ■ The classic reference on Bayesian statistics is Box and Tiao [47]. Tarantola's book [471] provides an extensive series of examples of applying Bayesian inference in a range of different inverse problems.

- ■ The notes at the end of Chapter 4 cite references that particularly address time series. Cressie's book [84] is the key reference for spatial statistics.

■ In the field of atmospheric science, the book by von Storch and Zwiers [455] is both comprehensive and accessible.

■ There is also a biennial conference of the American Meteorological Society on *Probability and Statistics in the Atmospheric Sciences*. The conference volumes of extended abstracts give an indication of the forefront of research in this field.

Exercises for Chapter 3

1. Prove the matrix-inversion lemma

$$[A + BC^{-1}B^T]^{-1} = A^{-1} - A^{-1}B[B^T A^{-1}B + C]^{-1}B^T A^{-1}$$

2. Show the equivalence of the general Bayesian estimate (3.3.12) to the result from the 'perturbation' form derived from (3.3.13c).

3. Show that, if a set of observations is split into two phases with independent data c_1 and c_2, then the parameter estimates $p_{inferred}(x|c_1, c_2)$ obtained using Bayesian estimation are equivalent to those obtained using Bayesian estimation with c_2 with priors given by the posterior estimates from Bayesian estimation applied to c_1.

Chapter 4

Time-series estimation

> The major problem [with time travel] is simply one of grammar ... for instance
> how to describe something that was about to happen to you in the past before you
> avoided it by time-jumping two days forward ...
>
> Douglas Adams: *The Restaurant at the End of the Universe.*

4.1   Time series

The discussion in the preceding chapter has considered vectors of observations, $\mathbf{c}$, and
vectors of model parameters, $\mathbf{x}$, without any consideration of the relations between
elements of these vectors. In practice many applications have such vector elements
representing sequences of a particular quantity distributed in space and/or time. In
studies of tracer inversions, the analysis of time series is of interest both in the analysis
of time-dependent problems and also as an example of the type of concepts that may
apply to spatial dependence.

For time-dependent problems we will want to consider the vectors $\mathbf{c}$ and $\mathbf{x}$ as a
sequence of subvectors, $\mathbf{c}'(n)$ and $\mathbf{x}'(n)$, associated with each time point, n. Thus we
put

$$\mathbf{c} = [\mathbf{c}'(1), \mathbf{c}'(2), \ldots, \mathbf{c}'(N)]^\mathsf{T} \tag{4.1.1a}$$

and

$$\mathbf{x} = [\mathbf{x}'(1), \mathbf{x}'(2), \ldots, \mathbf{x}'(N)]^\mathsf{T} \tag{4.1.1b}$$

and consider the statistical relations between the components. For the linear problem
described in Sections 3.3 and 3.4, the uncertainties in the observations, $\mathbf{c}$, will be

characterised by a covariance matrix, $\mathbf{X}^{-1}$, the prior estimates of $\mathbf{x}$ by $\mathbf{W}^{-1}$ and the posterior estimates of $\mathbf{x}$ by $[\mathbf{G}^T\mathbf{X}\mathbf{G} + \mathbf{W}]^{-1}$. The relevant questions are what sort of statistical models should be used for the sequence 'inputs' $\mathbf{c}'(.)$ and $\mathbf{x}'(.)$ (i.e. how do we specify $\mathbf{X}$ and $\mathbf{W}$) and how do we interpret the 'output', given by $[\mathbf{G}^T\mathbf{X}\mathbf{G} + \mathbf{W}]^{-1}$? (The same questions appear in more complicated forms if we go beyond the use of the multivariate normal distributions that lead to linear-estimation problems.)

This section starts with the simplest case of a single-variable time series. Thus we consider cases such as observations, $c_j = c(t_j)$, measurement errors, $\epsilon_j = \epsilon(t_j)$, and source parameters, $s_\mu = s(t_\mu)$. We further restrict consideration to equi-spaced time series: those where successive t_j are separated by a fixed interval Δt. For such equi-spaced time series there is an extensive statistical literature from which to obtain insights into the more complicated problems of vector time series and space–time variations. In working with equi-spaced time series, it is common to work in units such that $\Delta t = 1$ and to replace the time value, t, by an integer index $n = 1, 2, 3, \ldots, N$.

The most general statistical representation of a time series would be in terms of the joint probability distribution of all the $\mathbf{c}(n)$. In this representation, the time dimension has no special role. Time-series analysis and modelling applies to cases in which the time-ordering is a relevant part of the statistical description. One special class of time series consists of those defined as stationary. Such series have no preferred time origin. These have a constant mean, denoted $\bar{f}$ here, and the covariance between different times depends only on the time difference so that

$$E[f(n) - \bar{f})(f(j) - \bar{f})] = R'(n - j) \tag{4.1.2}$$

where $R'(.)$ is known as the autocovariance function. More precisely, stationary series are those for which there is no preferred time origin for any of the statistics of the series. A less restrictive condition is nth-order stationarity when there is no preferred time origin for any of the nth- or lower-order statistics. Equation (4.1.2) defines second-order stationarity, assuming that $\bar{f}$ is constant.

Stationary time series can be characterised by a spectral density, $h'(\theta)$, which is the Fourier transform of $R'(n)$, giving

$$R'(k) = \int_{-\pi}^{\pi} \cos(k\theta)\, h'(\theta)\, d\theta \tag{4.1.3a}$$

and

$$h'(\theta) = \frac{1}{2\pi} \sum_{k=-\infty}^{\infty} R'(k) \cos(k\theta) \tag{4.1.3b}$$

Since time is expressed in a dimensionless form, θ is a dimensionless frequency.

For purposes of comparing different time series, it is often convenient to regard $R'(k)$ as a discrete sample from a continuous function that we denote $R(t)$. Thus

$$R'(k) = R(k\,\Delta t) \tag{4.1.4}$$

We can express the autocovariance function, $R(.)$, and power spectrum, $h(.)$, as a pair of Fourier transforms:

$$R(t) = \int_{-\infty}^{\infty} \cos(\omega t)\, h(\omega)\, d\omega \qquad (4.1.5a)$$

and

$$h(\omega) = \frac{1}{2\pi} \int_{-\infty}^{\infty} \cos(\omega t)\, R(t)\, dt \qquad (4.1.5b)$$

subject to $\int_{-\infty}^{\infty} |R(t)|\, dt < \infty$. The (angular) frequency, ω, has the units of inverse time, unlike the dimensionless frequency, θ. They are related by

$$\theta = \omega\, \Delta t \qquad (4.1.6)$$

Box 4.1 describes the use of the relations above to define a standard normalisation for spectra and estimates of spectra. Several different normalisations exist in the literature. Our requirement is to have a standard that allows comparisons of spectra (or more usually estimated spectra) for time series that may differ in length and/or density of data.

As an alternative to characterising statistical models of time series using the frequency domain, there are various 'standard' time-domain models. They are constructed from a combination of random forcing, $\epsilon(n)$, and deterministic relations between the time-series values.

Autoregressive (AR) models. The general AR(k) process is defined [374: Section 3.5.4] by

$$f(n) = \sum_{m=1}^{k} a_m\, f(n - m) + \epsilon(n) \qquad (4.1.7a)$$

where the $\epsilon(n)$ are normally distributed random variables with zero mean and

$$\text{cov}[\epsilon_n, \epsilon_{n'}] = Q\delta_{nn'} \qquad (4.1.7b)$$

The power spectrum [374: equation (4.12.59)] is

$$h'(\theta) = \frac{Q}{2\pi} \frac{1}{\left|1 - \sum_{m=1}^{k} a_m \exp(-im\theta)\right|^2} \qquad (4.1.7c)$$

Fitting this spectral form is the basis of maximum-entropy spectral analysis (see Box 4.3).

A special case is the AR(1) model, which is specified by the parameters $a_1 = a$ and Q. Priestley [374: equation (3.5.16)] gives

$$R(0) = E[\epsilon(n)^2] = \frac{Q}{1 - a^2} \qquad (4.1.8a)$$

or, more generally,

$$R(n) = \frac{Q a^{|n|}}{1 - a^2} \qquad (4.1.8b)$$

Box 4.1. Spectral normalisation

Power-spectral densities can be defined with various normalisations. However, for quantitative studies, we need to define a standard normalisation for spectra.

- ■ Spectra from different sites need to be comparable, e.g. when the spectra have different proportions of high and low frequencies, it should be possible to assign the difference to an excess of high frequency in one or an excess of low frequency in the other (or both).

- ■ To achieve this, the normalisation should not change when the time series is lengthened – the only change should be an increase in frequency resolution and/or precision. Furthermore, apart from the inevitable effects of aliasing, the spectrum should not change if the spacing of data is changed.

- ■ The normalisation of the spectra should be consistent with equation (4.2.12).

A standard normalisation for spectral estimates of discrete time series is established by relating them to continuous functions via $R'(k) = R(k \Delta t)$ to give

$$\int_{-\pi}^{\pi} \cos(k\theta) \, h'(\theta) \, d\theta = \int_{-\infty}^{\infty} \cos(\omega k \, \Delta t) \, h(\omega) \, d\omega$$

$$= \frac{1}{\Delta t} \int_{-\infty}^{\infty} \cos(k\theta) \, h(\theta / \Delta t) \, d\theta$$

$$= \frac{1}{\Delta t} \int_{-\pi}^{\pi} \cos(k\theta) \sum_{m=-\infty}^{\infty} h\left(\frac{\theta + 2m\pi}{\Delta t}\right) \, d\theta \qquad (1)$$

whence

$$h'(\theta) = \frac{1}{\Delta t} \sum_{m=-\infty}^{\infty} h\left(\frac{\theta + 2m\pi}{\Delta t}\right) \qquad (2)$$

Therefore, $\Delta t \times h'(\omega \, \Delta t) = \Delta t \times h'(2\pi \nu \Delta t)$, plotted as a function of either ω or ν, is a standard normalisation that is equivalent to $h(.)$.

The $k = 0$ case of (4.1.3a) shows that this definition of $h'(\theta)$ is appropriate for use in (4.2.12). However, for comparison of spectra from time series with different sampling intervals, the quantity $h' \times \Delta t$ gives a standard, in that (apart from any aliasing effects) it is equivalent to h. (Note that (2) refers to the case where a discrete series is obtained by sampling individual points and this leads to aliasing of high frequencies. For discrete series obtained by averaging, such high frequencies will be suppressed to some degree and the aliasing will be less serious.)

Combining (4.1.8b) with equation (4.1.3b) gives

$$h'(\theta) = \frac{Q}{2\pi} \frac{1}{(1 - 2a\cos\theta + a^2)} \tag{4.1.8c}$$

A very special case is AR(1) with $a = 0$, which is uncorrelated 'white noise'. Another special limiting case is the random-walk model defined by $a = 1$. This model is non-stationary because the variance is unbounded.

Moving-average (MA) models. An alternative form of discrete-parameter time-series model is the moving-average model [374: Section 3.5.5] defined by

$$f(n) = \sum_{j=0}^{l} b_j \epsilon_{n-j} \tag{4.1.9}$$

These models have the property that the autocovariance, $R'(n)$, is zero for $n > l$.

Autoregressive moving-average (ARMA) models. The characteristics of the AR and MA models can be combined [374: Section 3.5.6] as

$$f(n) = \sum_{j=1}^{k} a_j f(n - j) + \sum_{j=0}^{l} b_j \epsilon_{n-j} \tag{4.1.10}$$

Autoregressive integrated moving-average (ARIMA) models. The ARMA models can be used as a basis for defining models of non-stationary processes [374: Section 10.2] as 'integrals' of stationary (ARMA) processes. The ARIMA model of order (k, d, l) for a process $f(n)$ is defined by

$$g(n) = \sum_{j=1}^{k} a_j g(n - j) + \sum_{j=0}^{l} b_j \epsilon_{n-j} \tag{4.1.11}$$

where $g(n)$ is the dth difference of $f(n)$.

ARIMA modelling, including the various special cases, was popularised by Box and Jenkins [46]. Autoregressive models are particularly convenient for use in the state-space modelling approach described in Section 4.4. Additional examples of this time-domain approach to time-series modelling are given by Young [525].

4.2   Digital filtering

Digital filtering is an important way of producing estimates of quantities expressed as time series. Our starting point is the general linear-estimation procedure described in Section 3.2. The statistical model is an additive-error model of the form

$$c_n = m_n + e_n \tag{4.2.1a}$$

where m_n is a 'signal' that is of interest and e_n is a residual 'noise' that is treated as random. The specific case for time series is

$$c(t) = m(t) + e(t) \tag{4.2.1b}$$

We use this model to examine the simplest case, which is that of linear estimation wherein we obtain estimates, $\hat{f}$, of some quantity, f, by taking linear combinations of the observations, c_n, with the chosen combination being the same regardless of the actual values of the c_n:

$$\hat{f}_j = \sum_n H_{jn} c_n \tag{4.2.2}$$

This includes both the case $f = m$ in which we wish to estimate m and the more general case in which we wish to estimate a function f that is linearly related to m. A measure of whether $\hat{f}$ is a 'good' estimate is given by

$$E[(\hat{f}_k - f_k)(\hat{f}_j - f_j)] = \left(f_k - \sum_p H_{kp} m_p \right) \left(f_j - \sum_q H_{jq} m_q \right)$$

$$+ \sum_{pq} H_{jp} H_{kq} E[e_p e_q] \tag{4.2.3a}$$

where $E[.]$ denotes a statistical expectation, calculated in terms of the distribution of the e_n. In particular, the mean-square error of component k is

$$E[(\hat{f}_k - f_k)^2] = \left(f_k - \sum_p H_{kp} m_p \right)^2 + \sum_{pq} H_{kp} H_{kq} E[e_p e_q] \tag{4.2.3b}$$

The second term in (4.2.3b) is the variance of the estimate and the first term is a squared bias. In special cases, it may be possible to *define* the relations so that the bias vanishes. For example if f is *defined* to be the seven-point running-mean of m then the seven-point running-mean filter will give an unbiased estimate of f. However, in many cases it may be desirable to accept biased estimates in order to reduce the variance (and the mean-square error). The difficulty is that, to do this effectively, one requires some knowledge of the unknown function, $f(.)$.

The previous section noted the importance of stationary processes: those whose statistical characteristics did not change with time. Digital filtering is a form of time-series analysis that preserves this stationarity.

We start by considering the case of continuous functions of time and use the notation $\langle . \rangle$ to denote a generalised time-averaging. In a linear stationary formalism that treats all times equivalently, such averages of a function, $a(t)$, can be written as convolutions with averaging kernels, $\Psi(.)$:

$$\langle a(t) \rangle = \int \Psi(t') a(t + t') \, dt' = \int \Psi(t - t'') a(t'') \, dt'' = \Psi * a \tag{4.2.4}$$

where $*$ denotes a convolution operation.

In this section, the integral form (equation (4.2.4) and its various special cases) will mainly be used as a conceptual tool. Actual time-averages (or, more precisely, estimates of such averages) will be obtained from sums over discrete data. Such digital filtering is denoted

$$\langle\langle z(n)\rangle\rangle = \sum_k \Psi(k) z(n+k) \tag{4.2.5}$$

Using digital filtering to construct estimates

$$\hat{f}(n) = \langle\langle c(n)\rangle\rangle \tag{4.2.6}$$

is a special case of the linear estimator (4.2.2) with

$$H_{jk} = \Psi(j-k) \tag{4.2.7}$$

The condition of stationarity allows analysis of the estimates in the frequency domain, using the filter response, $\psi(.)$, defined in terms of the dimensionless frequency θ as

$$\psi(\theta) = \sum_k \Psi(k)\, e^{ik\theta} \tag{4.2.8}$$

Considering estimates of $m(t)$ from (4.2.1b) of the form

$$\hat{m}(t) = \langle\langle c(t)\rangle\rangle \tag{4.2.9}$$

using a filter with coefficients $\Psi(k)$ and response $\psi(\theta)$ gives the mean-square error (MSE) as

$$E[(\hat{m}(t) - m(t))^2] = \int_{-\pi}^{\pi} \left[|1 - \psi(\omega)|^2 h'_m(\omega) + |\psi(\omega)|^2 h'_e(\omega)\right] d\omega \tag{4.2.10}$$

The optimal filter (defined as having the minimal MSE) has

$$\psi(\theta) = \frac{h'_m(\theta)}{h'_m(\theta) + h'_e(\theta)} \tag{4.2.11}$$

with MSE

$$E[(\hat{m}(t) - m(t))^2] = \int_{-\pi}^{\pi} \frac{h'_m(\theta) h'_e(\theta)}{h'_m(\theta) + h'_e(\theta)}\, d\theta \tag{4.2.12}$$

where $h'_m(\theta)$ and $h'_e(\theta)$ are the spectral densities of $m(t)$ and $e(t)$, respectively.

An important property of convolutions, whether over discrete or continuous time, is that convolution operations in the time domain transform into multiplication in the frequency domain. This means that digital filtering can be implemented by taking the Fourier transform of the time-series, multiplying by the desired frequency response and then performing the inverse transform on the product. This approach is used in many of the time-series analyses of CO_2 by the NOAA/CMDL group. An additional consequence of the relation between convolution operators and multiplications of the transforms of their kernels is that convolution operators, in particular those described below, commute. Thus, for example, if the estimates of derivatives are defined in terms of convolutions, then the derivative of a running mean is the running mean of the derivative, etc. Furthermore, if estimates are based on digital filters, then estimation (using a particular filter) commutes with averaging, differentiation, deconvolution and cycle estimation. This is discussed in more detail by Enting [131].

These operations of averaging, differentiation and other functionals are obtained by generalising the analysis to consider estimating a linear functional of the signal $m(t)$ rather than $m(t)$ itself. Four cases of particular importance are the following.

(i) **Averaging.** Averaging is one of the most direct applications of digital filtering. This makes it important to distinguish between digital filtering for defining averages and digital filtering for the purposes of estimation. In particular, we need to identify whether an estimate $\langle\langle z(t)\rangle\rangle$ that involves averaging is being treated as an estimate of $m(t)$ or of $\bar{m}(t)$ with the average being in turn defined by a convolution: $\bar{m}(t) = \langle\langle m(t)\rangle\rangle_\eta$.

(ii) **Numerical differentiation.** Differentiation is one of the most important of the numerical operations, since greenhouse-gas budgets represent a balance of sources, sinks and rates of change of concentration. As discussed in Chapter 8, numerical differentiation is also the archetypal ill-conditioned problem. In discrete-time-series analysis, differentiation needs to be approximated by some form of differencing.

(iii) **Deconvolution.** This is the case in which $m(t) = \int \rho(t - t') f(t') \, dt'$ and we wish to estimate $f(.)$. For the case in which $\rho(x) = 0$ for $x < 0$ (i.e. Volterra equations) there is a formal solution (see Section 6.6): $f(t) = \rho(0)\dot{m}(t) + \dot{\rho}(0)m(t)/\rho(0)^2 + \int \rho^*(t - t')m(t') \, dt'$, for an appropriate kernel $\rho^*(.)$.

(iv) **Cycle estimation.** Estimation of a time-varying seasonal cycle of the form $A(t)\cos(\omega t - \phi(t))$, with $A(t)$ and $\phi(t)$ slowly varying, can be performed by a technique known as complex demodulation [374: Section 11.2.2]. This has been applied to CO_2 data [474]. Enting [127] showed that the construction of the cycle by complex demodulation is equivalent to band-pass filtering of the original record. This was used as a basis for obtaining confidence intervals on the estimates $\hat{A}(t)$ and $\hat{\phi}(t)$ of amplitude and phase [131].

These four types of analysis correspond to obtaining stationary estimates, $\hat{f}$, of functionals of $m(t)$ (i.e. stationary cases of estimates described by (4.2.2), (4.2.3a) and (4.2.3b)) for $f = \bar{m}$, $f = \dot{m}$, $f(t) = \rho(0)\dot{m}(t) + \dot{\rho}(0)m(t)/\rho(0)^2 + \int \rho^*(t - t')m(t') \, dt'$ and $f = \Psi_{BP} * m$, where Ψ_{BP} is a band-pass filter.

The use of smoothing splines (see Box 4.2) has become quite popular in atmospheric-constituent studies. One of the reasons for this is that smoothing splines provide a means of handling non-equi-spaced data with the results being relatively insensitive to the density of data. As described in Box 4.2, fitting a smoothing spline acts like a digital filter. Splines have the advantage over digital filtering in that the spline fit is defined right to the end of the data record, but the degree of smoothing is greater near the ends than it is in the bulk of the record.

Box 4.2. Smoothing splines

Spline functions are defined as piecewise polynomials. As functions of t they are specified using a discrete set of points, t_j, known as nodes. The splines are mth-order polynomials between each pair of consecutive nodes. At the nodes, they are continuous, with $m - 1$ continuous derivatives. Smoothing splines are formally defined as the functions, $f(t)$, fitted to a data set, $c_j(t_j)$, so as to minimise

$$J_{\text{spline}}[f] = \sum_j (c_j - f(t_j))^2 + \Lambda \int_{t_1}^{t_N} \left(\frac{\mathrm{d}^2 f}{\mathrm{d}t^2} \right)^2 \mathrm{d}t \qquad (1)$$

Thus the smoothing spline is defined by a trade-off between the fit to the data (given by the first term) and the smoothness (given by the second term). The solution of this minimisation can be shown to be a cubic (i.e. $m = 3$) spline with nodes at each data point t_j. (Of course, actual spline-fitting calculations make use of this analytical result and minimise (1) over the set of cubic splines with nodes at the t_j.)

Smoothing splines have proved popular in atmospheric-constituent studies because they can produce a continuous curve fitted to unequally spaced data and can be extended to the end of the record. As well as being applied to time series, splines have been used to fit latitudinal variations for two-dimensional inversions [465].

To a good approximation, fitting a smoothing spline acts as a digital filter with response

$$\psi_{\text{spline}}(\omega) = \frac{1}{1 + (\omega/\omega_c)^4} \qquad (2a)$$

with 50% attenuation at frequency

$$\omega_c = (\Lambda \, \Delta t)^{-1/4} \qquad (2b)$$

corresponding to a period

$$T_c = 2\pi (\Lambda \, \Delta t)^{1/4} \qquad (2c)$$

It is this relative insensitivity to Δt (for reasonably large Λ) that means that the degree of smoothing is relatively insensitive to changes in density of data.

Enting [126, 131] has summarised these and other properties of smoothing splines, re-expressing results from the statistical literature [435, 436], in a notation consistent with the computer routines of de Boor [40].

4.3   Time-series analysis

In the analysis of stationary time series, the spectral densities play a central role. However, there is often little prior knowledge of these. Generally they need to be estimated from observations, c_n. For our generic model of 'observations = signal plus noise', this raises the immediate difficulty that such spectral estimation can at best produce the spectral density of the observations. In principle, such an estimate does not provide a basis for partitioning the spectrum into the separate contributions of 'signal' and 'noise'.

In some cases, however, the structure of the spectrum can suggest a partitioning into one or more 'signal' components and a residual irregular 'noise' contribution. A number of studies of CO_2 have used spectral estimation as a basis for such qualititative descriptive analyses, e.g. [474]. Manning [303] used spectral analyses to argue that there was a 'natural' decomposition of the variations in concentration of CO_2 into a long-term trend (periods over 5 years), a seasonal cycle (frequencies of 1 year or less) and an irregular component (periods of 1–5 years, plus high frequencies).

Priestley [374] gives an extensive discussion of the estimation of the spectral density. We summarise the key points, but restrict the analysis to series with zero mean.

We can define an estimate of $R'(k)$ by

$$\hat{R}'^*(k) = \frac{1}{N - |k|} \sum_{n=1}^{N-|k|} c_n c_{n+|k|}$$
(4.3.1)

This is termed the sample autocovariance function. It gives an unbiased estimate of $R'(k)$. The estimate $\hat{R}(k)$ is defined by

$$\hat{R}'(k) = \frac{1}{N} \sum_{n=1}^{N-|k|} c_n c_{n+|k|}$$
(4.3.2)

which is a biased estimate since

$$E[\hat{R}'(k)] = \left(1 - \frac{|k|}{N}\right) R(k)$$
(4.3.3)

Priestley [374: Section 5.3] quotes the sampling distributions for these estimates.

An initial estimate of $h'(\theta)$ can be obtained by taking the Fourier transform of $\hat{R}'(.)$. This is the periodogram, $I(\theta)$ (for which Priestley uses the notation I^*), given by

$$I(\theta) = \frac{1}{2\pi} \sum_{k=-n+1}^{N-1} \hat{R}'(k) \cos(k\theta) = \frac{1}{N\pi} \left| \sum_{n=1}^{N} c_n e^{-in\theta} \right|^2$$
(4.3.4)

The significance of using $\hat{R}'(.)$ rather than $\hat{R}'^*(.)$ is that, with the former, $I(\theta)$ is guaranteed to be positive definite. If $h(\theta)$ is continuous then we have

$$E[I(\theta)] = h'(\theta) + O(\log N/N)$$
(4.3.5)

where the bias occurs because of the use of $R'(.)$ rather than $\hat{R}'^*$. However, since the sampling variance is given by

$$\text{var}[I(\theta)] \sim [h'(\theta)]^2$$
(4.3.6)

$I(\theta)$ is not a consistent estimator of $h'(\theta)$ since the sampling variance does not go to zero as $N \to \infty$.

The solution is to use a 'window' characterised by a width, M, such that the window averages over a proportion M/N of the frequencies to reduce the variance. This

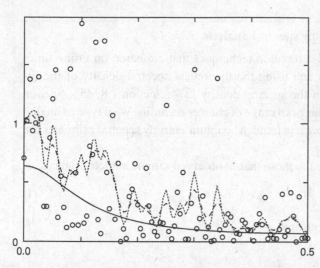

Figure 4.1 Illustrations of sample periodograms, for a synthetic AR(1) time series of 200 points, with lag-1 autocorrelation of 0.5. Open circles are the raw periodogram. Broken lines are estimates based on 13-point and 33-point binomial smoothing. The solid line is the theoretical spectrum.

smoothing will introduce a bias that is proportional to the second derivative of $h(\theta)$. Consequently, transforming the series to remove spectral peaks, or generally flattening the spectrum prior to spectral estimation, will reduce such bias. Figure 4.1 gives an example of a raw periodogram (shown by the open circles) together with two levels of smoothing.

Priestley [374: Table 6.1] lists a number of the more common windowing functions together with the variance and bias of the windowed estimate. For most windowing functions, the bias is proportional to $h''(\theta)/M^2$ with a (window-dependent) coefficient of order unity. The variance of the estimate behaves as

$$\mathrm{var}[\hat{h}'(\theta)] \propto [h'(\theta)]^2 \, M/N \tag{4.3.7}$$

where again the proportionality constant for most windowing functions is of order unity. For most purposes, a Daniell window (defined by the running mean over frequencies $\theta - M\pi/N \le \theta \le \theta + M\pi/N$) is adequate for spectral estimation.

An additional transformation that can be applied as part of spectral estimation is tapering. This involves multiplying the input series, c_n, by a taper function that is 1 in the middle of the series but which goes to 0 at the ends. The aim is to reduce the $O(\log N/N)$ bias in $\hat{h}'(.)$ at the expense of a small increase in the variance of the estimate. Generally, this is worthwhile only if there are sharp peaks in the spectrum.

The spectrum is often plotted on a logarithmic axis. Apart from the normal reason of using logarithmic plots to cover a large range of magnitudes, the confidence intervals on spectral estimates obtained from periodograms span a constant distance on such logarithmic plots (c.f. equation (4.3.6)).

Priestley [374: Section 6.1] also discusses the sampling properties of the periodogram for the case of discrete frequency components in the presence of additional noise. This is mainly of interest for cases in which the frequencies of the discrete components are unknown – discrete components at known frequencies can be analysed by regression analysis. In biogeochemical studies, discrete frequencies arise from annual

---

Box 4.3. Maximum-entropy spectral analysis

There is a class of spectral-estimation techniques that are based on fitting time-domain models to the data and using the theoretical spectral density of the fitted model as the estimate of the spectral density [374: Section 7.8; 455: Section 12.3.21]. Within this class there is a range of choices regarding what type of model, what order of model and how it is fitted. Maximum-entropy spectral estimates are a special case.

Formally, they are defined as those that fit observed sample autocovariances as

$$\int_{-\pi}^{\pi} \hat{h}'(\theta) e^{i\theta k} \, d\theta = \hat{R}(k) \qquad \text{for } k = 0 \text{ to } n \tag{1}$$

subject to maximising the entropy

$$\int_{-\pi}^{\pi} \log_e[\hat{h}'(\theta)] \, d\theta \tag{2}$$

It can be shown that the maximum-entropy method is equivalent to taking the spectrum based on fitting an AR(n) model using the Yule–Walker equations

$$\sum_{k=0}^{n} \hat{R}(|k - j|) a_j = \hat{R}(j) \tag{3}$$

using the sample autocovariances, $\hat{R}$. Priestley [374] discusses some of the limitations of the method.

---

and diurnal forcing – there is little need for analysis of signals containing frequency components that are discrete but unknown.

In the time domain, as opposed to the frequency domain, the main form of time-series analysis is the estimation of the parameters in the ARIMA models described above, including of course the various special cases. Box and Jenkins [46] describe the three stages of such time-series analysis as (i) identification of the model, (ii) estimation and (iii) diagnostic checking. For a known order, (k, d, l), of ARIMA model, the parameters can be estimated from the series by techniques such as maximum likelihood, including various approximations. However, the 'Identification' problem, determining the appropriate model order, is far less well-defined. The starting point recommended by Box and Jenkins is successive differencing until the time series appears stationary. This 'identifies' d, the order of differencing, which may of course be zero. The appropriate differencing reduces the estimation problem to that of estimating the order (k, l) of an ARMA model. The special case $(0, l)$, the moving-average model, can be identified by the covariances going to zero (i.e. the sample covariances dropping below some 'noise' level) for lags exceeding l. Similarly the case $(k, 0)$, which is the AR(k) model, can be identified by the partial autocovariances going to zero. The partial autocorrelations are defined as the solutions, $\phi(m)$, of the Yule–Walker equations

$$\sum_{m} R(|m - n|) \phi(m) = R(n) \tag{4.3.8}$$

For the general ARMA model, Box and Jenkins note that the autocorrelation decays as a mixture of exponentials and sine waves after the first $k - l$ lags and the partial autocorrelation decays in this way after $l - k$ lags. Chapter 5 of Priestley's book [374] is devoted to estimation in the time domain. Chapter 5 below gives examples of such time-series modelling applied to CO_2 data.

4.4   State-space estimation

The statistical model

An alternative form of digital filtering in the statistical analysis of time-dependent problems involves the use of a state-space representation. This formalism has been used in a number of tracer-inversion studies. The specifics of formulating tracer inversions in state-space form are discussed in Section 11.2. The present section reviews the general mathematical formalism. The state-space approach defines a *prior* statistical model in terms of a 'state-vector' describing the system which evolves from one time-point to the next as

$$\mathbf{g}(n + 1) = \mathbf{F}(n)\mathbf{g}(n) + \mathbf{u}(n) + \mathbf{w}(n) \tag{4.4.1a}$$

where $\mathbf{F}$ defines the unforced evolution of the state-vector $\mathbf{g}(n)$, $\mathbf{u}(n)$ is deterministic forcing and $\mathbf{w}(n)$ is a stochastic forcing with zero mean and covariance matrix $\mathbf{Q}(n)$. The $\mathbf{w}(n)$ at successive times are independent. The state-space model relates the state-vector to time series of observed quantities, $\mathbf{c}(n)$, by

$$\mathbf{c}(n) = \mathbf{H}(n)\mathbf{g}(n) + \mathbf{e}(n) \tag{4.4.1b}$$

where $\mathbf{H}$ defines the projection of the state onto the observations and $\mathbf{e}$ represents noise with zero mean and covariance matrix $\mathbf{R}$. As implied by the notation, the quantities $\mathbf{F}(n)$, $\mathbf{Q}(n)$, $\mathbf{H}(n)$ and $\mathbf{R}(n)$ defining the model can all depend on n – there is no restriction to stationary systems. Apart from the restriction to linear models, this is a very general formalism. The mixed stochastic–deterministic model (4.4.1a) can include either of the limiting cases of purely stochastic and purely deterministic modelling. Correlations in observational error can be incorporated by including an additional measurement-error model as part of the state $\mathbf{g}$.

Given a state-space model of the form (4.4.1a,b), the Kalman-filter formalism below (equations (4.4.2a)–(4.4.2e)) gives the optimal one-sided estimate of the state-vector $\mathbf{g}(n)$, i.e. the best estimate of $\mathbf{g}(n)$ that can be obtained using observations $\mathbf{c}(n')$ with $n' \leq n$. Truly optimal estimates, i.e. the best estimate using *all* the data, can be obtained by a technique known as *Kalman smoothing* [173] defined below.

Many different types of inversion can be accommodated within this general state-space-model/Kalman-filter formalism. Section 11.2 describes a number of these. Applications to studies of CFCs are described in Chapter 16. The state-space forms are less relevant to steady-state inversions because of the mismatch between the boundary conditions. State-space inversions start from an initial estimate whereas steady-state inversions apply to periodic fluxes. Mass-balance inversions can be expressed in a state-space form, but the estimation process does not correspond to a Kalman filter and so, even within the class of one-sided estimates, mass-balance inversions will be sub-optimal (see Section 11.3).

Estimation

The estimation procedure for time-point n starts with a prediction of the state-vector starting from the estimated state $\hat{g}(n-1)$ for the previous time-point and using the *prior* model (4.4.1a) to give a one-step prediction:

$$\tilde{g}(n) = F(n-1)\hat{g}(n-1) + u(n-1) \tag{4.4.2a}$$

Given the assumptions of the state-space model, the prediction $\tilde{g}(n)$ is the best estimate of $g(n)$ that can be obtained using observations $c(n')$ with $n' \leq n-1$. This prediction of the state-vector has covariance matrix

$$\tilde{P}(n) = F(n-1)\hat{P}(n-1)F(n-1)^{\mathsf{T}} + Q(n-1) \tag{4.4.2b}$$

where $P(n)$ is the covariance of the estimate $\hat{g}(n)$. From the prediction, the observational data are used to provide an updated estimate, $\hat{g}(n)$, of the state-vector:

$$\hat{g}(n) = \tilde{g}(n) + K(n)[c(n) - H(n)\tilde{g}(n)] \tag{4.4.2c}$$

These estimates differ from the projection, $\tilde{g}(n)$, by a linear combination of the mismatch between the observations and the predictions, $H(n)\tilde{g}(n)$, of the observations.

These estimates have the covariance matrix

$$\hat{P}(n) = [I - K(n)H(n)]\tilde{P}(n)[I - K(n)H(n)]^{\mathsf{T}} + K(n)R(n)K(n)^{\mathsf{T}} \tag{4.4.2d}$$

The optimal estimate is obtained when K is given by the Kalman gain matrix:

$$K(n) = \tilde{P}(n)H(n)^{\mathsf{T}}[H(n)\tilde{P}(n)H(n)^{\mathsf{T}} + R(n)]^{-1} \tag{4.4.2e}$$

The procedure is started from an initial estimate, $\hat{g}(0)$, with covariance $\hat{P}(0)$.

While these equations (4.4.2a)–(4.4.2e) fully specify the Kalman-filter procedure, in deriving the equations and possible generalisations it is useful to identify the vector of 'innovations' $c(n) - H(n)\tilde{g}(n)$ and to denote the covariance of this vector

$$P_I(n) = [H(n)\tilde{P}(n)H(n)^{\mathsf{T}} + R(n)] \tag{4.4.2f}$$

In fact (4.4.2a), (4.4.2b), (4.4.2c) and (4.4.2d) are valid for arbitrary K, but the optimal one-sided estimate requires that $K(n)$ be given by (4.4.2e). There is an alternative expression for $\hat{P}(n)$ that has a simpler structure (see problem 3), but it applies only for the optimal estimate based on (4.4.2e) and may be less stable numerically than (4.4.2d) [173].

Box 4.4. Recursive regression

The regression problem is one of fitting a set of observed data, $z(n)$, as a linear combination of M specified functions $h_\mu(n)$, using the statistical model

$$z(n) = \sum_\mu \alpha_\mu h_\mu(n) + e(n)$$

In the Kalman-filter formalism it is necessary for the state to include the quantity to be estimated. Therefore we take the state as a constant specified by

$$\mathbf{x}(n)^\mathsf{T} = [\alpha_1, \alpha_2, \ldots, \alpha_M] \qquad \text{for all } n \tag{1}$$

Thus, its evolution is defined by $\mathbf{F}(n) = \mathbf{I}$, $\mathbf{Q}(n) = \mathbf{0}$ and $\mathbf{u}(n) = \mathbf{0}$.

The observation equation is specified for the case of scalar observations by

$$\mathbf{H}(n) = [h_1(n), h_2(n), \ldots, h_M(n)] \tag{2}$$

The Kalman-filter formalism readily generalises to multi-component regressions

$$\mathbf{z}(n) = \sum_\mu \alpha_\mu \mathbf{h}_\mu(n) + \mathbf{e}(n) \tag{3a}$$

using

$$\mathbf{H}(n) = [\mathbf{h}_1(n), \mathbf{h}_2(n), \ldots, \mathbf{h}_M(n)] \tag{3b}$$

with the measurement variance given by

$$\mathbf{R}(n) = \mathrm{cov}(\mathbf{e}(n), \mathbf{e}(n)) \tag{3c}$$

For the Kalman-filter formalism we also need to specify the initial estimate of $\hat{\mathbf{g}}(0)$ and its variance, $\hat{\mathbf{P}}(0)$. Non-Bayesian estimation corresponds to the limit of arbitrarily large diagonal elements in $\hat{\mathbf{P}}(0)$.

The properties $\mathbf{F}(n) = \mathbf{I}$, $\mathbf{Q}(n) = \mathbf{0}$ and $\mathbf{u}(n) = \mathbf{0}$ give a significant simplification of the Kalman-filtering formalism, giving $\tilde{\mathbf{P}}(n) = \hat{\mathbf{P}}(n-1)$ and $\tilde{\mathbf{x}}(n) = \hat{\mathbf{x}}(n-1)$. Consequently, the estimation equations reduce to

$$\hat{\mathbf{x}}(n) = \hat{\mathbf{x}}(n-1) + \mathbf{K}(n)[\mathbf{c}(n) - \mathbf{H}(n)\hat{\mathbf{x}}(n-1)] \tag{4a}$$

with

$$\hat{\mathbf{P}}(n) = \hat{\mathbf{P}}(n-1) - \mathbf{K}(n)\mathbf{P}_\mathrm{I}(n)\mathbf{K}(n)^\mathsf{T} \tag{4b}$$

where

$$\mathbf{K}(n) = \hat{\mathbf{P}}(n-1)\mathbf{H}(n)^\mathsf{T}\mathbf{P}_\mathrm{I}(n)^{-1} \tag{4c}$$

with

$$\mathbf{P}_\mathrm{I}(n) = \left[\mathbf{H}(n)\hat{\mathbf{P}}(n-1)\mathbf{H}(n)^\mathsf{T} + \mathbf{R}(n)\right] \tag{4d}$$

To summarise, the application of the Kalman-filtering formalism requires a definition of a state, $\mathbf{g}$, and specific values at each time point, n, for $\mathbf{F}(n)$, $\mathbf{H}(n)$, $\mathbf{u}(n)$, $\mathbf{Q}(n)$ and a set of observations, $\mathbf{c}(n)$, and their covariance, $\mathbf{R}(n)$. A starting value, $\hat{\mathbf{g}}(0)$, and its covariance, $\hat{\mathbf{P}}(0)$, are also required.

Two-sided estimation

The formalism for two-sided estimation (technically known as Kalman smoothing) is given by Gelb [173]. Generally this gives estimates with lower variance than the Kalman filter. Actually, several different forms of smoothing can be defined. If we consider observations $\mathbf{c}(n)$ for $n = 1, 2, \ldots, N$, then we have

- Kalman filtering, as defined by (4.4.2a)–(4.4.2e) above, whereby we estimate $\mathbf{g}(n)$ using $\mathbf{c}(n')$ for $n' \leq n$;
- fixed-lag smoothing, whereby we estimate $\mathbf{g}(n)$ using $\mathbf{c}(n')$ for $n' \leq n + m$ for a fixed m;
- fixed-interval smoothing, whereby we estimate $\mathbf{g}(n)$ using $\mathbf{c}(n')$ for all $n' \in [1, N]$; and
- fixed-point smoothing, to estimate $\mathbf{g}(n)$ for fixed n using $\mathbf{c}(n')$ for all $n' \in [1, N]$ as N increases.

Fixed-lag smoothing can be implemented by defining an augmented state, $\mathbf{g}^*(n) = [\mathbf{g}(n), \mathbf{g}(n-1), \ldots, \mathbf{g}(n-m)]^\mathsf{T}$ and applying the Kalman-filtering formalism using $\mathbf{g}^*$ as the state.

There are several different ways of implementing fixed-interval smoothing [173]. Two of these are the following.

- The 'two-filter' form whereby the forward estimate, denoted $\hat{\mathbf{g}}_\mathrm{F}(n)$, is calculated as defined by (4.4.2a)–(4.4.2e) above, with covariance matrix $\hat{\mathbf{P}}_\mathrm{F}(n)$, and a backward estimate $\hat{\mathbf{g}}_\mathrm{B}(n)$, with covariance matrix $\hat{\mathbf{P}}_\mathrm{B}(n)$, is calculated by running the Kalman filter in reverse order (with $\hat{\mathbf{P}}_\mathrm{B}(N)^{-1} = \mathbf{0}$). The optimal estimate is $\hat{\mathbf{g}}(n) = \hat{\mathbf{P}}(n)[\hat{\mathbf{P}}_\mathrm{F}(n)^{-1}\hat{\mathbf{g}}_\mathrm{F}(n) + \hat{\mathbf{P}}_\mathrm{B}(n)^{-1}\hat{\mathbf{g}}_\mathrm{B}(n)]$, where $\hat{\mathbf{P}}(n)^{-1} = [\hat{\mathbf{P}}_\mathrm{F}(n)^{-1} + \hat{\mathbf{P}}_\mathrm{B}(n)^{-1}]^{-1}$.
- The Rauch–Tung–Streibel form in which $\hat{\mathbf{g}}(n) = \hat{\mathbf{g}}_\mathrm{F}(n) + \mathbf{A}(n)[\hat{\mathbf{g}}(n+1) - \tilde{\mathbf{g}}_\mathrm{F}(n+1)]$, with $\mathbf{A}(n) = \hat{\mathbf{P}}(n)\mathbf{F}^\mathsf{T}[\tilde{\mathbf{P}}(n+1)]^{-1}$ initiated with $\hat{\mathbf{g}}(N) = \hat{\mathbf{g}}_\mathrm{F}(N)$ and with the covariance evolving as $\hat{\mathbf{P}}(n) = \hat{\mathbf{P}}_\mathrm{F}(n) + \mathbf{A}(n)[\hat{\mathbf{P}}(n+1) - \tilde{\mathbf{P}}_\mathrm{F}(n+1)][\mathbf{A}(n)]^\mathsf{T}$ with $\hat{\mathbf{P}}(N) = \hat{\mathbf{P}}_\mathrm{F}(N)$.

The smoothing techniques differ in the types of numerical problems that may be encountered. It should also be noted that there is a special class of state-space model, termed 'non-smoothable', for which the one-sided estimates are optimal. Applying the smoothing procedure gives no further reduction in uncertainty. One important case of this occurs when the state does not evolve in time ($\mathbf{Q} = \mathbf{0}$ and $\mathbf{F} = \mathbf{I}$). This will occur if the state comprises a fixed set of parameters that are to be estimated from the time series, **c**. This case corresponds to the recursive implementation of linear-regression analysis and the estimation equations (equations (11.2.3a)–(11.2.3c)), which are simplified forms of (4.4.2a)–(4.4.2c) are given in Chapter 11 and Box 4.4. Some tracer-inversion studies have used this formalism.

Overview

Some of the advantages of state-space modelling and estimation are the following.

- The recursive form of the estimation is computationally compact.
- The technique can handle non-stationary processes. In particular, it can handle missing data, either as $\mathbf{R}(n) \to \infty$ or $\mathbf{H}(n) = 0$. Either choice leads to $\mathbf{K}(n) = \mathbf{0}$.
- The technique can handle multiple time series and interpret them in the light of model relations between the series.
- The technique can be extended to apply to weakly non-linear problems. Gelb [173: Chapter 6] describes the 'extended Kalman filter'. The time-evolution is expressed using a non-linear relation, $\mathbf{x}(n + 1) = \mathbf{F}(\mathbf{x}(n))$, while a linearisation of $\mathbf{F}(.)$ is used to propagate the covariance matrix. A non-linear relation $\mathbf{H}(\mathbf{x})$ can be used to relate the state to observations, while the gain matrix is defined in terms of a linearisation of $\mathbf{H}(.)$. This procedure has become increasingly widespread in assimilation of meteorological data.

The main disadvantages of the state-space approach are as follows.

- The state-space estimation is non-optimal unless the rather complicated two-sided Kalman-smoothing formalism is used.
- The covariance matrices $\hat{\mathbf{P}}(n)$ apply to $\hat{\mathbf{g}}(n)$, i.e. to the state estimates at a single time-point. Information about correlations between $\hat{\mathbf{g}}(n)$ and $\hat{\mathbf{g}}(n')$ is not generally produced by state-space estimation. Of course, if an augmented state $\mathbf{g}^*$ is defined, as in the implementation of fixed-lag smoothing, then such correlations will be calculated within the lag interval.

There are many ways in which tracer-inversion problems can be expressed in terms of the state-space formalism defined here.

- Section 11.3 shows how a number of inverse problems of deducing trace-gas sources can be expressed as state-space estimation problems.
- Section 11.2 considers the technique of mass-balance inversion in a state-space form, noting that the inversion technique differs from the Kalman filter and so will in general be sub-optimal (even within the class of one-sided estimates).
- The relation between Green-function forms of inversion and state-space forms is noted in Section 11.2.

■ In the second part of this book, covering recent applications, the Kalman filter appears mainly in connection with studies of CFCs [34, 197, 187] and the regional inversion of CH_4 sources [450].

Further reading

■ Priestley's book [374] has been my main reference for time-series analysis.

■ Gelb [173] covers the Kalman filter and its extensions.

■ Box and Jenkins [46] is the classic account of ARIMA modelling.

■ Draper and Smith [114] give a comprehensive account of regression analysis.

■ Some of the key references on splines are the analytical studies by Silverman [435, 436], the book by de Boor [40], which includes listings of a number of Fortran routines, and the book by Wahba [495].

Exercises for Chapter 4

1. Given an AR(1) process for points 0, 1, 2, 3, 4, . . . , with parameters a and Q, express the subprocess for points 0, 2, 4, . . . , as an AR(1) process with parameters a' and Q', giving a' and Q' as functions of a and Q.

2. From the previous problem, show that the spectral density for the subprocess is as expected from aliasing of the spectral density of the original process.

3. Equation (4.4.2d) gives the prediction error-covariance matrix for arbitrary gain, $\mathbf{K}$. If the optimal gain, given by (4.4.2e), is used, then the prediction error covariance can be reduced to the simpler form $\hat{\mathbf{P}} = (\mathbf{I} - \mathbf{KH})\tilde{\mathbf{P}}$. Prove that this follows from (4.4.2e) and (4.4.2d).

4. Show how the case of AR(1) correlated errors in scalar measurements can be incorporated by an expansion of the state-vector. Show that this is equivalent to the standard form when the correlation goes to zero.

Chapter 5

Observations of atmospheric composition

> High-quality stratospheric measurements such as those for carbon 14, retain their
> value indefinitely, but high-quality stratospheric models need to be modified every
> year or so.
>
> H. S. Johnson: Response to Revelle medal citation [76].

5.1   Measurement of trace constituents

Trace atmospheric constituents exist in the atmosphere in one of three phases: gases,
liquids (where minor constituents dissolve in water droplets) and solids (as components
of aerosol particles). These different forms have a wide range of physical and chemical
properties, leading to a range of atmospheric lifetimes ranging from milliseconds to
millennia.

The main aspects of trace-gas measurement that need to be considered are

  (i) the sampling technique;
 (ii) the physico-chemical basis of the measurement technique; and
(iii) the laboratory requirements for achieving the requisite accuracy and precision.

5.1.1   Sampling

Some of the main divisions in the type of sampling are the following.

Flask sampling. One of the simplest strategies is that in which air is collected in
containers and returned to a central laboratory for analysis. This has the
advantage of avoiding differences in the measurement calibration between

different laboratories and the additional advantage of requiring only simple technology in the field. The disadvantages of flask sampling are the low frequency of data and, for many compounds, the possibility of changes in composition either during the trapping process or during the period in transit from the sampling site to the measurement laboratory. The primary requirement is for flasks with inert surfaces, generally glass or stainless steel. However, other flask characteristics such as the valves and their lubrication have also proved to be important for some compounds.

Pre-concentration. In some cases it may be both possible and desirable to concentrate the relevant species prior to measurement. This is only practical if the required measurement is a ratio of two constituents and if the trapping acts on both constituents equally. (This is most reliably achieved by ensuring that complete trapping occurs.) One of the most common cases of such ratio measurement is that of isotopic ratios, measured by mass spectrometry or by measurements of radioactivity in the case of radionuclides. Cryo-trapping is a convenient form of pre-concentration for many species. Chemical absorption is useful in some cases such as absorption of CO_2 in NaOH solution for measurements of ^{14}C levels. Absorption onto filters is applicable for aerosols and some reactive species.

In situ **measurement.** This involves having an instrument at the sampling site so that the measurement is made *in situ*. Conventionally the term '*in situ*' is applied when an instrument measures the composition of air pumped from the atmosphere into a laboratory, aircraft or field instrument. This is to be distinguished from the 'free-atmosphere' measurements described in the following point. The primary difficulty with *in situ* methods is costs, both for the instrument itself and for operators to run it. The need for cross-calibration between different sites introduces an additional difficulty.

Free-atmosphere measurement. An alternative is to make measurements of the gas genuinely '*in situ*' in the free atmosphere. This implies some form of remote sensing and can be achieved by measurement techniques based on radiative absorption. Satellite-based measurements are of particular importance because they can provide extensive (often global) coverage. However, the primary information is often in terms of vertically integrated concentrations, with little information about the vertical distribution. An exception is 'limb-sounding', whereby radiative absorption of sunlight is measured for a range of paths with different vertical distributions, as a satellite moves to or from the shadow of the earth. This special geometry restricts the data coverage compared with forms of remote sensing that acquire data from all parts (or all daylight parts) of each orbit. Recovery of concentration data from satellite-derived radiative data involves an additional class of inverse problem that goes beyond the scope of this book [488, 492]. Burrows [60] has reviewed current techniques for remote sensing of atmospheric composition. A summary

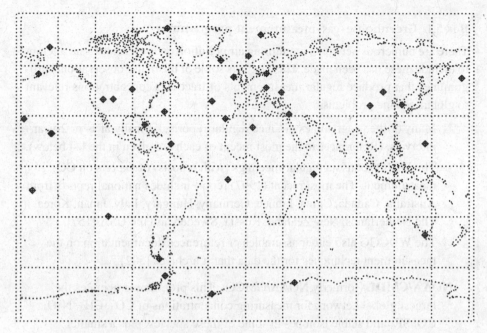

Figure 5.1 The NOAA/CMDL CO_2-observing network (fixed sites with data coverage for 1998 and 1999).

of the increasing scope for satellite observations of tropospheric chemistry is given by Singh and Jacob [437]. A ground-based form of 'free-atmosphere' measurement is the use of reflectors to give long (multi-kilometre) absorption paths to measure concentrations of weakly absorbing species such as OH, either through long baselines or using multiple reflections [199].

Each form of sampling can be done at fixed sites (fixed laboratories, fixed flask-sampling sites or long-term satellites) or on a short-term basis during intensive special-purpose measurement campaigns. Most aircraft data come from targeted campaigns, although a small but increasing number of programmes maintain long-term sampling from aircraft. Aircraft programmes can be particularly important for assessing spatial structure such as that of plumes or layers, e.g. [477]. The space shuttle has also provided a platform for short-term measurement programmes [404]. Figure 5.1 shows the NOAA/CMDL CO_2 flask-sampling network, excluding ship-based sampling, but including those sites with sufficient data to construct annual means for 1998 and 1999.

5.1.2 Measurement techniques

A wide range of techniques is used for measuring the concentrations of atmospheric constituents. They are grouped here in terms of the physical or chemical properties underlying the measurement technique.

Box 5.1. Greenhouse-gas-measurement programmes

There are numerous networks monitoring air pollution on local scales – no attempt to include these has been made. This box lists some of the major observational programmes that produce regular measurements of trace-gas concentrations relevant to global-change problems.

- Many of the programmes produce regular reports, typically at 1- or 2-year intervals. (References to the most recent of each are given in the list below).
- The WMO sponsors regular meetings (with reports) of experts on CO_2 measurement. The most recent (1997) report includes national reports from Australia, Canada, China, France, Germany, Hungary, Italy, Japan, Korea, The Netherlands, New Zealand, Russia, Sweden and the USA [157].
- The WDCGG also either assembles or references documentation on the measurement techniques for the data that it archives [502].

NOAA/CMDL (formerly NOAA/GMCC). This programme operates the largest global network for measuring concentrations of CO_2, CH_4, N_2O, CO, SF_6 and stable isotopes in some of these species, plus a smaller halocarbon network [427].

CSIRO. The CSIRO operates the Global Atmospheric Sampling Laboratory (GASLAB) with the second-largest flask network [160], emphasising high southern latitudes. GASLAB provides calibration and quality assurance for Cape Grim Baseline Air Pollution Station [479], where measurement programmes are operated by a number of groups.

SIO. The CO_2-measurement programme of the Scripps Institution of Oceanography (SIO) has been expanded from Keeling's original measurements at Mauna Loa and the South Pole to a network of Pacific stations measuring concentrations both of CO_2 and of $^{13}CO_2$ [248].

AGAGE. The AGAGE programme measures concentrations of CFCs and their replacements at five sites at frequencies of several measurements per hour. Concentrations of additional compounds such as CO and N_2O and CH_4 have also been measured. The history of the objectives, measurement techniques and results has been reviewed by Prinn *et al.* [382].

Japan. Several groups in Japan operate CO_2-measurement programmes and report their results to the WDCGG.

Radiative absorption. A number of trace gases can be detected through their radiative absorption. Of course, for many of them it is their radiative properties that are the reason for environmental concern. Concentrations of CO_2 have been measured mainly by infra-red absorption; concentrations of O_3 are measured mainly by ultra-violet absorption. Concentrations of the OH radical are measured by long-path (multi-kilometre) absorption techniques.

Fourier-transform infra-red (FTIR) instruments capture an entire infra-red absorption spectrum and so provide measurements of all species that have absorption features that are sufficiently strong and distinct from those of other constituents.

Refractive index. Keeling [251] recently used the refractive index of air, measured by an interferometer, to determine the oxygen : nitrogen ratio.

Radioactivity. There are several radioactive nuclides that characterise biogeochemical processes and/or atmospheric transport and chemistry. Measurement involves counting the decay of either the species of interest or one of its decay products.

^{222}Rn is of importance due to its contribution to background environmental radioactivity. An additional interest in ^{222}Rn is its role as a diagnostic for the recent land origin of air masses. This is because ^{222}Rn has a half-life of 3.8 days and the sources are primarily from land. ^{222}Rn has also been used as a diagnostic of vertical mixing in the troposphere.

In the period of atmospheric testing of nuclear weapons, concentrations of many artificial radionuclides were measured. Among the results of such measurements were detection of tests, information about atmospheric circulation and characterisation of devices via isotopic signatures.

One of the most important radionuclides in biogeochemical studies is ^{14}C. This has a half-life of 5730 years and is produced naturally in the stratosphere. Therefore ^{14}C acted as a 'clock' for processes on time-scales of centuries to millennia, with ^{14}C levels indicating the time since the carbon was part of the atmospheric reservoir. Nuclear-weapon testing, particularly in the period 1961–2, almost doubled the amount of ^{14}C in the atmosphere. The progression of this ^{14}C pulse through the biota and oceans gives additional information on the time-scales of biogeochemical processes. Because of the small proportions and the long half-life, measurement of concentrations of ^{14}C has required long counting periods and relatively large samples, as well as extreme care in avoiding contamination. More recently, accelerator mass spectrometry (AMS) has allowed ^{14}C measurements on much smaller samples than had previously been possible.

Chemistry. An important class of techniques involves chemical reaction of the target constituent followed by detection of the product, often by optical or spectroscopic means. One example is gas chromatography, in which the peaks for CO_2, CO and CH_4 are separated and then CO_2 and CO are converted to CH_4 and detected quantitatively by FID.

Gas chromatography. Chromatography is actually used as a separation and pre-concentration technique. Once peaks of different gases have been separated, detection uses flame-ionisation detectors or electron-capture detectors. In GCMS instruments, a mass spectrometer is used for quantitative determination after initial chromatographic separation.

Mass spectrometry. Molecules can be ionised and then separated by mass using a combination of electric and magnetic fields. The principles are described by Criss [85]. Mass spectrometers used in biogeochemical studies can generally only separate masses rounded to integers (relative to ^{12}C). This means that, for example, $^{13}C^{16}O_2$ cannot be distinguished from $^{12}C^{17}O^{16}O$. Generally mass spectrometers are used for isotopic studies, but they have also been used to determine oxygen : nitrogen ratios [27, 270] as an alternative to interferometric techniques.

Paramagnetism. Manning *et al.* [302] recently described how a paramagnetic-oxygen analyser can be configured in a form suitable for continuous monitoring of oxygen levels at a precision of 0.2 ppm.

5.1.3   Measurement issues

Other issues associated with measurement are the following.

Drying. There are two main reasons why drying of air samples can be required. The first is to prevent interference with the actual measurement process. An example is the need to dry air prior to measuring the concentration of CO_2 by infra-red absorption. Secondly, drying of air may be required in order to prevent further chemical reaction (or isotopic exchange) in a flask sample. The main techniques for drying air samples are chemical drying with hygroscopic compounds such as magnesium perchlorate [156] and cryogenic drying for compounds with very low boiling points.

Determining the calibration. Many of the measurement techniques described above compare a sample with a reference gas in order to obtain the requisite precision. These include the NDIR instruments used to measure concentrations of CO_2, mass spectrometers and gas chromatographs. To produce concentration values, it is first necessary to establish the relation between the signal ratio and the concentration ratio (there may be significant non-linearity, particularly with some older CO_2-measuring instruments). It is then necessary to link the concentration of the reference gas back to an international standard. For CO_2, what is required is to define standards to an accuracy of 0.1 ppm by volume or better for concentrations of over 300 ppm by volume. The calibration has to be maintained across a series of local calibration gases, through to local standards and ultimately tied back to international standards. This degree of accuracy needs to be maintained over periods of many decades.

As well as the difficulties of maintaining standards under laboratory conditions, additional problems arise when one is propagating these standards to multiple sites, often in remote locations with associated logistical difficulties. Achieving comparability among different observational networks involves an extra degree of difficulty. However, it is extremely important for the reliability of the type of inverse calculations described in this book, since it is *differences* in concentration that provide the information. Cross-comparisons based on circulating high-pressure cylinders between laboratories indicate that discrepancies persist. For details of the historical developments see Pearman [353], Keeling *et al.* [246] and WMO [511]. The task of achieving the requisite inter-comparability is addressed by the WMO Experts Group on CO_2 Measurement and is described in their reports, most recently [157]. Masarie *et al.* [310] describe an on-going project aimed at assessing and improving the degree of inter-comparibility between the CSIRO and NOAA/CMDL networks. The process is one of timely detection (and removal through use of independent information) of systematic operational differences. Da Costa and Steele [92] describe new analyser developments that should reduce calibration problems by achieving greatly improved precision through instrument stabilisation. It also greatly reduces the amounts of reference gas because of the low gas-flow through the instrument, thereby reducing the number of reference tanks (and associated calibrations) that are needed. It can also greatly facilitate inter-network comparisons by removing the requirement to ship high-pressure cylinders.

Sensitivity to conditions. Instruments vary in the extent to which they are sensitive to environmental conditions such as temperature, pressure, humidity and vibration. Highly sensitive instruments require carefully controlled laboratory conditions. The 'carrier-gas effect' [352] in CO_2-concentration measurements is another example of a confounding influence on trace-gas measurement.

Size and portability of instrument. There are several physical and operational characteristics of instruments that are important. Size and weight will be important restrictions for aircraft or satellite use and also at remote locations. In addition, requirements for working reference gases, electrical power and intervention by the operator can impose restrictions on the locations at which measurements are practical.

Note that, in this analysis, it is important to distinguish between the precision, which reflects the degree of reproducibility of the measurements, and the accuracy, which reflects the correctness of the values obtained. In interpreting trace-gas concentrations, most of the information lies in differences in concentration in space and/or time. Inter-network precision is the most important requirement for inversion studies, although absolute accuracy tied back to fundamental physical standards is important for ensuring confidence when one is comparing measurements taken over very long time periods.

5.2 Data processing

The raw data from measurements are generally subjected to a number of transformations before being used in interpreting atmospheric-constituent budgets. Enting and Pearman [141] identified the main processes as

editing, to remove faulty data related to known problems such as malfunction of instruments, routine maintenance and local contamination;

calibration, to convert the data to an international standard concentration scale;

selection, to remove data that are believed not to represent baseline conditions; and

averaging, to reduce the effects of 'noise' and to reduce the volume of data for publication.

The 'editing process' is a vital part of an operational trace-gas measurement programme, but lies outside the scope of this book, as do the issues of calibration noted in the previous section. Averaging and similar processing of data are discussed in the following section. For now, we present a discussion of selection of data, updated from [141]. This description of selection procedures that have been used operationally or analysed in investigative studies provides a reference point for the discussion in Section 9.3 on the role of selection of data in matching observations to models.

Even at the best of surface sites, it is necessary to perform some degree of selection of data in order to remove local influences on the concentrations. Often, the aim has been to produce so-called 'baseline data' that are, conceptually, representative of large-scale air-masses. In practice, this conceptual definition has to be replaced by operational definitions tailored to each site. For example, Figure 5.2 shows hourly mean CO_2 data from Cape Grim, Tasmania. It shows two types of behaviour: relatively steady 'baseline' levels of around 360 ppm, from air-masses from the southern ocean, and the 'pollution events', in which the air-mass has recently passed over land.

The criteria for selection need to be based on extensive studies of the variability at the site. Some of the main criteria that are used (or could be used) are the following.

Wind direction. At many sites, particularly coastal sites, a 'baseline sector', usually with a large ocean fetch, is defined so that only data with winds from that sector are taken as the baseline. For example, Baring Head in New Zealand uses a southern sector, whereas Cape Grim in Tasmania, Australia uses a south-westerly sector. Examples of studies of selection based on wind direction are for Mt Cimone [70] and Izaña [426].

Variability of data. The degree of variability within short runs of continuous analyser data is used as a selection criterion at several sites. For example, at Izaña in the Azores the variability within each half-hour period is used. The Cape Grim selection criteria are applied to hourly means. An hourly mean is selected as 'baseline CO_2' if it is part of a run of 5 h during which none of the other four hourly means differs from it by more than 0.3 ppm by volume. The

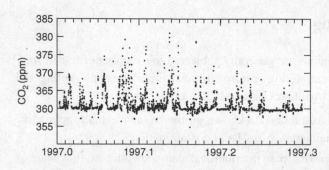

Figure 5.2 Several months of hourly mean CO_2 data from Cape Grim, Tasmania [L. P. Steele, personal communication, 2001].

NOAA continuous-analyser programmes use both within-hour variability and (except at the South Pole) between-hour variations as selection criteria.

Wind-speed. In order to reduce contamination from local sources and sinks, a minimum wind-speed is often specified. For Amsterdam Island, the minimum that is used depends on the sector. Early reports of provisional data from Cape Grim used wind-speed and direction as the baseline-selection criteria.

Wind consistency. The persistence of the wind direction is another criterion that can indicate large-scale air-masses.

Particle count. Particulate counts (cloud-condensation nucleii) are used at Cape Grim as part of the sampling criteria for flask programmes, with high particle counts indicating land-contact to be excluded.

Flask consistency. Another approach, used in the NOAA/CMDL flask-sampling programme, is to take pairs of samples in the same sampling session and reject data when concentrations from the flasks disagree, assuming this to be due to highly variable local contamination and/or operator or handling error.

Air-mass trajectories. Computed air-mass trajectories are seldom available for routine use in selection of data but have been used in a number of exploratory studies.

^{14}C. Levin [281] has described the use of ^{14}C data to remove data affected by local fossil-fuel sources in the CO_2 record from Schauinsland (a mountain above the Rhine valley).

Radon. As noted above, ^{222}Rn (with a half-life of 3.8 days) provides a way of distinguishing air-masses that have recently passed over land since the radon sources from land are much greater than those from the ocean. For example, exploratory studies using radon data have been undertaken for CO_2 records from Amsterdam Island [172], Schauinsland [281] and Cape Grim.

Other constituents. Anomalous concentrations of other atmospheric constituents in flasks can be indicative of contamination. In some CSIRO programmes anomalous CO_2 data have been used as a criterion for rejecting other data from the same flask.

The determination of appropriate selection criteria for a particular site requires comprehensive studies of the variability in concentration of CO_2 at the site. While some of these studies can be undertaken while assessing the *a priori* suitability of the site, a full evaluation requires the data from an operational period of several years.

Beardsmore and Pearman [26] have reviewed the reasons for the choice of the current operational definition of Cape Grim baseline data. Results of later studies [128 and references therein] have suggested that a narrower sector of wind directions would reduce the variability in the Cape Grim data. For Mauna Loa, Keeling *et al.* [247] have discussed local influences on the record from the Scripps Institution of Oceanography (SIO) programme. Thoning [475] describes the selection criteria for the continuous-analyser programme of the National Oceanic and Atmospheric Administration (NOAA).

With flask sampling, the limited amount of data restricts the possibilities for detailed studies of the selection criteria. Tans *et al.* [466, 468] addressed this problem by looking at the statistics of flask samples from the NOAA sites where continuous analysers are operated. They obtained estimates of the statistical errors expected from the limited sampling rate by studying the statistics of sets of pseudo-flask data, sampled from the continuous-analyser record. They then estimated the experimental errors by comparing the real flask data with the continuous data. They concluded that, while long-term averages were in good agreement, interannual differences could be as large as 0.4 ppm by volume.

Apart from the greater frequency of continuous-analyser data compared with flask data, an important advantage of the continuous data is the ability to go back over the record retrospectively and change the selection criteria. With flask sampling, most of the selection takes place in the definition of sampling conditions – once the sample has been taken, these sampling conditions cannot be changed retrospectively. The continuous data also contain information about more local details of the source–sink distribution and, as described in Chapter 17, increasingly many studies are making use of such information.

In the past, global-scale modelling has concentrated on interpreting the baseline concentrations. More recently studies on the continental and sub-continental scale have been based on interpreting the perturbations about the baseline, as described in Chapter 17. With improvements in resolution of global models, a convergence of scales can be expected, with global models making use of the full concentration records.

5.3   Data analysis

As well as the editing and selection processes noted above, many of the reported concentrations of greenhouse gases have been subject to additional processing/analysis of data. The vast majority of data sets on greenhouse-gas concentrations come in the form of time series: sequences of measurements from particular sites. Therefore analyses have emphasised time-series techniques of the type discussed in Chapter 4.

Data analysis is performed for two main reasons:

to report data in a compact standard summary form; and

for interpretative studies.

In interpretative studies, the form of the study will tend to prescribe appropriate forms of statistical analysis. However, for reporting of data, the expectation is that the data will later be used in interpretative studies whose form is unknown at the time of reporting the original data and so there is little scope for optimally targeted forms of statistical analysis.

Some of the summary statistics that are commonly reported are the following.

Monthly (or annual or other) means. These are often the starting point for
    further processing into one of the forms listed below, or for input into

modelling studies. If there is an error model for the original data and if the averaging process is known, then the error statistics for such means can be calculated. More commonly, the monthly means will be used to derive an empirical 'noise' model as discussed in the following sections.

Linear or polynomial trends. These can be fitted by regression analysis. This is really only suitable for short records.

Long-term trends. These can be estimated without assuming specific functional forms by applying low-pass filters to the records. A closely related technique is the use of smoothing splines (see Box 4.2). These act as low-pass filters with a relatively broad transition band. Advantages of smoothing splines are that they do not require equi-spaced data and that the filtering characteristics are relatively insensitive to variations in density of data.

Mean cycles. These can be determined by regression fits of periodic functions, which are often characterised by a small number of Fourier components. An alternative characterisation of an annual cycle is as mean monthly offsets from a long-term trend.

Varying cycles. Estimates of the time-variation of a seasonal (or other) cycle can be obtained using the technique of complex demodulation (e.g. [474]).

Other. Additional possibilities include using results of covariance/correlation analyses, spectral estimation and cross-correlation/cross-spectra.

This book emphasises that any statistical analysis requires, whether explicitly or implicitly, a statistical model. Implicit in most statistical analyses of trace-gas concentrations is the characterisation of the records as consisting of the sum of one or more 'signals' plus 'noise'. Such 'additive' models are usually appropriate given the additive nature of the principle of conservation of mass.

Starting from monthly means, each of the other types of data processing listed above can be related to the digital filtering described in Section 4.2. (Polynomial fits can be considered as the limit of a running polynomial fit when the averaging length equals the length of data and similarly estimates of fixed cycles can be regarded as the totally smooth limit of complex demodulation.) Digital filtering is particularly appropriate for stationary time series and the procedures correspond to estimating a signal $x(t)$ with spectral density $h'_x(\theta)$ in the presence of additive noise with spectral density $h'_n(\theta)$. As noted in Section 4.2, for any filter with response $\psi(\theta)$ we have the mean-square error of the estimate given by

$$E[(\hat{x}(t) - x(t))^2] = \int_{-\pi}^{\pi} \left[|1 - \psi(\omega)|^2 h'_x(\omega) + |\psi(\omega)|^2 h'_n(\omega) \right] d\omega \qquad (5.3.1)$$

In all of the types of data processing listed above, the accuracy with which these components can be extracted from the observations depends on the statistical characteristics of the 'noise' term. In particular, how likely is the 'noise' to simulate an apparent signal? This needs to be analysed in terms of (5.3.1). In contrast, the standard formulae for estimates of uncertainty for techniques such as regression fits are usually based on the assumption that the errors are independent.

One extensive form of data processing of trace-gas data is the use of NOAA CO_2 records to produce an extended data set with a spatial coverage uniform in time – see Box 5.2. This has been an important data set for mass-balance inversions. However, the extent of the processing makes the statistical characteristics of this data set difficult to ascertain.

Box 5.2. GLOBALVIEW

The NOAA/CMDL programme has the most wide-ranging of the networks for sampling greenhouse gases. The development of the network is on-going and there have been several periods of major expansion in coverage. In a few cases, sampling at particular locations has been discontinued. A consequence of these developments is that the coverage of the network is not uniform. This presents problems for a number of techniques for analysing these data.

In order to address the problem of variations in data coverage over time, Masarie and Tans [309] have undertaken an activity described as 'data extension'.

Missing data are replaced by adding an offset to reference values based on nearby sites. The reference is defined from LOESS (robust locally weighted regression [73]) fitted to marine boundary sites as a function of latitude each 1/48 year. The offset between a particular site and the reference curve is defined as a mean plus 12-month and 6-month Fourier components, fitted for periods when data were available.

The initial construction of an 'extended' CO_2 data set has been followed by a similar data extension of the NOAA/CMDL CH_4 data set and other species are being added.

The construction of a data set that is uniform in time is particularly helpful in mass-balance inversions. The specific form of the 'GLOBALVIEW' extended data sets is less suitable for inversion techniques that are treated as statistical estimation. Essentially the GLOBALVIEW approach is making some assumptions regarding smoothness about the sources. Ideally, analyses should, if required, make such assumptions explicitly rather than have them implicit in the data extension. However, for many inversion techniques, particularly those involving one-sided estimation, the data-extension approach seems definitely preferable to ignoring the issue of changes in the sampling network. Furthermore, the overall activity of integrating data sets is highly desirable, particularly when it is extended to the integration of data sets from several institutions.

As an example of the difficulties involved in uncertainty analysis, the bootstrap sensitivity analysis presented by Masarie and Tans generated synthetic samples by selecting 100 subsets of network sites and performing a separate data extension on each of the 100 sets and then using these in the inversion.

5.4 Error modelling

5.4.1 Classification

The attempt to achieve a comprehensive uncertainty analysis is the major theme of this book. A valid uncertainty analysis is essential for the credibility of policy decisions based on greenhouse-gas science. The use of uncertainty analysis as the basis for experimental design, as described in Chapter 13, provides an additional reason for a careful uncertainty analysis. Morgan *et al.* [324] present a very general overview of the origins of uncertainty; some of their key points are summarised in Box 5.3. Of the causes that they identify, those most relevant to trace-gas measurements are statistical variation/inherent randomness, variability and approximation.

Box 5.3. Sources of uncertainty

Chapter 4 of *Uncertainty* by Morgan *et al.* [324] covers 'The nature and sources of uncertainty'. Although their scope is wider than seems relevant to the inversions discussed in this book, some of these wider considerations are relevant to policy-related applications of results from studies of greenhouse gases and so some key points are summarised here, together with the recommendation to read the book by Morgan *et al.*

They identify different types of variable for which different sources and treatments of uncertainty will apply:

- ▪ empirical quantities, generally requiring a probabilistic treatment of uncertainty;
- ▪ defined constants, which are taken as certain by definition;
- ▪ decision variables, treated parametrically;
- ▪ value parameters, treated parametrically;
- ▪ index variables (e.g. position and time) taken as certain by definition;
- ▪ model-domain parameters, treated parametrically;
- ▪ state quantities, whose treatment will depend on the form of the inputs; and
- ▪ outcome criteria, whose treatment will depend on the form of the inputs.

The parametric treatment involves a sensitivity analysis but not the definition of a probability distribution. The argument is that these quantities have no 'true value' to be represented by a probability distribution.

For the empirical quantities a probabilistic treatment is recommended. Generally this probabilistic treatment will propagate to state variables of the model and the outcomes. The sources of uncertainty in empirical quantities were identified as

- ▪ statistical variation (of measurements) and inherent randomness (of the empirical quantity), both of which can be treated by conventional statistics;

- variability of the quantity, noting that the frequency distribution (describing the variability) is generally not the same as the probability distribution describing the uncertainty;

- approximation, whereby the error can often be assessed only by analysing some cases without the approximation;

- subjective judgment, noting that estimates of systematic bias seem to be frequently subjective under-estimates; and

- disagreement (often from differences in perspective) and linguistic imprecision (especially in quantifying qualitative information).

The definition of 'error' is dependent on the context. As noted in the introduction, the statement by Enting and Pearman [141] that "Any variability that cannot be modelled, whether synoptic variation, interannual variation or longitudinal variation (if using a two-dimensional model) must be treated as part of the 'noise', acting to reduce the signal-to-noise ratio" should be reworded as *any variability that cannot be modelled deterministically needs to be modelled statistically*. Enting and Pearman further noted that "much of the so-called 'noise' in the observations represents contributions from sources that are not properly treated, either due to model error or due to use of a truncated representation of the source distribution".

The basis of much of the uncertainty analysis is treating the observations in the form of a signal plus noise and then propagating the noise distribution through the estimation procedure. In Chapter 3, we noted that $p(\mathbf{c}|\mathbf{x})$, which describes the statistics of data error, could include both measurement error and model error. In many cases this distinction can become blurred.

The issues can be clarified by considering the main contributions to the spread in $p(\mathbf{c}|\mathbf{x})$ as

(i)   measurement error associated with the laboratory procedure;

(ii) sampling error relative to the inherent variability (in both space and time) of the concentrations;

(iii) model-truncation error, including both quantitative limitations on the resolution of a model (either in transport or in source discretisation) and quantitative changes when processes are ignored or highly parameterised; and

(iv) transport-model error.

Clearly (i) should be classified as a data error and (iv) as a model error. Cases (ii) and (iii) could be classified either way. The categories shade into one another. It is a matter of choice whether components of the data that cannot be represented by a model should be treated as due to limitations of the model (truncation error in the general sense used above) or as model error. One particularly important case to note is that the use of GCM winds means that synoptic variation must be treated as some form of noise or error. This is discussed further in Section 9.3.

A number of points follow.

- The amount of the 'error' terms depends on how much of the variation in concentration is modelled. The more that can be modelled the smaller the error component. As a trivial example, the use of transport modelling reduces (and ideally removes) the representativeness error that occurs when one is estimating global trends from a small number (e.g. one) of time series.

- Because of the ambiguities in the classification, there is a need to avoid double counting when model error is considered explicitly.

5.4.2   Time-domain models

There have been several studies that have attempted to classify the variability in the CO_2 records as signal plus noise. One approach is to construct a statistical description (or model) of the total CO_2 record, recognising that the division into signal and noise will depend greatly on the modelling context. In some cases the form of the statistical model (e.g. breaks in the power spectrum) can suggest an empirical signal/noise classification that could apply across a wide range of modelling contexts. Martín and Díaz [307] produced a model based solely on the CO_2 time series that took the form of an ARIMA (autoregressive integrated moving-average) model:

$$[1 - \phi_1 \mathcal{B} - \phi_2 \mathcal{B}^2 - \phi_3 \mathcal{B}^3]\nabla_{12}\,\nabla_1[c(t) - \bar{c}] = [1 - \alpha_1 \mathcal{B}^{12}][1 - \alpha_2 \mathcal{B}]\epsilon(t)$$

$$(5.4.1)$$

where $\mathcal{B}$ is the backward shift operator, ∇_1 is the one-step difference operator and ∇_{12} is the 12-step difference operator. Tunnicliffe-Wilson (1989) modelled the monthly mean Mauna Loa CO_2 data series from NOAA as an ARMA (autoregressive moving-average) model

$$[1 - \mathcal{B}][1 - \mathcal{B}^{12}]c(t) = [1 - 0.49\mathcal{B}][1 - 0.86\mathcal{B}]\epsilon(t) \qquad (5.4.2)$$

with var $\epsilon = 0.09$ but this was a model for the series as a whole, not just the irregular component.

As a variation on this approach, authors of some studies have performed decompositions of CO_2 time series into 'regular' and 'irregular' components and produced time-series models of the 'irregular' component. Table 5.1 summarises some time-series models that have been proposed for the 'irregular' or 'noise' component of monthly mean CO_2 data at Mauna Loa and, in one case, for the South Pole. These models are to some extent based on different assumptions about what constitutes the 'regular' component(s). Thus some of the 'noise' variability captured by some of the error models will represent variations that other studies would regard as 'signal'.

Cleveland et al. [74] represented the irregular components of the SIO Mauna Loa and South Pole records as first-order autoregressive (AR(1)) processes, as indicated in Table 5.1. They noted that this model did not capture the ENSO-scale variations in the long-term growth.

Table 5.1. *Models of the irregular or noise component, n_j, of various monthly CO_2 time series, expressed in terms of white-noise forcing ϵ_j.*

Site	Model	var ϵ	Reference
Mauna Loa	$n_{j+1} = 0.63n_j + \epsilon_j$	0.066	[74]
South Pole	$n_{j+1} = 0.72n_j + \epsilon_j$	0.049	[74]
Mauna Loa	ϵ_j	0.23	[458]
Mauna Loa	$n_{j+1} = 0.48n_j + 0.18n_{j-1} - 0.12n_{j-2} + \epsilon_j$	0.063	[307]

Surendran and Mulholland [458] analysed the SIO Mauna Loa time series by fitting an exponential and 12-month and 6-month sinusoids. They then used an AR(2) model to capture the variability associated with the ENSO phenomenon. They suggested that the residuals could be represented as white noise with a variance of 0.2303.

Martín and Díaz [307] modelled the CO_2 data in terms of fossil fuel and sea-surface temperature and fitted the residuals as an AR(3) model:

$$[1 - 0.48\mathcal{B} - 0.18\mathcal{B}^2 + 0.12\mathcal{B}^3][n(t) - \bar{n}] = \epsilon(t) \tag{5.4.3}$$

with var $\epsilon = 0.063$.

Manning [303] used estimated power spectra for a range of sites to argue in favour of decomposing CO_2 time series into a 'trend' representing time-scales over 5 years, a slowly varying seasonal cycle and an 'irregular' component that includes both inter-annual variability and high-frequency components.

Thoning *et al.* [476] constructed an AR(1) model of the irregular component of *daily* mean data from the NOAA/GMCC Mauna Loa record. They did not quote the autocorrelation but, using the theoretical AR(1) power spectrum (equation (5.4.4a) below), it can be estimated from their Figure 6 as $a \approx 0.8$. This would imply an autocorrelation of order 0.001 for monthly data. Clearly the synoptic and subseasonal variation modelled by Thoning *et al.* is only a very small part of the variability identified by Cleveland *et al.*

Figure 5.3 shows some of the power spectra for these error models, plotted as a function of frequency $\nu = \theta/(2\pi \Delta t)$. These use (see, for example, Priestley [374: Section 4.12.3])

$$h'(\theta) = \frac{\text{var } \epsilon}{2\pi} \frac{1}{1 - 2a\cos\theta + a^2} \tag{5.4.4a}$$

for AR(1) models, or the more general expression

$$h'(\theta) = \frac{\text{var } \epsilon}{2\pi} \frac{1}{\left|1 - \sum_{j=1}^{N} a_j e^{ij\theta}\right|^2} \tag{5.4.4b}$$

for AR(N). Also shown in Figure 5.3 is the spectrum of the MA(1) model obtained from the singular-value analysis described below.

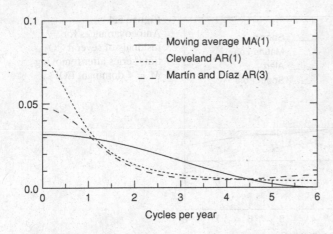

Figure 5.3 Power spectra for error models of monthly mean Mauna Loa time series, as described in Table 5.1 by Cleveland *et al.* [74] and Martín and Díaz [307] and the moving-average model suggested by the EOF analysis of equations (5.4.4a) and (5.4.4b).

5.4.3 Relations between time series

An alternative approach to separating the signal and noise components of CO_2 time series is to identify common components, regarded as signals, of multiple time series. Those components that have a sufficiently large spatial scale to appear at multiple sites are aspects for which regional interpretation could reasonably be sought.

The computational procedure is to write the time series $c_x(t)$ at a set of locations, x, as a combination of spatial distributions, u_{xq}, with time series $v_q(t)$ for $q = 1, 2, \ldots$:

$$c_x(t) = \sum_q u_{xq} \lambda_q v_q(t) \tag{5.4.5a}$$

which is the singular-value decomposition of $c_x(t)$ (see Box 3.2). This technique is commonly used in the earth sciences under the name of 'empirical orthogonal functions' (EOFs). In analysing monthly mean CO_2 data, it was found that initially the singular values decreased rapidly, followed by a slow decline as q increased. The relatively slow decrease is characteristic of the singular values for independent time series. From these relations, residual series $r_x(t)$ were constructed by removing a small number of the largest modes:

$$r_x(t) = c_x(t) - \sum_{q=1}^{M} u_{xq} \lambda_q v_q(t) \tag{5.4.5b}$$

for $M = 4, 5$ and 6. Figure 5.4 shows the sample autocovariances for several of the residual series. Most of them (and the others not shown) exhibit a rapid drop to near zero autocorrelation after lags of 2 months or more. This suggests a MA(1) model as a common model of this residual variation. Problem 2 of this chapter suggests how this form might arise as a result of constructing monthly means of autocorrelated weekly data. The residuals for Samoa were anomalous, with very slowly dropping autocorrelation. This is most plausibly interpreted as an effect of interannual variation in transport, changing the extent to which Samoa sees northern- versus southern-hemisphere air. Since this is a characteristic that affects other sites very

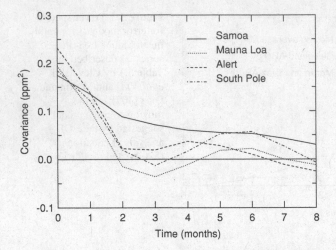

Figure 5.4
Autocovariances for residuals of several CO_2 time series after removing $M = 4$ dominant EOFs.

much less, the EOF analysis left this variability in the residuals. When one is interpreting CO_2 data using a model with a realistic representation of interannual variation in transport, this long-term CO_2 correlation should be treated as part of the signal. However, in a model without realistic interannual variations, an appropriate 'noise' model of CO_2 at Samoa would include long-term correlations of the type shown in Figure 5.4.

There have also been studies that sought relations between the statistical characteristics of different gases. The classic relation is that of Junge [236], which related the relative variability to atmospheric lifetime as

$$\sqrt{\text{var}[c_\eta]}/\bar{c}_\eta \approx 0.14\lambda_\eta \qquad (5.4.6a)$$

where λ_η is the inverse of the lifetime, in years, of species η. This relation was fitted by a range of compounds with lifetimes from 0.01 to 100 years.

A more extensive study was presented by Jobson *et al.* [232] who found coherent behaviour of the form

$$\sqrt{\text{var}[\ln c_\eta]} \propto (\lambda_\eta)^\beta \qquad (5.4.6b)$$

where generally β was smaller than the value unity corresponding to the Junge relation. In studies spanning a range of conditions, both in the troposphere and in the stratosphere, relation (5.4.6b) was found to fit a range of compounds. However, different values of β were required for different conditions.

Because of the nature of the available data sets, this chapter has concentrated on statistics in the time domain. Detailed spatial statistics of trace-gas concentrations are rarely available. Aircraft data provide one way of obtaining such statistics, e.g. [477, 65].

Other issues of error modelling are treated in the context of transport-model error, specifically resolution and source truncation in Section 8.3 and selection and representativeness of data in Chapter 9.

To summarise the current situation of error modelling in the context of tracer inversions:

- correlations in observations are important;
- the approach of working in the frequency domain [145], giving independent errors at each frequency if the original time series is stationary, is really applicable only to so-called cyclo-stationary problems;
- in time-dependent inversions a realistic error model is required; and
- for inversions that interpret the type of non-baseline data shown in Figure 5.2, stationary-time-series models are unlikely to be adequate.

Further reading

- Ayers and Gillett [17] have reviewed many of the aspects of experimental techniques involved in measurements of atmospheric composition.
- A more recent review is given by Roscoe and Clemitshaw [414]. Additional discussion is given by Mankin *et al.* [301].
- The intersociety volume [287] aims at comprehensive coverage of sampling and analysis.
- For CO_2, the reports from the WMO experts meetings (most recently [157]) communicate the 'state of the art' in issues of CO_2 measurement.

Exercises for Chapter 5

1. Assuming that one has equi-spaced data and working in the frequency domain, derive the approximation (equation (2a) of Box 4.2) for the response of the smoothing spline regarded as a digital filter.

2. If monthly average data are obtained by averaging four 'weekly' values within the month, with an AR(1) noise, what is the autocorrelation for the averages, at lags of 1 and 2 months. What value of the 'weekly' autocorrelation gives an autocorrelation of 0.5 for successive monthly averages?

Chapter 6

The sources and sinks

When in April the sweet showers fall
And pierce the drought of March to the root and all
The veins are bathed in liquor of such power
As brings about the engendering of the flower,
When also Zephyrus with his sweet breath
Exhales an air in every grove and heath . . .
  G. Chaucer: *The Canterbury Tales (prologue)*
  (*circa* 1400). Trans. N. Coghill (1951).

6.1   Classification

The principle underlying diagnostic use of modelling of tracer transport is that the
space–time distributions of trace-gas concentrations in the atmosphere reflect the
space–time distributions of their sources and sinks. Observations of these distributions can in turn provide information about the processes involved. Tracer-inversion
calculations use the relation between fluxes and concentrations in order to make specific estimates of fluxes. The possibility of using atmospheric-concentration data to
make more direct inferences about processes is addressed in Chapter 12.

In this chapter, we review the general characteristics of sources that will influence the effectiveness of such inversions. In practice, the question of which source
characteristics are important depends on the objectives of the inversion.

In considering the sources, there are two questions that we need to ask.

■ What do we wish to learn about the sources and sinks?

■ What prior information is available for Bayesian estimation?

The first question is addressed in very broad terms in the following section. To put the discussion in context, we consider some of the generic ways of classifying sources and sinks of greenhouse gases.

Location. This is the primary information required by forward modelling of atmospheric transport of trace constituents. Conversely, information about source strengths at specific locations is the most direct output from inverse modelling. The inverse problem is subject to loss of detail at small scales, as noted in Chapter 1 (see Figure 1.3) and discussed in more detail in Chapter 8. Therefore the locational information obtained from tracer inversions is generally in a 'degraded' low-resolution form.

Process. The description of sources and sinks in terms of processes can occur at many levels of complexity. A simple case is that of characterising anthropogenic compounds such as CFCs in terms of a source (described as anthropogenic, with the details outside the concern of atmospheric science) and a sink (due to photo-dissociation in the stratosphere). For purposes of the 'national inventories' required by the Framework Convention on Climate Change and Kyoto Protocol, a classification of anthropogenic sources according to prescribed categories (see Box 6.1) is required. The classification of processes becomes far more complicated for gases that also have natural sources and sinks. The most important of these are CO_2, CH_4 and N_2O. Some halocarbons such as CH_3Cl and CH_3Br also have natural sources, as do various compounds, such as CO, that have indirect effects on greenhouse-gas budgets.

In parallel with the anthropogenic categories prescribed by the Kyoto Protocol, the natural fluxes can be partitioned according to the type of system: most obviously into land versus ocean and then in further detail, especially when one is classifying terrestrial fluxes in terms of the type of ecosystem. Classification of sources according to processes is of particular importance when it allows the use of information from process models. At the simplest level, this can be in the form of relations between fluxes of different tracers, paving the way for multi-tracer inversions (see Section 10.5). Many cases of such multi-tracer inversions have been based on different isotopic species of a single compound.

Functional role. Classifying sources according to their functional role implies imposing a model-based 'conceptual' framework on the fluxes. Hence, there is a need to exercise special caution, to ensure that the concepts are consistent with the real-world system. The initial classification that we propose is to classify fluxes as 'deterministic' versus 'random' (Figure 6.1). This distinction depends on how we model the system. The discussion in Section 5.4 of the context-dependence of error-modelling of concentration data is equally applicable to fluxes. The 'random' component is whatever we cannot model with whatever deterministic model is being used. Variations in flux due to climatic variability will often be treated as random. They can, however, be

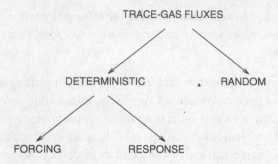

Figure 6.1 A schematic diagram of the functional classification of trace-gas fluxes.

regarded as deterministic if one has a model of climatic forcing of fluxes, driven by observed climate data, e.g. [93]. We further divide the deterministic component into 'forcing' versus 'response'. This classification is defined from the point of view of the atmosphere. The 'forcing' is the set of fluxes that do not depend (causally) on the amount of tracer in the atmosphere. The 'response' is the set of fluxes that do depend on the atmospheric concentration of tracer. Examples of 'responses' are chemical destruction (with loss rates proportional to concentrations of reactants when single molecules are involved) and dissolution of CO_2 in the oceans.

In the classification above, the deterministic components of atmospheric budgets define the (non-random component of the) rate of increase of mass of tracer as the difference between an anthropogenic forcing (generally an input) and a natural response (generally a loss). If the system behaves linearly then contributions from different times can be added. In this case, the global-mean natural responses can be characterised by response functions, $\rho_\eta(t)$, which are the proportion of a unit input of tracer, η, remaining in the atmosphere after a time t. Thus

$$M_\eta(t) = \int_{t_0}^{t} \rho_\eta(t - t')s_\eta(t')\,\mathrm{d}t' \tag{6.1.1}$$

By defining $\rho_\eta(.)$ as a proportion (i.e. $\rho(0) = 1$), we have imposed a requirement that $M_\eta(.)$ and $s_\eta(.)$ be in consistent units – $s_\eta(.)$ must have the units of $M_\eta(.)$ divided by time. Some studies incorporate conversions of units into the definition of $\rho(.)$. The inverse problem of deducing $s_\eta(.)$ given $M_\eta(.)$ is discussed in Section 6.6.

The representation in (6.1.1) can be transformed into a flux form by differentiating (6.1.1) to give

$$\dot{M}_\eta(t) = s_\eta(t) + \int_{t_0}^{t} \dot{\rho}_\eta(t - t')s_\eta(t')\,\mathrm{d}t' \tag{6.1.2a}$$

where $\dot{\rho}_\eta(.)$ is the derivative of $\rho_\eta(.)$.

Recalling that we work in units such that $\rho(0) = 1$, this can be expressed as

$$\dot{M}_\eta(t) = s_\eta(t) + \Phi_{\mathrm{resp}:\eta}(t) \tag{6.1.2b}$$

with

$$\Phi_{\mathrm{resp}:\eta}(t) = \int_{t_0}^{t} \dot{\rho}_\eta(t - t')s_\eta(t')\,\mathrm{d}t' \tag{6.1.2c}$$

Box 6.1. The Kyoto Protocol

The Kyoto Protocol was negotiated and signed in December 1997. It is defined with respect to the Framework Convention on Climate Change (FCCC). The FCCC specifies general objectives for nations – the Kyoto Protocol imposes stronger and more specific obligations.

The substances controlled by the Kyoto Protocol are carbon dioxide, CO_2; methane, CH_4; nitrous oxide, N_2O; hydrofluorocarbons (HFCs); perfluorocarbons, (PFCs); and sulfur hexafluoride, SF_6.

The Kyoto Protocol is specifically intended to exclude substances controlled under the Montreal Protocol. Furthermore, the controls apply to specific categories of anthropogenic emissions, classified into the main groups of *energy*, *industrial processes*, *solvent and other product use*, *agriculture*, and *waste*.

Some of the main operational (as opposed to procedural) features of the Convention and Protocol are the following.

- ■ The restrictions are in terms of emission targets for the period 2008–2012 for developed nations, specifically those listed in Annex B of the Kyoto Protocol.

- ■ The targets are in terms of CO_2 or equivalent CO_2. For the first commitment period, the definition of equivalence uses 100-year global-warming potentials (see Box 6.2) using the values estimated for the IPCC's Second Assessment Report. The definition could be changed for any later commitment period.

- ■ There is a requirement for parties to submit national communications including national inventories of greenhouse-gas emissions.

The use of the 100-year GWP as the definition of CO_2-equivalence implies that the greenhouse-gas cycles are being viewed in terms of a 'forcing' (anthropogenic) whose effect is reduced by the 'response' of natural systems.

The natural response is in the form of sinks. The FCCC requires parties to protect such sinks. This is of limited significance for those gases for which the main sinks are through chemical destruction in the free atmosphere, or for uptake of CO_2 by the oceans. However, for terrestrial uptake of CO_2 the sinks are in areas under national control and in many cases are subject to threat.

The Kyoto Protocol also has the potential for crediting nations with enhancement of terrestrial CO_2 sinks. The details are still being negotiated at the time of writing.

Here $\Phi_{resp:\eta}$ is the trace-gas flux from the 'response' processes, with the positive signs indicating fluxes *to* the atmosphere. Generally we will have $\dot{\rho}(t) < 0$ so that $\Phi_{resp:\eta} < 0$. More complicated non-decaying responses may occur if one considers multi-stage processes and generalises (6.1.1) to describe the response of trace gas η to the sources of another trace gas, η'.

Box 6.2. Global-warming potentials

Equations (6.1.1), (6.1.2a) and (6.1.2b) characterise atmospheric trace-gas fluxes in terms of forcing and response. This type of characterisation has assumed major policy significance because it forms the basis of the global-warming potentials (GWPs). These define the measures of equivalence between gases for use in specifying targets under the Kyoto Protocol.

The GWP aims to combine two characteristics in which the greenhouse gases differ: their ability to absorb infra-red radiation and the time that they remain in the atmosphere. Since the various gases are lost from the atmosphere on different time-scales, there is no natural time-scale over which to make the comparison. It is necessary to define a time horizon, T, over which the radiative impact is evaluated.

The starting point is the absolute global-warming potentials (AGWPs). These are defined as

$$\text{AGWP}_\eta = a_\eta \int_0^T \rho_\eta(t) \, dt \tag{1}$$

where a_η is the radiative absorption of gas η and ρ_η is the response function that describes how a perturbation to the concentration of gas η disappears from the atmosphere. This can also be written as

$$\text{AGWP}_\eta = a_\eta T \bar{\rho}_\eta(t) \tag{2}$$

where $\bar{\rho}_\eta(t)$ is the average amount in the atmosphere over the time period T.

For a specified time horizon, the GWP of a gas η is defined as the AGWP of η, relative to the AGWP of a reference gas. From (2) it follows that

$$\text{GWP}_\eta = \frac{a_\eta \bar{\rho}_\eta(t)}{a_{\text{ref}} \bar{\rho}_{\text{ref}}(t)} \tag{3}$$

The original IPCC definition used CO_2 as a reference [433]. This had the disadvantages that first, the response of CO_2 was (and is) subject to significant uncertainty; and secondly, the radiative effectiveness of CO_2 decreases with increasing concentration. The revised IPCC definition [2] used a specific model calculation of CO_2 as the reference case. This means that real CO_2 will not necessarily have a GWP of 1, although the Kyoto Protocol adopts GWPs with 100-year time horizons and a GWP of 1 for CO_2 for the first commitment period (2008–2012).

For gases with natural cycles, using equation (6.1.1) to analyse anthropogenic inputs requires that it refers to the anthropogenic perturbation to the natural cycle. For these gases, M_η is the excess above pre-industrial levels. There will often be an additional degree of natural variability that will complicate diagnostic studies.

The classification of sources as 'anthropogenic' versus 'natural' reflects the requirements of policy applications, particularly for the Framework Convention on Climate Change and Kyoto Protocol. In the terminology defined here $s_{\text{fossil:CO}_2}(t)$ then is clearly a 'forcing' of 'anthropogenic' origin. The CO_2 from this source is lost from the atmosphere by the 'natural' processes of uptake by the oceans and biota. Nevertheless, as represented by equation (6.1.2c), this uptake flux is an explicit consequence of the forcing flux. My own view is that this 'response' to an anthropogenic forcing is clearly an 'anthropogenic effect'. However, I would suggest that it is appropriate to term such fluxes 'indirect anthropogenic effects' and to distinguish them from the 'direct anthropogenic fluxes'.

The important point to appreciate is that, for many scientific or policy purposes, what is required is a 'process' or 'functional-role' description of the trace-gas fluxes, but the information contained in trace-gas distributions primarily reflects the location of the fluxes.

6.2   What do we want to know?

The study of the various gases considered in this book takes place in a context of unprecedented levels of human-induced global change. Figure 6.2 gives a schematic representation of some of the human and natural components involved. This sort of representation has formed the basis of a class of model known as 'integrated assessment models'. These models aim to capture the interactions of human influences on natural systems and the effects of such global changes on human systems. The schematic representation emphasises the role of the atmospheric components with a carbon-cycle component describing transfers of carbon and a climate component describing transfers of energy, momentum and water. However, the integrated assessment models focus attention on the causal chain, represented by the horizontal arrows. This runs from

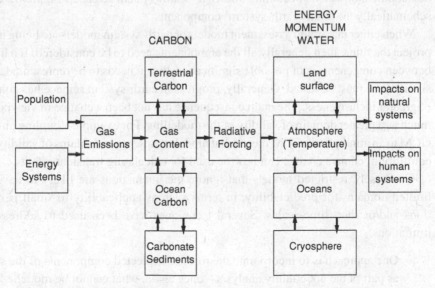

Figure 6.2 A schematic representation of an integrated-assessment model.

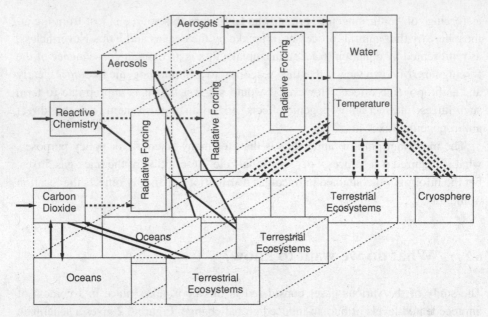

Figure 6.3 A schematic representation of the earth-system model, showing carbon, chemistry and climate systems acting in parallel through components of the earth system (represented as staggered planes). Dashed arrows represent transfers of energy, chain arrows are for exchanges of water and solid arrows are for other mass transfers.

human choices and actions through to emissions, concentrations, radiative forcing, climate change and impacts.

Increasing sophistication in global-scale modelling has revealed the need to consider complex linkages even within the natural components. This realisation has led to the development of what is becoming known as 'earth-system science'. Figure 6.3 shows schematically the main 'earth-system' components.

When either integrated assessment models or earth-system models are being used to project the future then generally all the components need to be considered. If a linkage between components is of possible significance, then it needs to be represented and its significance must be tested. Generally, progress in earth-system science has consisted of greater inclusiveness. The main consequence has not been a change in viewpoint so much as a greater domain of validity of the modelling. For example, coupling an ocean GCM to an atmospheric GCM means that models whose main domains of validity were equilibrium climate change can now be used for calculating transient effects.

Conversely, restricted models that ignore the interactions are likely to have only limited domains for predictability, in terms such as applicability to small perturbations and/or short time-scales. Several techniques have been used to address such limitations.

■ One approach is to incorporate the roles of neglected components of the system as part of the uncertainty analysis – once again, what cannot be modelled deterministically needs to be modelled statistically.

■ In particular, the uncertainties in the human inputs have often been addressed by using, as inputs to the earth-system components, a suite of *scenarios* that capture a range of different but self-consistent possibilities for human influence.

■ In some cases the influence of other earth-system components can be addressed through a parameterisation that gives an 'effective behaviour'. Representing the behaviour of greenhouse gases in terms of response functions is a simple example. The response-function representation (6.1.1) has assumed great importance in greenhouse-gas studies because it forms the basis of the definition of *global-warming potentials* (GWPs) (see Box 6.2). GWPs have been adopted by the Kyoto Protocol as the basis for converting emissions of various gases into the common scale of 'CO$_2$-equivalents'. Although the literal definition of the GWPs is as a measure of integrated radiative forcing, in times of growing emissions GWPs can give a rough proxy for actual warming. In other words, the concept gives a measure of the effective behaviour of the earth system with somewhat greater validity than the literal definition. The limits of the GWP concept are explored by Smith and Wigley [438].

The situation is rather different when we turn to diagnostic modelling in general and inverse modelling in particular. In such studies we can model one or more components of the earth system and address interactions with other components by using the *observed* behaviours of these other components rather than trying to model them. For example, atmospheric CO$_2$ can be modelled with air–sea fluxes expressed in terms of observed ocean p_{CO_2} rather than using an ocean-carbon model. Of course, such effective decoupling does not provide any tests of the neglected components. As an example, for many years the majority of carbon-cycle models specified the input of ^{14}C from nuclear tests by running models that tracked atmospheric observations without checking for the consistency with feasible release patterns until discrepancies were noted by Hesshaimer *et al.* [211].

This ability to deal with components in 'isolation' would suggest that we can expect inverse modelling to play an increasingly important role in the development of earth-system science. It would also suggest that the types of inversion required will become more complex, with greater likelihood of non-linearity and simultaneous inversion of many disparate types of data. However, the utility of such studies is conditional on the requirement, seldom addressed in any detail, of converting the results of diagnostic inversion modelling into reduction of uncertainty in prognostic modelling.

6.3   Types of information

There are many different types of information available about the exchanges of greenhouse gases directed to and from the atmosphere. These will reflect different aspects of the classification described in the previous section. We identify four broad groupings as 'inventories', 'local flux measurements', 'process models' and 'inversions'.

FCCC inventories. The UN Framework Convention on Climate Change (FCCC) and the Kyoto Protocol commit the signatories to produce national inventories of greenhouse gases, according to a specified set of methodologies and in terms of specified sectors (see Box 6.1). These inventories are prepared according to default IPCC methodologies [490] modified for local conditions when local expertise is available. These national inventories are supplemented by 'country studies' [48] for nations with limited capabilities for producing inventories. A recent IPCC report covers issues of 'best practice' in uncertainty analysis of the process of producing an inventory [223]. There are several difficulties in relating these inventories to trace-gas concentrations, particularly for CO_2: the inventories are for anthropogenic emissions only, with special treatment of anthropogenic sinks. An example is that the Kyoto Protocol counts only the reafforestation sinks from forests established after 1990.

Other inventories. Prior to the development of the FCCC national inventories described above, various inventories of trace-gas emissions had been compiled, mainly for research purposes. Some of these are summarised in Table 6.1. These inventories are compiled from a range of sources: economic statistics for many of the anthropogenic fluxes and remote sensing, related to more-direct 'ground truth' for many of the natural fluxes. Bouwman *et al.* [44] have reviewed many of the issues involved in scaling small-scale information to global-scale inventories, especially for natural systems, emphasising in particular the great mis-match in temporal scales between inventories (years or more) and atmospheric-transport models (hours). A major inventory activity, the Global Emissions Inventory Activity (GEIA), has been established as part

Table 6.1. *Some global inventories of trace-gas emissions.*

Gas	Class	Period	Grouping	Reference
CO_2	Fossil	1950–95	Annual by nations	[305, 12]
CO_2	Fossil	1950, 1960, 1970, 1980, 1990	$1° \times 1°$	[11]
CH_4	Wetlands	(1980s)	$1° \times 1°$	[311]
CH_4	Animals	1984	$1° \times 1°$	[280]
CH_4	Rice growing	(1980s)	$1° \times 1°$	[312]
GHG	Anthropogenic	—	$1° \times 1°$	[345]
GHG	Various	Various	Various	[183]
NH_3	All	1990	$1° \times 1°$	[43]

of the IGAC IGBP project. The initial scoping paper by Graedel *et al.* [183]
explores many of the issues involved in inventory development. Current
electronic access to GEIA information is described in Appendix B. The
EDGAR data base [345] from the RIVM (National Institute of Public Health
and the Environment – The Netherlands) covers direct and indirect greenhouse
gases. Differences between national inventories and the EDGAR data base
have been reviewed by van Amstel *et al.* [7].

Local flux measurements. There are various ways of measuring fluxes from
points on the earth's surface. At the smallest scales, a closed chamber can be
placed over a location of interest and the flux determined by monitoring the
change of trace-gas concentration in the chamber. Although it has the greatest
precision, this type of technique has the disadvantage of potential bias through
disturbing the system by shielding the surface from the normal influences of
wind, radiation and precipitation [286]. In the free atmosphere,
micrometeorological techniques can be used, determining the flux as the
integral of the instantaneous concentration times the instantaneous velocity due
to eddies (see, for example, [263: p. 172]). On a larger scale, techniques that
effectively use the boundary layer as a very large chamber have been developed
[103 and Box 17.2].

    All of these techniques produce results on scales much smaller than those
considered in global trace-gas modelling. Therefore, use of such results in
global studies requires a way of scaling the local information. The most
plausible way of using this type of detailed information is by using process
models as intermediaries. For regional scales, however, it may be possible to
incorporate results from boundary-layer-accumulation techniques directly into
the regional transport modelling.

Process models. There is a wide range of 'process models' for surface fluxes of
the various greenhouse gases, particularly those with large natural cycles: CO_2,
CH_4 and N_2O. Some of these models are summarised in Section 6.4.
Section 6.2 has noted these process models as components of more-inclusive
earth-system models and/or integrated assessment models and suggested an
important role for inversion studies in this wider context.

Inversions. Inversion of tracer distributions is the main focus of this book. There
are several ways in which such inversions can relate to the other types of
information about fluxes. First, the other information can be combined with
that produced by the inversion calculation, to produce a combined 'best
estimate'. Essentially, this is what is being done with the Bayesian formulation
of the inversion process. An alternative is to use the inversion as independent
verification of the other types of information. It is likely that techniques for
such independent verification will become increasingly important if and when
the Kyoto Protocol comes into effect.

6.4   Process models

The term 'process model' is used here mainly to describe a model that calculates surface fluxes for trace gases. In inversion calculations, the main roles of process models have been in providing (a) prior values for the quantities that are being estimated, (b) spatial and temporal structures of synthesis components and (c) error models characterising the uncertainty in the prior values. In addition, process models, of various degrees of sophistication, can provide linkages between fluxes of different trace gases. One of the simplest examples of this is relations linking the fluxes that differ only in their isotopic composition (e.g. $^{12}CO_2$ versus $^{13}CO_2$). In this case the 'model' consists of the specification that the isotopic fluxes are in the ratio equal to (or otherwise related to) the isotopic ratio of the source reservoir.

An emerging development is the use of inverse tracer modelling to deduce parameters in process models rather than simply fluxes [397]. This is described in Section 12.4, which also notes a few early studies that are precursors of this type of calculation [164, 467]. In addition, inversion studies estimating atmospheric lifetimes of CFCs (see Section 16.2) are also a form of parameter estimation. Section 6.2 described earth-system models and integrated assessment models (e.g. [381]), noting the potential role of inverse studies as diagnostic tools for evaluating earth-system models that have been developed for prognostic studies.

In discussing process models, we first identify those that are sufficiently self-contained to be used in prognostic mode with appropriate forcing and have the potential to be incorporated as components of an earth-system model, apart from possible problems due to size and complexity. Secondly, we identify a number of 'models' that would seem to be confined to diagnostic roles. As well as this division, there is the possibility of a range of models from different parts of the deterministic–stochastic spectrum of models. Examples of statistically based models are those based on statistical relations between the fluxes and more readily observable proxy quantities – estimating spatial distributions of soil-carbon fluxes from maps of soil type [392] is an example.

Some types of models for use in prognostic mode

Atmospheric transport. Atmospheric-transport models have been described in Chapter 2. As a component of an earth-system model, transport would probably be treated in the on-line mode, as an addition to the AGCM climate component.

Atmospheric chemistry. Section 15.3 discusses models of atmospheric chemistry. See also [51, 501].

Ocean. Models of ocean biogeochemistry, particularly those representing the carbon cycle of the ocean, have evolved from simple parameterisations through to a summary of a range of models used in evaluating emissions for the second assessment by the IPCC given by Enting *et al.* [144]. There has been a recent inter-comparison of three-dimensional ocean-carbon models [421].

Terrestrial systems. Modelling of terrestrial carbon systems has developed considerably in recent years. These models include dependences on factors such as climate and nutrients [82 and references therein].

Industrial systems. As discussed above, representation of industrial emissions has generally been through inventories for past periods or as suites of scenarios for the future. Modelling industrial systems can play a role in scenario development by capturing inter-dependences between emissions of different gases. Other aspects of modelling of emissions can include relating emissions (which is what atmospheric models require as input) to production (which is generally what is known from economic data). This is important for gases such as the CFCs, for which much of the emission represented 'leakage' during and after use of the product containing the CFC. The IPCC Special Report on Emission Scenarios (SRES) [333] synthesises a large amount of information on scenario development.

Changes in use of land. Prognostic modelling of emissions from changes in use of land can be achieved only as a component of modelling human systems. However, in earth-system modelling, prognostic modelling of the processes involved in changes in use of land can be important in accounting for effects additional to those following directly from the emissions. Some of these issues are addressed in the IPCC Report on Land Use, Land Use Change and Forestry (LULUCF) [500].

Examples of models with primarily diagnostic roles

Atmospheric transport. A case in which the representation of atmospheric transport is reduced to diagnostic relations is the 'tracer-ratio' method described in Chapter 12.

Atmospheric chemistry. Production/loss chemistry, as described in Section 15.3.

Ocean. Air–sea exchange of gases can be treated by specifying the oceanic content of the particular tracer and a gas-exchange relation. An important aspect of air–sea exchange of gases in diagnostic studies is the specification of the isotopic ratios of gas fluxes as fractionation causing offsets from the ratios of the source reservoir.

Terrestrial systems and changes in use of land. In recent decades, remote sensing has provided additional data [82]. Again, isotopic relations provide additional constraints.

Human systems. In diagnostic modes, emissions from human systems are often specified by inventories for which the underlying 'model' takes the form emission = activity × intensity. Often population is used as a proxy for levels of activity.

The use of indirect or 'proxy' variables is a common feature of models used in diagnostic mode. In particular, remote-sensing data can be used to give detailed spatial coverage and then linked to fluxes through structural relations having varying degrees of statistical and deterministic modelling.

6.5   Statistical characteristics

The linear Bayesian-estimation procedure described in Chapter 3 expresses the estimates of source parameters as $\hat{x}_{BLS} = [G^T X G + W]^{-1}[G^T X c + W z]$, where G is the discretised Green function of a transport operator, c is the set of observations with error covariance and z is a set of prior estimates of the source parameters with covariance W. Applying this formalism requires a characterisation of the sources in terms of the z and W.

There are several variations on this approach: the various recursive synthesis techniques described in Section 11.2 require essentially the same information as that contained in z and W, but in a factored form consistent with the defining equations for state-space modelling. Regularisation techniques (see Box 8.1) correspond to special cases of Bayesian estimation such as $z = 0$ and $W = \Lambda I$. Inversion calculations lacking formal criteria for regularisation have often used the judgment of the modellers in place of formal specification of z and W. Of course, the specification of the probability distribution in terms of z and W applies only to simple cases in which linear estimation follows from multivariate normal distributions for $p_{prior}(x)$. Inversions based on more complicated models of the source statistics will require additional information about the distribution.

The specification of W needs to cover both the uncertainty in each of the components as well as the amount of correlation: in time, in space and among multiple tracers.

Some of the main requirements for statistical modelling of the source/sink distributions are the following.

- One must characterise the role of source variability. Box 5.3 has noted variability as a potential contributor to uncertainty. For sources, variability in space and time will be significant in assessing the uncertainty in prior estimates, especially those based on some sort of sampling. Source variability will also be important for the data/model uncertainty in determining how representatively the wind-fields 'sample' the source distribution.

- One must distinguish between urban and rural sources, to some extent distinguishing anthropogenic from natural fluxes. This is particularly significant in our own work. Australia has an unusual pattern of urbanisation with a very high proportion of the population living in a small number of relatively large cities. Seasonal variations provide a distinction that is

particularly relevant to the northern-hemisphere land areas, with a great
reduction in many natural sources in winter.

Some common approaches to obtaining space–time statistics are as follows.

- The use of a 'proxy' distribution. In particular, distributions of use and/or
  generation of electricity are often used in modelling studies as a substitute for
  distributions of other forms of industrial activity and their associated
  emissions. This introduces two compounding levels of uncertainty. More
  generally, the 'activity × intensity' representation of many anthropogenic
  fluxes introduces a double level of uncertainty and a possible breakdown in the
  validity of the multivariate normal assumption. In particular, uncertainties in
  activities and uncertainties in intensities are likely to have very different
  correlations in space and time.
- Remote sensing, especially from satellites. This introduces new classes of error
  statistics, including those characterising long-term drifts in calibration and
  variability associated with orbital geometry. Of course, this is really a special
  example of using proxy data – in this case radiances.

In the absence of detailed statistical modelling of source/sink distributions, the most
plausible approach is to be relatively conservative by assigning relatively large uncertainties to the prior estimates. Generally there is little scope for testing the statistical
assumptions because the amount of detail in the source specifications is too small and
the source components are too disparate to give usable numbers of cases for statistical
testing.

The synthesis-inversion techniques described in Sections 10.3 and 11.2 estimate
trace-gas fluxes in terms of scaling factors for pre-defined regional distributions. This
raises the question of determining the best ways to choose the components and the
statistical implications of the choice. Wunsch and Minster [522] recommend using a
large number of components; only a small number of components will be resolved,
but these will be determined by the data rather than by an arbitrary decision of the
modeller. Putting this philosophy into practice may involve significant computational
demands unless adjoint modelling (see Section 10.6) is used. Furthermore, actually
asking which components are resolved is an ambiguous question in the absence of some
normalisation of the two vector spaces defining the data and the source parameters.
The most meaningful way of specifying the normalisations is in terms of the spread
in the observations and prior estimates.

Section 8.3 discusses the selection of components for synthesis inversion in terms
of the related issues of resolution. However, going beyond the general principles
developed in Section 8.3 will require that at least the specification of the uncertainties
is determined on a more detailed scale than that used in the synthesis components.
Remotely sensed data from satellites can provide useful measures of the scales of
variability in surface properties, e.g. [481, 482, 99].

6.6 Global constraints

Section 6.1 introduced the concept of using response functions to characterise the dissipation of perturbations in concentration due to perturbing sources. The response functions $\rho_\eta(t)$ are the proportion of a unit input of tracer η remaining in the atmosphere after a time t. The notation was

$$M_\eta(t) = \int_{t_0}^{t} \rho_\eta(t - t') s_\eta(t') \, \mathrm{d}t' \tag{6.6.1}$$

Mathematically, this is a convolution equation; the process of inverting (6.6.1) is known as deconvolution.

In practice, the response functions, $\rho(.)$, are generally derived from models, although a number of workers [506, 233] have then constructed numerical models based on the use of such response functions from other models. The formulation (6.6.1) leads to three distinct mathematical problems.

 (i) Determine $M(.)$ given $\rho(.)$ and $s(.)$. This is the forward problem and is usually done with a numerical model of the processes.
 (ii) Determine $s(.)$ given $\rho(.)$ and $M(.)$. This is the 'standard' deconvolution problem. An important example is the estimation of sources from gas concentrations measured in ice cores.
 (iii) Determine $\rho(.)$ given $M(.)$ and $s(.)$. This problem is not often considered in atmospheric trace-gas studies. One case is the estimation of $\rho(.)$ for CO_2 as an AR(2) model by Young $et\ al.$ [527], using a spline fit to ice-core data and fossil-carbon emissions. Because of the neglect of autocorrelation and the omission of fluxes from changes in use of land, the results are primarily illustrative rather than being a definitive estimate of $\rho_{CO_2}(.)$.

Another important application, related to case (iii), is the estimation of atmospheric lifetimes, for which the response is taken as $\exp(-\lambda t)$ and the lifetime λ^{-1} is estimated from a combination of source and concentration data. This estimation problem is discussed in Chapter 16.

The deconvolution of $s(.)$ from $M(.)$ is of importance in atmospheric-transport inversions because it may provide additional information, particularly when (6.6.1) is generalised to describe isotopic variations. However, it is necessary to ensure that, when such additional information is treated as independent, the inversion formulation does not implicitly incorporate (6.6.1), i.e. that double counting is avoided.

It is also possible to consider hybrids of cases (i) and (ii). Some of the CO_2 modelling calculations [144] used in the IPCC Radiative Forcing Report were carried out by deducing $s(.)$ from $C(.)$ for past times and then calculating $C(.)$ from future emission scenarios $s(.)$. Hybrid calculations with $\rho(.)$ unspecified are also possible, particularly if additional constraints such as $\rho(.) > 0$, $\dot{\rho}(.) < 0$ and $\ddot{\rho}(.) > 0$ are included [134, 129].

In view of the importance of the general form of response function (6.6.1), it is worth noting that there is a generic form of the inverse, as a problem in solving for $s(.)$, expressed (for the case normalised with respect to $\rho(0) = 1$) as

$$s(t) = \dot{M}_\eta(t) + \dot{\rho}_\eta(0)M_\eta + \int_{t_0}^t \rho_\eta^*(t - t')M_\eta(t')\,dt' \tag{6.6.2}$$

for an appropriately defined kernel $\rho_\eta^*(.)$. If ρ_η is represented as the sum of N exponentials, then $\rho_\eta^*(.)$ can be represented as the sum of $N - 1$ exponentials. This relation was described by Enting and Mansbridge [135] using an analysis based on Laplace transforms. In Section 8.2, inverse problems are classified in terms of the equivalent order of differentiation. Equation (6.6.2) explicitly shows that the inversion of (6.6.1) involves differentiation of $M_\eta(t)$.

An alternative form of the inversion relation, more suitable for use in a numerical model, can be obtained by rewriting the flux relation (6.1.2a) as

$$s_\eta(t) = \dot{M}_\eta(t) - \int_{t_0}^t \dot{\rho}_\eta(t - t')s_\eta(t')\,dt' \tag{6.6.3}$$

The source depends on the present rate of change of concentration and the past sources, so source estimates can be calculated by integrating forwards in time.

Transforming this relation into a finite-difference form gives

$$\Delta t \times s(t + \Delta t)$$

$$\approx M(t + \Delta t) - M(t) - \int_0^t [R(t + \Delta t - t') - R(t - t')]s(t')\,dt'$$

$$= M(t + \Delta t) - \int_0^t R(t + \Delta t - t')s(t')\,dt' \tag{6.6.4}$$

This can be termed a 'mass-balance' form of deconvolution. A schematic representation based on (6.6.3) is given in Figure 6.4, for the fossil-plus-terrestrial CO_2 flux (using approximate data and an approximate carbon-uptake response function for the ocean, exaggerated to enhance the illustration). The source is represented as a series of pulses. As shown in the upper panel, the source estimate at each time point (shown in the lower panel) is derived from the difference between the specified concentration (dashed line) and the model calculation of the concentration due to all previous pulses. Of course, at any time, the model is working with the carbon distribution from the combined past source rather than with separate pulses. The most common form of operation is to have this response to past sources calculated with a model that represents specific processes rather than using a response-function representation. Exceptions that are based on the use of response functions occur in the work of Wigley [506], who used a response-function representation of uptake by the ocean, and Trudinger [487], who represented response functions within a state-space formalism. Using the state-space formalism means that the Kalman-filter procedure can be used to obtain estimates of terrestrial CO_2 fluxes and associated uncertainties.

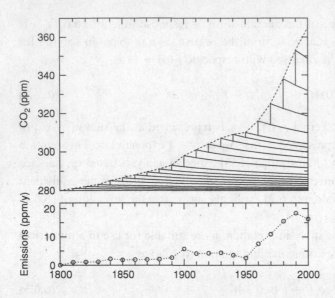

Figure 6.4 A schematic representation of the mass-balance form of deconvolution. The source is represented as a series of pulses. The source estimate at each time point is derived from the difference between the specified concentration (dashed line) and the model calculation of the concentration due to all previous pulses. (These are approximate illustrative values only.)

Some of the main examples of deducing sources from concentration records apply to the estimation of CO_2 sources based on the concentration record from ice-cores [e.g. 434, 487]. This form of deconvolution can be extended to joint inversion of ice-core records for CO_2 and $^{13}CO_2$, a process that has become known as double deconvolution [e.g. 248, 487].

Further reading

■ The national inventories prescribed under the FCCC provide for developed nations to report their emissions. There are additional reports covering various developing nations, produced in cooperation with developed nations. Further details are available on the UNFCCC web-site www.unfccc.int. The methodology for taking an inventory [490] and guidelines for good practice and uncertainty management [223] are essential references for those involved in producing or analysing these inventories.

■ The IPCC's Special Report on Emissions Scenarios (SRES) [333] and that on Land Use, Land-Use Change and Forestry (LULUCF) [500] contain extensive discussions of anthropogenic sources.

■ Volume 7 of the EUROTRAC series [116] describes estimates of emissions for Europe.

■ The remote-sensing literature is an on-going source of information about source distributions.

■ Cressie's book [84] is a comprehensive reference on spatial statistics.

Exercises for Chapter 6

1. For the toy model of Box 1.5, derive the case of the limit of stability:
   $2S_{CH_4} > (\xi S_{OH} - S_{CO}) > 0$ for the assumptions, $\zeta = 1$ and $\lambda_{OH} = \lambda_{CO} = 0$,
   made by Guthrie [186]. Generalise it to $0 < \zeta \leq 1$. See Appendix B for values
   for use in numerical calculations using the toy chemistry model.

2. Derive the deconvolution relation (6.6.2). (Hint: use Laplace transforms.)

Chapter 7

Problem formulation

In solving a problem of this sort, the grand thing is to be able to reason backwards.

A. Conan Doyle: *A Study in Scarlet.*

7.1   **Problems and tools**

The previous chapters have covered the main components required for tracer-inversion calculations to be able to deduce trace-gas fluxes from concentrations. These are

(i) atmospheric-transport models;

(ii) estimation techniques;

(iii) measurements of atmospheric composition; and

(iv) independent estimates of source/sink distributions.

Before going into descriptions of specific computational techniques, this chapter gives an overview of the range of different ways in which the pieces can be fitted together. This expands on the schematic description given in Figure 3.1.

As a starting point, we need to recall the discussion in the introduction addressing the use of the word 'model'. We distinguished the terms 'transport model' and 'statistical model', where the statistical model was called upon to cover any incompleteness in our deterministic understanding. The general case could be regarded as a single model with both deterministic and stochastic aspects; the state-space-modelling formulation exemplifies such a combination of modelling, with deterministic and stochastic components closely integrated (see Sections 4.4 and 11.2). However, in many cases some

Figure 7.1 A representation of a black-box atmospheric-transport model, calculating $\mathbf{m} = \mathbf{Gs}$.

degree of separation between the transport model (or other deterministic model) and the statistical model is both possible and useful.

Various relations are possible: we can have a deterministic model in which the stochastic components are random functions of the outputs of the deterministic component (e.g. observations = model calculation + random noise), deterministic models with random inputs, or a wide range of hybrid cases. It is also important to distinguish 'stochastic modelling' from techniques of 'stochastic analysis'. Examples of the latter are the Monte Carlo techniques noted in Chapter 3, Section 7.2 and Box 7.1.

Weizenbaum [503] attributes to psychologist Abraham Maslow the observation that 'to a person who has only a hammer, the whole world looks like a nail'. So how do the world's problems appear to someone whose primary tool is a semi-Lagrangian tracer-transport model?

Numerical models of atmospheric tracer transport are specified in 'procedural' form: DO THIS, THEN DO THIS, THEN DO THE NEXT THING ..., using procedural computer languages, most notably Fortran (see Chapter 2). Such steps are combined, generally in a hierarchical manner, to produce a procedure for numerical integration of the transport equation (1.1.1). Usually the model's most common (and perhaps only) mode of operation will be that of numerical integration of this forward problem: given a time history of the sources at each point, integrate the equations to determine concentrations. The model takes an initial concentration field $c(\mathbf{r}, t_1)$ and a source specification $s(\mathbf{r}, t)$ for the time interval $[t_1, t_2]$ and is integrated to give $c(\mathbf{r}, t)$ for $t \in [t_1, t_2]$. Figure 7.1 gives a schematic representation of this calculation, treating the transport model as a 'black box'. The following section describes how such a model can be used for inversion calculations (estimating sources, $\mathbf{s}(\mathbf{r}, t)$), either by surrounding the forward model with an outer level of computational 'shell' or by transforming the model in the case of mass-balance inversions.

In these descriptions, the statistical model appears as a 'shell' around the deterministic model which is represented by the relation $\mathbf{m}(\mathbf{s})$ giving the concentrations (actually one or more functions of time: $m_x(t)$) in terms of the sources (which are also functions of time). The uncertainties in the observations, $\mathbf{c}$, are modelled in terms of a distribution $p(\mathbf{c}|\mathbf{m})$. The prior uncertainties in the sources are represented by $p_{\text{prior}}(\mathbf{s})$. What is needed are effective techniques for obtaining $p_{\text{inferred}}(\mathbf{s}|\mathbf{c})$ as given by the Bayes solution (3.2.2e).

Section 7.2 illustrates a number of different formulations of the inverse problem of deducing fluxes from concentrations, building on the schematic picture of Figure 7.1.

Box 7.1. Some Monte Carlo algorithms

Monte Carlo sensitivity. This applies when one has an estimator $\hat{\mathbf{x}}_X(\mathbf{c})$, but lacks the means to propagate the errors associated with $\mathbf{c}$ through the estimation procedure. The Monte Carlo sensitivity analysis constructs an ensemble of estimates, $\hat{\mathbf{x}}_{X:j}$, as $\hat{\mathbf{x}}_{X:j} = \hat{\mathbf{x}}_X(\mathbf{c} + \mathbf{e}_j)$, where $\mathbf{e}_j$ is a realisation from a simulation of the error distribution of $\mathbf{c}$.

Monte Carlo integration. This evaluates $\Omega = \int dx_1 \cdots \int dx_M f(\mathbf{x})$. This is done by defining $g(\mathbf{x}) = f(\mathbf{x})/p(\mathbf{x})$ for a known, readily simulated, probability distribution, $p(\mathbf{x})$. Thus $\Omega = \int dx_1 \cdots \int dx_M p(\mathbf{x})g(\mathbf{x})$, giving $\Omega \approx \sum_j g(\mathbf{x}_j)$ summed over a sample from the distribution $p(\mathbf{x})$. This technique can be used to evaluate functionals of the posterior distribution $\int h(\mathbf{x}) p_{\text{inferred}}(\mathbf{x}|\mathbf{c}) \, d\mathbf{x}$ in cases in which either $h(x)$ or $p_{\text{inferred}}(.)$ is too complicated to be readily integrable and/or the dimensionality is large [471: Box 3.1].

Ensemble sampling. This is Monte Carlo integration with $p_{\text{inferred}}(\mathbf{x}|\mathbf{c}) = p_{\text{prior}}(\mathbf{c})p_{\text{theory}}(\mathbf{c}|\mathbf{x})$ taking the role of $f(\mathbf{x})$ above and using $p_{\text{prior}}(\mathbf{x})$ in the role of $p(\mathbf{x})$. Any functional $h(.)$ of $p_{\text{inferred}}(\mathbf{x}|\mathbf{c})$ is approximated by $\int h(\mathbf{x}) p_{\text{inferred}}(\mathbf{x}|\mathbf{c}) \approx \sum_j h(\mathbf{x}_j) p_{\text{theory}}(\mathbf{c}|\mathbf{x}_j)$. Krol et al. [266] describe the variant for the case in which one knows $p_{\text{prior}}^*(\mathbf{c}|\mathbf{x}) = A p_{\text{prior}}(\mathbf{c}|\mathbf{x})$ with an unknown normalisation, A. In this case $\int h(\mathbf{x}) p_{\text{inferred}}(\mathbf{x}|\mathbf{c}) \approx \sum_j h(\mathbf{x}_j) p_{\text{theory}}(\mathbf{c}|\mathbf{x}_j) / \sum_j p_{\text{theory}}(\mathbf{c}|\mathbf{x}_j)$.

Monte Carlo inversion. This is described by Press et al. [373]. Samples $\mathbf{x}_j$ are generated within some prior domain and the results, $\mathbf{m}(\mathbf{x})$, are checked for agreement with the data. The distribution of the solution is taken as being given by the ensemble of cases for which agreement is judged acceptable.

Monte Carlo optimisation. Arsham [15] describes a number of stochastic search algorithms, generally variations on deterministic searches, for finding extrema of objective functions. The stochastic nature helps avoid convergence to local extrema. In *simulated annealing*, the search characteristics are systematically changed over time, decreasing the size of the random steps in the parameter space. Initial large steps are designed to find the region of the true extremum and are followed by smaller steps once the extremal region has been located. Monte Carlo optimisation can be applied to an inversion problem in which estimates are defined by extrema of objective functions.

Some of these algorithms are described further by Tarantola [471: Chapter 3].

To obtain a perspective from related problems, Section 7.3 discusses assimilation of meteorological data and Section 7.4 extends this to assimilation of chemical data. Section 7.5 discusses the formulation of inverse problems in other areas of science related to atmospheric composition.

7.2 Studying sources

Figure 7.1 gives a schematic representation of a tracer-transport model: a 'black box' whose input is a specification of tracer sources (as functions of space and time) and whose output is a set of calculated tracer concentrations. We also express this in mathematical terms as calculating a concentration vector $\mathbf{m}$ from a source vector $\mathbf{s}$, in each case as a function of time.

We can define a number of approaches to the inversion problem of deducing trace-gas sources given concentration data. Each involves successively greater modifications to the structure of the transport model defined for time-integration of the 'forward' problem. Figures 7.2–7.6 give schematic representations of how the modelling framework is adapted for the various problems.

Forward modelling. To use the 'standard' model form, numerically integrating (1.1.1) requires a pre-specified set of sources, possibly from some form of process model. The results are then compared with observations. This uses a numerical model unchanged from the form described above. However, such calculations are not (in our terminology) inversion calculations. The schematic diagram in Figure 7.2 shows how such usage requires no insight (formally) into the internal structure of the model. In terms of the linear model relation, the forward model calculates $\mathbf{m} = \mathbf{G}\mathbf{s}$ for a specific input, $\mathbf{s}$, or more generally $\mathbf{m} = \mathbf{G}(\mathbf{s})$ if non-linear processes are involved.

Synthesis. Synthesis inversions are performed by defining a set of basis sources, $\mathbf{s}_\mu(t)$, and calculating the corresponding tracer distributions $\mathbf{m}_\mu(t) = \mathbf{G}\mathbf{s}_\mu$. The sources are estimated by fitting $\mathbf{m}_{\mathrm{obs}} \approx \sum_\mu \alpha_\mu \mathbf{m}_\mu$, whence the estimated source is $\hat{\mathbf{s}} = \sum_\mu \alpha_\mu \mathbf{s}_\mu(t)$. The schematic diagram in Figure 7.3 shows that multiple runs of the forward model are needed, but that again no access to the

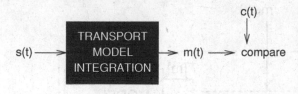

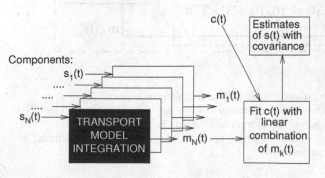

Figure 7.2 Forward modelling with a black-box atmospheric-transport model, calculating $\mathbf{m} = \mathbf{G}\mathbf{s}$ for comparison with observations, $\mathbf{c}$.

Figure 7.3 The use of forward modelling with a black-box atmospheric-transport model, calculating $\mathbf{m}_k = \mathbf{G}\mathbf{s}_k$ for matching to observations, $\mathbf{c}$, by least-squares fitting $\mathbf{c} \approx \sum_j \alpha_j \mathbf{m}_j$ to produce $\hat{\mathbf{s}} = \sum_j \alpha_j \mathbf{s}_j$.

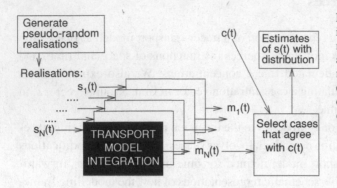

Figure 7.4 Monte Carlo inversion: forward modelling with a black-box atmospheric-transport model, calculating an ensemble of $\mathbf{m}_j = \mathbf{G}\mathbf{s}_j$ for comparison with observations, $\mathbf{m}_{obs}$.

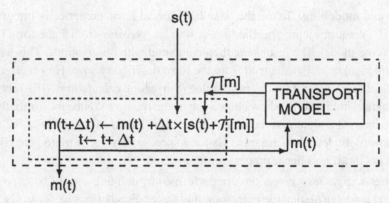

Figure 7.5 A representation of an atmospheric-transport model as successive loops invoking a 'black-box' transport-model component, calculating $\dot{\mathbf{m}} = \mathcal{T}[\mathbf{m}] + \mathbf{s}$. (The use of Euler time-stepping is purely for illustrative purposes.)

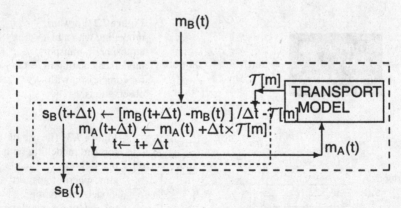

Figure 7.6 A schematic representation of the transformation of an atmospheric-transport model by integration as successive loops invoking a 'black-box' transport-model component, to perform mass-balance inversion calculating $\dot{\mathbf{m}}_A = \mathcal{T}[\mathbf{m}] + \mathbf{s}$ and $\mathbf{s}_B = \dot{\mathbf{m}}_B - \mathcal{T}[\mathbf{m}]$. (Again, the use of Euler time-stepping is purely for illustrative purposes.)

internal structure of the model is required. In Bayesian synthesis, prior information is used in the fitting process and again this requires no modification to the forward model. The formalism is described in Section 10.3.

Monte Carlo inversions. Monte Carlo inversions have a long history in seismology problems [253, 372] but have been little used in tracer-transport studies. An example of the use of a Monte Carlo approach in greenhouse-gas studies was provided by Parkinson and Young [349], who used a combination of a deterministic model, $\mathbf{m}(\mathbf{x})$, and a prior distribution $p(\mathbf{x})$ to define $p(\mathbf{m}, \mathbf{x})$. (They modified the deterministic model to include a small stochastic component, but their discussion largely ignores this aspect.) They explored the distribution $p(\mathbf{m}, \mathbf{x})$ by Monte Carlo simulation. Kandlikar [242] has described a comparable Monte Carlo investigation of the global methane budget. Figure 7.4 gives a schematic representation of this type of procedure as a 'shell' surrounding a 'black-box' deterministic-transport model.

Box 7.1 gives a brief mathematical summary of other applications of Monte Carlo techniques relevant to inversion problems.

Mass-balance. Mass-balance inversions require a transformation of the forward transport model that is represented as the 'black box' in the previous cases. Figure 7.5 shows the transport model expressed as a time-integration loop that repeatedly calls a routine (the inner black box) to evaluate the transport terms and calculate changes in concentration as the sum of transport and sources. Figure 7.6 shows how a slight modification of the loop, given the same ability to calculate transport, gives the ability to calculate sources as the difference between the rate of change of concentration and the transport term. The mass-balance-inversion technique is described in Section 11.1.

Analysis of sensitivity and uncertainty. There are several ways in which these various uses of transport modelling can be extended to incorporate analyses of sensitivity and/or uncertainty. The simplest is simply to implement deterministic perturbations, i.e. repeated calculation with slightly different values of the sources or other parameters. A differential form of this is to integrate equations for the derivatives with respect to one or more parameters. This is known as the tangent linear model (see Section 10.6) and automatic code-generation techniques for transforming computer routines into routines that implement the tangent linear model are available. The Monte Carlo inversion techniques produce a measure of uncertainty, from the spread of realisations that give acceptable fits to the data. As indicated in Figure 7.3, synthesis inversions also produce covariances for the estimates. Figure 7.7 represents this as an 'outer-level' box producing posterior covariances as its output. This can be used as the basis of an experimental design process, changing the type of data and its specified precision to determine how best to improve the estimates. This role is discussed in Chapter 13. Figure 7.8 shows this process schematically, indicating that the 'experimental-design' aspects act as a shell around a synthesis-inversion calculation.

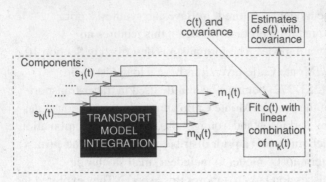

Figure 7.7 A representation of synthesis inversion as a 'black box', taking inputs of data and producing estimates of sources and variances of these estimates.

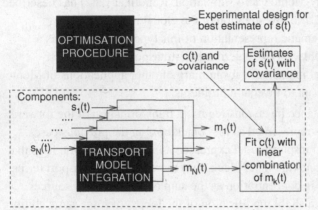

Figure 7.8 A representation of the use of the synthesis-inversion 'black box' for experimental design, minimising the variances of the source estimates by adjusting the data set symbolised by $c(t)$.

Adjoint. The adjoint method permits the calculation of sensitivities, including sensitivities to source components (see Section 10.6). If the Green-function matrix is expressed as $G_{j\mu} = \partial m_j/\partial s_\mu$ then the forward model calculates $G_{j\mu}$ for all j for a single μ while the adjoint model calculates $G_{j\mu}$ for all μ and a single j. The calculation requires a program that is distinct from that for the basic forward-transport model. In particular, the adjoint code steps backwards through time. The term 'adjoint' refers to its being the adjoint operator for the tangent linear model described above. There are automatic adjoint compilers that transform the code for a numerical model into the code for its adjoint. Section 10.6 describes the formalism and notes a number of applications in biogeochemical studies in general and atmospheric-tracer inversions in particular.

7.3   Data assimilation

Chapter 2 noted that atmospheric GCMs used for climate modelling differed in their usage from those used for operational forecasting. In the absence of infinitely precise data at infinitely fine resolution as a starting point in a model of arbitrarily high resolution, forecasts become increasingly inaccurate as time progresses, until they are at best correct only as climatological averages. In other words, atmospheric GCMs have sensitive dependences on initial conditions. Some of the issues arising from this are discussed in Box 7.2. For the present section, the important point is that, to achieve any reliability of forecasting, the models must be continually updated with new observational data in order to prevent the model state drifting away from reality. This task is known as *data assimilation*. Lorenc [288] identified the distinctive characteristics of data assimilation, relative to other inverse problems in the earth sciences, as

  ■ a large pre-defined parameter basis, i.e. the variables of the forecast model;
  ■ a set of observations that were far too few to determine these basis components; and
  ■ a large amount of prior information from on-going operation of the forecasting system.

An additional issue is that generally the observed quantities differ from the dynamical variables in the models.

The procedures for data assimilation have been described in the book *Atmospheric Data Analysis* by Roger Daley [94]. The present section notes some of the key features of that presentation and identifies parallels with tracer inversions. Data assimilation is presented as a problem that overlaps tracer-inversion problems in the areas of process inversion (see Section 12.5) and assimilation of chemical data (see Section 7.4) and a mass-balance inversion based on data assimilation [97]. Data assimilation is also of interest here as the origin of the 'analysed' transport fields that can be used in atmospheric-transport modelling (see Chapter 2).

Daley identifies four steps in what has become known as the 'data-assimilation cycle':

  (i) quality control (checking of data);
  (ii) objective analysis;
  (iii) initialisation; and
  (iv) a short forecast to prepare the next background field (i.e. the prior estimates).

The way in which this cycle is used is that the output of step (iii) produces 'analysed' meteorological fields on a specified grid at specified time intervals, generally 6 or 12 h. The fields at each time-step differ from those at the previous time-step due to a combination of three changes:

  (a) the time-evolution calculated from a short-term forecast, i.e. from stage (iv) of the previous iteration of the data-analysis cycle (these forecast fields are termed the 'background field');
  (b) a correction to bring the background field into agreement with the objective analysis of the data produced in stage (ii); and
  (c) a final correction to apply constraints of dynamical consistency to the fields, i.e. the initialisation step.

Daley [94] notes that, in modern practice, the largest changes come from the forecast model, followed by changes implied by the objective analysis followed in turn by changes implied by the initialisation step. In part, this simply reflects the order in which the changes are applied.

The analysed fields produced at stage (iii) are used as the basis for operational multi-day forecasts. The model used for these forecasts is not necessarily the same model as that which is used for the short-term prediction involved in producing the background field in step (iv) of the data-assimilation cycle.

Daley's presentation includes a historical review starting from the early analyses of constructing weather maps by LeVerrier in France and Fitzroy in the UK in the nineteenth century. This work and operational practice in the first half of the twentieth century is now known as 'subjective analysis' (due to the large amount of judgment required by the analyst) in contrast to 'objective analysis' based on specific mathematical procedures. The techniques of 'objective analysis' involve various forms of 'filtering' that need to suppress 'noise' and interpolate the data onto a regular grid.

Formally, the optimal estimate of any field will be the weighted average of the observations nearby and the 'background' field produced by the short-term forecasts. (In the early applications of data assimilation, background fields based on mean climatology were all that was available.) Determining the optimal weighting requires an error model for each of the data streams that are being combined and, since data from different locations are being combined, the error model must included a specification of its spatial dependence.

The feature in which 'data assimilation' differs most from the majority of the inverse problems addressed in this book is step (iii) in the cycle described above: the initialisation step. The requirement is that the fitted fields are consistent with the atmospheric dynamics. An important issue is that the atmosphere has two modes of variability: longer-scale variability known as Rossby modes and shorter-scale variability known as inertial-gravity modes. In the atmosphere, the inertial-gravity modes are only weakly excited, i.e. they account for only a small part of the kinetic energy. However, if the initial state of a model contains a large proportion of such waves then they will continue to propagate.

A number of early models addressed the problem by using numerical schemes that suppressed the propagation of inertial-gravity waves. However, in modern models, the loss of accuracy involved in this process is regarded as unacceptable.

A widespread formulation of the data-assimilation problem is based on minimising various objective functions – this has become known as a variational approach since, when it is expressed in terms of specifying continuous functions of space (and possibly time), the formalism is that of the calculus of variations. Specifying the problem as one of minimising an objective function provides a generic approach to fitting a set of non-linear equations. The objective is defined as the weighted sum of squares of deviations from the desired equations.

Some of the prior information is contained in the time-evolution of the meteorological fields. Therefore improved estimates should be obtained by making the observations consistent with the time-evolution. This process is known as four-dimensional assimilation [28]. Lorenc [288] shows that, for linear models, this is equivalent to a sequence of three-dimensional (i.e. single-time) assimilations based on the Kalman filter. Todling [480] analyses data assimilation in terms of the Kalman filter, concentrating on procedures that give computationally feasible approximations to the ideal formalism. One of the most serious computational problems is the propagation of the uncertainties. Formally this is expressed by the term $\mathbf{FPF}^\mathsf{T}$ in the evolution equation for uncertainties (see equation (4.4.2b)). In practice this expression needs to be approximated. Todling relates four-dimensional assimilation to the fixed-lag Kalman smoother, noting that going from the filter to the smoother exacerbates the computational demands. The variational form of four-dimensional assimilation takes the form of minimising (7.3.1a) subject to (7.3.1b), where

$$J = \int_{t_0}^{t_1} \sum_j [c_j(t) - m_j(t)]^2 \, dt \qquad (7.3.1a)$$

and the m_j are the model variables to be fitted and the $c_j(t)$ are the observations. The model equations are

$$\dot{m}_j(t) - f_j(\mathbf{m}) \qquad (7.3.1b)$$

This is done by minimising

$$J' = \int_{t_0}^{t_1} \sum_j [c_j(t) - m_j(t)]^2 + \int_{t_0}^{t_1} \sum_j \Lambda_j(t)[\dot{m}_j(t) - f_j(\mathbf{m})] \, dt \qquad (7.3.1c)$$

where the Lagrange multipliers $\Lambda_j(t)$ are introduced in order to force the model equations (7.3.1b) as exact constraints.

In the Kalman-filter formalism, update equations are written as

$$\hat{\mathbf{g}}(n) = \tilde{\mathbf{g}}(n) + \tilde{\mathbf{P}}(n)\mathbf{H}(n)\mathbf{P}_\mathrm{I}^{-1}[\mathbf{c}(n) - \mathbf{H}(n)\tilde{\mathbf{x}}(n)] \qquad (7.3.2a)$$

Then one can calculate

$$\mathbf{h}(n) = \mathbf{P}_\mathrm{I}^{-1}[\mathbf{c}(n) - \mathbf{H}(n)\tilde{\mathbf{x}}(n)] \qquad (7.3.2b)$$

by numerical solution of the minimisation of

$$J(\mathbf{h}) = \tfrac{1}{2}\mathbf{h}(n)^\mathsf{T}\mathbf{P}_\mathrm{I}\mathbf{h}(n) - \mathbf{h}(n)^\mathsf{T} \cdot [\mathbf{c}(n) - \mathbf{H}(n)\tilde{\mathbf{x}}(n)] \qquad (7.3.2c)$$

without the need for explicit inversion of $\mathbf{P}_\mathrm{I}$, the covariance matrix for the innovations. These techniques require the use of pre-specified prescribed analytical functions as approximations to the covariance matrices. This is because implementation of the Kalman-filter procedure for updating the covariance matrices is computationally not feasible for large problems.

Daley [94] identified a number of 'future directions', many of which have indeed become important since 1991 when his book was written. These include four-dimensional variational methods (using the time dimension to ensure greater consistency between the analysis and the model equations) and assimilation of data on surface properties. In the latter applications Daley identified a 'spin-up' problem due to the longer time-scales of surface processes relative to those in the atmosphere.

The techniques of data assimilation have made a number of direct contributions to the practice of tracer inversion. Law [274] has used objective interpolation techniques to extend the small number of trace-gas records into a field specified for all surface grid points as required by the mass-balance-inversion method described in Section 11.1. Dargaville [96, 97] has used a more direct form of data assimilation in making the transport model follow the observed data only around the measurement sites [274]. A greater overlap between trace-gas inversion and meteorological data assimilation can be expected in the future. Forecasting models are increasingly using assimilated data on surface biology and hydrology [117] and a convergence with tracer studies formulated as process inversions (see Section 12.4) seems highly probable. Chemical-data assimilation [155, 118, 439] is outlined briefly in the following section. Most applications have been for distributions of ozone, which, since the radiation budget will be improved, contribute improvements to forecasting models.

Box 7.2. **Chaos and predictability**

■ The detailed time-evolution of the atmosphere exhibits a sensitive dependence on initial conditions. Differences between states with similar initial conditions tend to grow exponentially in time.

■ Often this is called the 'butterfly effect': the flapping wings of a butterfly in Brazil/China etc. can 'cause' a hurricane etc. in the North Atlantic.

■ This limits the time-scale on which useful *weather* predictions, i.e. predictions that are better than just assuming the average climate, are feasible.

These properties reflect what is known as 'chaos', which, in the 1970s, exploded from a few isolated results into an entirely new field of mathematical science. Two notable accounts are by Gleik [178] for the non-technical reader and by Lorenz [290] with technical detail and comments on his own work. It includes the text of a talk called 'Predictability: does the flap of a butterfly's wings in Brazil set off a tornado in Texas?', which was apparently the origin of the 'butterfly' example noted above. There are also textbooks, e.g. [354].

On many occasions, chaos has been mis-represented in attempts to discredit GCM modelling that implies a need for restrictions on emissions of greenhouse gases. To clarify this, we note the following points.

■ A sensitive dependence on initial conditions is a property both of models and (almost certainly) of the real atmosphere. A sensitive dependence on initial conditions is not, as is sometimes implied, a failing of the models.

■ The exponential growth of perturbations as $e(t) = e(0)\alpha^t$ is not of unrestricted applicability. Having a 0.1-K perturbation double in 2 days does not mean that it will become a 102.4-K error in 20 days. Exponential growth may imply that a 0.0001-K perturbation grows to 0.1 K after 20 days, but ultimately the growth of perturbations is limited by the scale of the model climatology. Fortunately for climate modellers, climate models do not exhibit unbounded growth of *finite* perturbations. Even more fortunately, for climate modellers and everyone else, neither does the real atmosphere.

■ Any climate-model prediction of the temperature on 1 June 2100 at a particular point will be no better than the average climate of the model at that time. What climate models should be able to predict is averages.

■ Since the only possible valid long-term predictions from climate models are climatological averages, the issue for any particular model is the following: how realistic is the climate in the model? A sensitive dependence on initial conditions of the atmosphere is largely irrelevant to this question. A dependence on initial conditions of the oceans is more problematic.

Box 7.3. Toy transport model A.3. The four-zone atmosphere

This model has been the basis of a large number of applications through its use in the ALE/GAGE/AGAGE studies as the main method for interpreting their data, generally with use of the Kalman filter as the estimation procedure.

This model represents the atmosphere using four equal-area latitude zones. For each of these zones the troposphere is divided into two layers with upper boundaries at 500 and 200 hPa. The first version of the model that was employed in ALE/GAGE studies [90] used a single stratospheric reservoir, giving nine boxes, with a stratospheric turnover time of 4 years. A more recent version has extended the zonal division into the stratosphere, giving a total of 12 boxes. The use of these models has been summarised by Prinn *et al.* [382].

The transport coefficients were represented as four-term Fourier series in time. In the original use of the model, the estimation process included an additional scale factor that was used to multiply the horizontal diffusion coefficients.

Cunnold *et al.* [90: Table 2] report the behaviour of the model in terms of the matrix of partial derivatives (**H** in the notation of this book) that are required when the model is used in the extended Kalman-filter formalism to estimate lifetimes and calibration factors.

7.4   Chemical–data assimilation

The tracer-inversion problems for CO_2, CH_4 and the halocarbons, which were introduced in the previous chapters, have mainly involved relations between surface concentrations and surface fluxes. Because the relation between sources and concentrations is linear, each of these is formally necessary and sufficient for specifying the other and, if only one of these data sets is available, the information will be consistent with any transport model.

There is, however, a more complicated class of problems, termed chemical-data assimilation, in which non-linearity may allow multiple solutions for a given data set or alternatively preclude any solution.

Some of the main characteristics of the chemical-data-assimilation problem are that,

- as with assimilation of meteorological data, there can be a large amount of information in the prior estimate (the forecast); and
- complex constraints apply to relations between tracers due to the non-linearity of chemical reactions and the diversity of time-scales involved.

Ménard [316] has reviewed techniques for the case of a single tracer, to which the issue of relations between tracers does not apply. The analysis is couched in terms of the Kalman filter and addresses the issues involved in achieving practical algorithms for large problems. Some of the key aspects are comparable to the application of the Kalman filter in assimilation of meteorological data, c.f. Todling [480].

Ménard quotes examples from a study of assimilation of CH_4 data [318, 317] testing the statistical model used for the assumed covariance matrices used in the variational form (7.3.2a)–(7.3.2c). He also notes a formalism that ensures that assimilated concentrations are positive by working with lognormal distributions. This work has been extended by Khattatov *et al.* [259]. Elbern *et al.* [118] describe a four-dimensional variational formalism for assimilating ozone data into an air-quality model. This involves using an adjoint model to define the variational changes in their objective function.

Segers *et al.* [429] present model studies that compare several techniques for dealing with non-linearities. These include

(i) ensemble methods, using multiple realisations from the distribution of possible states;
(ii) linearisations of the Kalman-filter equations, approximating the extended Kalman filter; and
(iii) second-order linearisations to provide a closer approximation to the extended Kalman filter.

Franssens *et al.* [161] describe the use of four-dimensional assimilation of aerosol data from satellite data. The time-variation is that due to atmospheric transport, which is used to provide the additional information to counteract the sparseness of the data – the data are from solar occultation and so only two transects per orbit are obtained. The adjoint equations are used to calculate the gradient of the penalty function. A special aspect of this problem is the need to convert from optical extinction, the quantity that is measured, to aerosol number density, which is the variable required by the transport model.

A range of chemical-data-assimilation studies has been presented in a recent workshop on assimilation of ozone data [439], covering a range of scales from urban-scale to global. Fisher and Lary [155] present global-scale analyses from their four-dimensional chemical assimilation scheme [271].

7.5   Related inverse problems

There are many other inverse problems that are closely related to the atmospheric-tracer-inversion problem, whether through mathematical similarity or similarity in terms of area of application. Problems that are mathematically similar are discussed in Section 8, where we review a systematic classification of the degree of difficulty of these problems.

Looking at inverse problems in the area of atmospheric science and/or biogeo-chemical cycles, some important cases (in addition to the data-assimilation problem described in the previous section) are the following.

Estimating atmospheric transport. Inverse problems involving estimation of atmospheric transport are discussed in Chapter 18. Generally the quantities that are estimated are a small number of characteristic time-scales.

Remote sensing. Chapter 5 noted remote sensing, especially from satellites, as a means of measuring atmospheric composition. An important technical issue is that it is generally necessary to obtain a simultaneous inversion for the vertical profiles of concentration and temperature. An early description of some of the mathematical issues is given by Twomey [488]. The work of Rodgers on remote sensing has made major contributions to the more general study of inverse problems [411].

Deconvolution of time series. In Section 6.1, we noted the characterisation of atmospheric budgets of greenhouse gases in terms of forcing and response and the associated convolution representation

$$m(t) = m_0 + \int_{t_0}^{t} \rho(t - t')s(t') \, dt' \tag{7.5.1}$$

Section 6.6 describes the inverse problem of deducing $s(.)$ given $c(.)$.

Ocean tracer inversions (physical). The 'classic' ocean inversion problem is that of deducing the circulation from data on temperature and salinity. The book by Wunsch [521] deals extensively with this problem.

Ocean tracer inversions (biogeochemical). A more complicated class of ocean tracer inversion is those that analyse biologically active tracers and seek simultaneously to invert ocean-circulation and biological-process rates. Typically the tracers included C, ^{14}C, O, N and P. Early studies were presented

by Bolin *et al.* using a 12-box ocean model [39]. This was addressed as a
linear-estimation problem, subject to linear constraints. Examples of more
recent work in biogeochemical inversions are [351] and [432]. Applications
based on adjoint modelling are given by [510] and [425].

Ocean inversion. More recently, the data from the World Ocean Circulation
Experiment (WOCE) have been analysed in an inversion that combined
biogeochemical information and the traditional data used in physical
oceanography [167].

Further reading

■ The lack of material on atmospheric-tracer-inversion problems was the reason
for writing this book. Various chapters from the Heraklion conference [244]
cover particular aspects of tracer inversion and a number of related inverse
problems.

■ For assimilation of meteorological data Daley's book [94] remains a key
reference. For recent developments a number of reviews need to be consulted
[480]. Chemical-data assimilation is a rapidly evolving field but there are some
useful reviews [318, 317, 439]. Similarly, surface-data assimilation is being
increasingly recognised as important. The ECMWF and WCRP/GEWEX
workshops [117] covered a wide range of relevant topics and should provide a
starting point for identifying more recent work.

Exercises for Chapter 7

1. Use Laplace transforms to obtain explicit forms of the deconvolution relation
   (6.6.2) when $\rho(t) = Ae^{-\alpha t} + (1 - a)e^{-\beta t}$.

2. For the model $c_n = G_n s + \epsilon_n$ show that the recursive solution is equivalent to
   the direct matrix inversion presented in Box 3.3.

Chapter 8

Ill-conditioning

The data may be contaminated by random errors, insufficient to determine the unknowns, redundant, or all of the above.

D. D. Jackson [224]

8.1   Characteristics of inverse problems

In Chapter 1, inverse problems were characterised as those in which the chain of calculation or inference was in the opposite direction to real-world causality. In the real world, dissipative processes often lead to a loss of information about details of real-world causes. In such cases, attempts to reconstruct the small-scale structure of the 'cause' require an amplification of attenuated details in the 'effect', but this will generally lead to amplification of errors (see Figure 1.3). Commonly, such ill-conditioning, rather than the cause–effect relation, is used as the definition of an 'inverse problem'.

The study of such ill-conditioned inverse problems has been long-established in seismology and in a wide range of other earth sciences. Inversions of composition data are important in oceanography [521]. The present chapter addresses some of the underlying mathematical issues involved in working with these problems. The analysis draws on accounts by Rodgers [411] and Twomey [488] and in particular that of Jackson [224].

The linear inverse problem

Many of the important characteristics of inverse problems can be studied in the linear inverse problem which, following previous chapters, we express as finding the solutions $\mathbf{x}$ of

$$\mathbf{c} \approx \mathbf{m} = \mathbf{Gx} \tag{8.1.1a}$$

The overall approach in this book is to regard such 'solutions' as estimates, $\hat{\mathbf{x}}$, based on the information in relation (8.1.1a), obtained using the concepts and techniques of statistical estimation described in Section 3.3. For the purposes of discussion in this section, we follow Jackson [224] and consider linear estimates:

$$\hat{\mathbf{x}} = \mathbf{Hc} \tag{8.1.1b}$$

where the $\mathbf{c}$ are the observations corresponding to the model predictions, $\mathbf{m}$, and $\mathbf{H}$ is regarded as an approximate inverse of $\mathbf{G}$.

The analysis of (8.1.1a) and (8.1.1b) is characterised by three numbers:

- N, the number of observations;
- M, the number of parameters; and
- P, the rank of $\mathbf{G}$, i.e. the number of non-zero singular values, λ_q, in the singular-value decomposition (SVD) of $\mathbf{G}$ (see Box 3.2).

Note that $P \leq M$ and $P \leq N$.

The problem of estimating parameters, $\mathbf{x}$, from a set of observations, $\mathbf{c}$, using model predictions, $\mathbf{m}(\mathbf{x})$ (given as $\mathbf{Gx}$), is termed 'underdetermined' if $M > P$ and 'overdetermined' if $N > P$; it may be both. For underdetermined problems, there will be a null space of dimension $M - P$ (see Box 3.1). Any vector from this null space can be added to a solution of (8.1.1a). For overdetermined problems we can expect to obtain only a solution for $\mathbf{x}$ that approximately satisfies (8.1.1a).

In the analysis of ill-conditioned problems, we have to address the case in which the singular values may be very small, even though they are not strictly zero. In particular, for ill-conditioned problems, as we increase the resolution (i.e. increasing M and N) we find that, while there are additional non-zero singular values, λ_q, these may rapidly drop to zero – increasing the resolution ceases to add useful information.

For most of this chapter, 'resolution' is discussed in a rather abstract manner. In some of the examples, it is taken as resolution in the time domain. In much of the discussion of tracer inversions it is spatial resolution that is of most interest. In other contexts, the question of interest could be the ability to 'resolve' two or more different processes in the sense of being able to distinguish between them.

Regularisation using the SVD

The singular-value decomposition (SVD) (see Box 3.2) can be used to clarify the possibilities. The SVD allows us to write

$$G_{j\mu} = \sum_{q=1}^{P} U_{jq} \lambda_q V_{\mu q} \tag{8.1.2a}$$

where $\mathbf{U}$ and $\mathbf{V}$ are orthogonal matrices and $\lambda_q > 0$.

Box 8.1. **Regularisation**

The ubiquity of ill-conditioned inverse problems has led to generic approaches for obtaining stable solutions. The class of approach known as regularisation originated by Tikhonov [478] (and independently by other workers) uses additional constraints to stabilise the solution. An ill-posed problem of the form

$$\mathcal{G}f(\mathbf{y}) = s(\mathbf{y}) \tag{1}$$

is solved subject to a constraint of the form

$$||\mathcal{R}f(\mathbf{y})|| \le c \tag{2}$$

where $||.||$ denotes a norm (see Box 12.2), such as the mean square, and $\mathcal{R}$ is an operator such as differentiation.

The standard solution technique for constrained optimisation is to introduce a Lagrange multiplier, Λ, and minimise

$$J = ||\mathcal{G}f(\mathbf{y}) - s(\mathbf{y})|| + \Lambda||\mathcal{R}f(\mathbf{y})|| \tag{3}$$

with respect to $\mathbf{y}$ and use the constraint equation to determine the value of Λ.

Within this general class of technique, there is a range of choices for the appropriate norm and a wide range of choices for the type of constraint. Some of the main choices are constraints on the function or one of its derivatives, especially the second derivative, implying a smoothness constraint.

Craig and Brown [81], on which the above summary is based, also note the class of regularisation technique based on statistical arguments – essentially the Bayesian approach advocated in this book.

Anderssen [8] has noted that occasionally the actual solution of an ill-posed problem is not required, rather it is some functional of the solution that is needed. In such cases, he recommends the use of computational techniques that evaluate the functional directly.

Some empirical studies of the applicabilities of various types of regularisation in CO_2 inversions have been given by Fan et al. [151], who compared bounds on sums of squares of sources, Bayesian constraints and truncated SVD, and by Baker [20], who compared truncated SVD, smoothing in time and aggregation in space.

We can consider a class of approximate inverses of **G** of the form

$$H_{j\mu} = \sum_q U_{jq} f_{\text{reg}}(\lambda_q) V_{\mu q} \tag{8.1.2b}$$

which are used to construct estimates:

$$\hat{x}_\mu = \sum_j H_{j\mu} c_j \tag{8.1.2c}$$

In order to reproduce the observations we require $f_{\text{reg}}(\lambda) \approx \lambda^{-1}$ but, to avoid having a large variance in the estimates, we need to avoid large values of $f_{\text{reg}}(\lambda)$. Choosing between these conflicting requirements involves some sort of 'trade-off' and has become known as regularisation (see Box 8.1) – hence the notation $f_{\text{reg}}(.)$.

A particular case of (8.1.2b) is the truncated singular-value expansion which uses a cut-off, λ_c, for the lower bound on the 'useful' singular value and defines the approximate inverse through (8.1.2b) with

$$f_{\text{reg}}(\lambda) = \lambda^{-1} \qquad \text{if } \lambda > \lambda_c \tag{8.1.3a}$$

or

$$f_{\text{reg}}(\lambda) = 0 \qquad \text{if } \lambda \leq \lambda_c \tag{8.1.3b}$$

The pseudo-inverse (see Box 8.2) can be regarded as a special case of (8.1.3a) and (8.1.3b) with $\lambda_c = 0$. Choosing $\mathbf{H}$ to be the pseudo-inverse gives a solution that is a least-squares fit (if the problem is overdetermined) and a minimum-length solution (if the problem is underdetermined) – the proof of this is left as an exercise – problem 1 below.

Box 8.2. The pseudo-inverse

The Moore–Penrose pseudo-inverse (or simply pseudo-inverse) of a matrix $\mathbf{A}$ is a matrix denoted $\mathbf{A}^{\#}$ with the properties [22: Appendix A] that

$$\mathbf{AA}^{\#}\mathbf{A} = \mathbf{A} \tag{1a}$$

$$\mathbf{A}^{\#}\mathbf{AA}^{\#} = \mathbf{A}^{\#} \tag{1b}$$

and that $\mathbf{A}^{\#}\mathbf{A}$ and $\mathbf{AA}^{\#}$ are symmetric. The pseudo-inverse is unique. If $\mathbf{A}$ is square and non-singular, then $\mathbf{A}^{\#} = \mathbf{A}^{-1}$.

An explicit expression for the pseudo-inverse can be obtained using the singular-value decomposition (see Box 3.2). If we write

$$A_{j\mu} = \sum_q U_{jq} \lambda_q V_{\mu q} \tag{2a}$$

then

$$A^{\#}_{\mu j} = \sum_q U_{jq}\, f_{\text{reg}}(\lambda_q)\, V_{\mu q} \tag{2b}$$

with

$$f_{\text{reg}}(\lambda) = \lambda^{-1} \qquad \text{if } \lambda \neq 0 \tag{2c}$$

or

$$f_{\text{reg}}(\lambda) = 0 \qquad \text{if } \lambda = 0 \tag{2d}$$

For the linear problem

$$\mathbf{Ax} = \mathbf{c} \tag{3}$$

the solution

$$\hat{\mathbf{x}} = \mathbf{A}^{\#}\mathbf{c}$$

has the smallest solution (smallest $\mathbf{x}^{\mathsf{T}}\mathbf{x}$) if (3) is underdetermined. It is also the best fit (smallest $[\mathbf{A}\hat{\mathbf{x}} - \mathbf{c}]^{\mathsf{T}}[\mathbf{A}\hat{\mathbf{x}} - \mathbf{c}]$) if (3) is overdetermined. The proof of these properties is left as an exercise.

With an appropriate normalisation of $\mathbf{G}$, the Bayesian estimation described in Section 3.3 corresponds to

$$f_{\text{reg}}(\lambda) = \frac{\lambda}{\lambda^2 + 1} \tag{8.1.4}$$

The use of (8.1.4) avoids unbounded amplification of errors by using the prior estimates for those components that are poorly resolved by the data. In the absence of a systematic Bayesian analysis, techniques such as ridge regression and other forms of regularisation can achieve a similar effect, but in a less systematic way. In a non-Bayesian context, regularisation of the form (8.1.2b) is a form of biased estimation. The nature of the 'Bayesian' form (8.1.4) emphasises an important aspect of using approximate inverses based on the SVD: the SVD is not invariant under changes in the relative normalisation of the parameter space or the observation space. In other words, the SVD will generally change as the units of the parameters and measurements are changed. The Bayesian form (8.1.4) corresponds to choosing units (and possibly combinations of observations and parameters) that make the covariance matrices for the observations and the prior estimates of parameters equal to the identity matrix.

Some of the key questions, whose answers determine the mathematical characteristics of an inverse problem, are the following.

- How many of the λ_q are non-zero?
- How does P, the number of non-zero singular values, change with increasing resolution, i.e. as $M, N \to \infty$?
- How rapidly do the non-zero λ_q go to zero as q increases?
- How does this change with increasing density of data?

Generally, in geophysical inverse problems the first two of the questions above are of lesser importance. Data are collected with the aim of giving information about the causal parameters in the system. The case in which some causal components have absolutely no effect on the observations ($M > P < N$) is unlikely. Similarly, apart from measurements that appear as exact duplicates (including the same error), each measurement will usually be giving some information about the system. Far more likely are (i) causal components that have very little influence on observable quantities and/or (ii) observations that are nearly, but not completely, redundant. These cases correspond to singular values that are small, but not precisely zero.

Estimation of surface fluxes

The development of tracer-inversion techniques that reflect the ill-conditioned nature of the problem has occurred in a rather *ad hoc* manner. Early mass-balance-inversion studies showed the need to smooth the concentration data both in space and in time in order to prevent instability, even with two-dimensional models. Formal analysis of the ill-conditioned nature of the problem of deducing surface sources came from analytical studies of the 'toy model' using diffusion [138; see also Box 2.2], combined with numerical studies based on operational transport models [136, 132]. Plumb and Zheng [361] used a 'toy model' with transport based on diffusion plus solid-body rotation and performed the SVD for the Green function relating observations to sources for the case of a five-site observational network. Although much of their interest was in the underdetermined nature of this problem, they noted the increase in error-amplification factors, λ_q^{-1}, for estimating higher-order modes. They determined relative error-amplification factors of 1, 3.33, 7.36 and 10.04 (in their 'fast-advection' case) for the four modes of source variation that could be detected by the ALE/GAGE sampling network. Fan *et al.* [151] plotted the singular values of their Green-function matrix, showing that there was a rapid decrease in the first four singular values and then a relatively slow decrease followed by a more rapid decrease for the last few values. Using a two-dimensional model, Brown [57] performed a singular-value analysis of the problem of deducing surface sources from surface-concentration data. The source discretisation was 18 latitude bands × four Fourier components in time. The resolution of the data was 18 locations × 12 monthly values, with a small number of missing data. The singular values, λ_q, exhibited a generally $1/q$ decay. This SVD analysis was used to analyse the signal-to-noise characteristics of the CH_4 data set used by Brown and to identify λ_c for use in (8.1.3a) and (8.1.3b).

The following section considers a number of specific tracer-inversion problems in terms of the form of the decrease in the λ_q. Most of these analyses are based on 'toy models', in some cases supplemented by tests with numerical models.

In Section 8.3, the issues of resolution are analysed in a more abstract form, not tied to specific problems, but in a way that brings together the multiple meanings of the terms 'resolution'.

8.2   Classification of inverse problems

Numerical differentiation

Numerical differentiation is often regarded as an archetypal inverse problem. The order of differentiation can be used as a measure of the difficulty of the problem. This measure can be extended to other problems, with a q^{-n} rate of attenuation for singular values in the forward problem meaning that the inverse problem is comparable in difficulty to that of nth-order numerical differentiation. This analogy can be extended to inverse problems with non-integer n (such as problems from stereology), which can be related to the generalised theory of fractional-order differentiation (see Box 8.3). Inverse problems in radiative transfer, e.g. those involved in remote sensing of the atmosphere, may have an exponential (i.e. α^q) rather than power law (i.e. q^α) attenuation rate.

Following Anderssen and Bloomfield [9], we analyse numerical differentiation as an ill-conditioned problem by working in the frequency domain, using Figure 8.1 for illustration. The solid curve in the top panel of Figure 8.1 gives an example of $h_f(\omega)$, the spectral density of a signal. We consider data-spacings of $\Delta t = \frac{1}{2}, \frac{1}{4}$ and $\frac{1}{8}$, i.e.

cut-off frequencies of $\nu = 1, 2$ and 4. In the consistent spectral normalisation described in Box 4.1, halving the data-spacing halves the spectral density. The dashed curves in the top panel are the spectral density for the same white-noise process for the three different data-spacings. The centre panel shows filters that can produce estimates of the derivatives (with $\psi(-\omega) = -\psi(\omega)$). The solid curves are the optimal filters for the three data-spacings, defined by taking the derivatives of functions $\hat{f} = \Psi * c$, estimated using

$$\psi_{\text{opt}}(\omega) = \frac{h'_f(\omega)}{h'_f(\omega) + h'_e(\omega)}$$

The dashed curves show the filters defined by differencing a second-order binomial smoothing at the three data spacings. In these cases, changing the data-spacing re-scales the filter both in the magnitude and in the spread of frequencies. The lower panel shows $h'_e(\omega)\psi(\omega)^2$, the frequency-dependence of contributions to the variance of the estimated derivative $\hat{\dot{f}}$.

Figure 8.1 shows limits on how well derivatives can be estimated. The optimal estimates simply ignore the high-frequency components once the power in the signal drops below the noise level. Decreasing the data-spacing decreases the noise power at any one frequency, but the power in the function will drop more rapidly – if it doesn't then the derivative doesn't exist. Numerical differentiation that uses a specific differencing scheme, regardless of the signal-to-noise ratio, incorporates more and more of the noise as the data-spacing decreases, as indicated by the dashed curves in Figure 8.1(c).

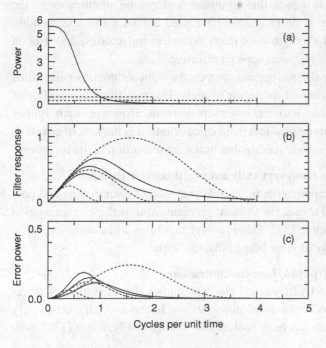

Figure 8.1 Characterising numerical differentiation. (a) The spectral density of an illustrative function (solid curve) and noise (dashed curves) for data spacings of 0.5, 0.25 and 0.125 time units; (b) responses, $\psi(\nu)$, of filters for estimating the derivative (solid curves are optimal filters for a specified signal and noise power, dashed curves are differenced binomial filters); and (c) contributions to the variance in estimates of the derivative, $\dot{f}$, from the six filters.

Classification

Using numerical differentiation as a reference case, we can list a number of tracer-inversion problems in which the degree of ill-conditioning has been identified. This builds on the classification by Enting [122]. The description is based on the spatial aspects, characterised by wavenumber, k.

Estimating surface sources from surface concentrations

When one is using atmospheric transport to relate surface sources to surface concentrations, the forward problem has a k^{-1} attenuation rate for the dependence on latitude. The early study by Bolin and Keeling [37] implied a k^{-2} attenuation, because they failed to distinguish between surface concentrations and vertical averages. Enting and Mansbridge [136] identified the k^{-1} attenuation in a two-dimensional numerical model, but incorrectly attributed the difference from Bolin and Keeling by suggesting that it was due to the difference between advection and diffusion. Newsam and Enting [339] noted that the degree of attenuation will be determined by the most dispersive process and that the distinction between k^{-1} and k^{-2} was due to the formulation of the problem rather than the way in which transport was modelled. The k^{-1} attenuation in diffusive models is illustrated by the toy models B.1 and B.2 and a number of the exercises based on these models.

The k^{-1} attenuation has also been confirmed in a three-dimensional model [132]. Prinn [375] has queried the applicability of these results, noting that, apart from molecular diffusion, all atmospheric-transport processes are due to some form of advection. The use of diffusive parameterisations in models is merely a substitute for unresolved sub-grid-scale advection and in practice additional unphysical numerical diffusion may be present in models. However, this argument neglects the significance of the time-scales involved. Global-scale inversions have used annual mean and monthly mean data, which represent averages over many advective trajectories; and it is on these scales that atmospheric transport appears diffusive.

However, for problems involving regional sources, the highest signal-to-noise ratios are for time-scales too short for such averaging to apply. This may introduce a different form of asymptotic behaviour into the inversion problem. However, such studies remain to be done. In current regional inversion calculations, it is the lack of coverage of data rather than any atmospheric mixing that limits the resolution of the inversions.

Estimating surface sources from vertically averaged concentrations

This problem has potential application in the interpretation of aircraft data and possibly satellite data. As noted above, the forward problem exhibits the k^{-2} attenuation calculated by Bolin and Keeling [37]. This is a very rapid loss of information, but it also decreases the probability of there being 'aliasing' error.

Free-atmosphere sources from surface concentrations

This problem is severely underdetermined since it implies the derivation of three-dimensional fields from two-dimensional observations. In this case, there is a very large null space. The problem has been analysed by Enting and Newsam [138], who suggested that what could be estimated from surface observations would be weighted averages of the vertical distribution of sources. Furthermore, the greater the degree of spatial variability in the surface data the more the averaging would be weighted towards the earth's surface. For the purely diffusive model, Enting and Newsam found that the vertical averages were attenuated as $k^{-3/2}$ at latitudinal wavenumber k.

Incorporating extra constraints
Owing to the ill-conditioning of the 'pure' tracer-inversion problems, many studies incorporate additional information. This raises the question of whether these data change the nature of the inverse problem, or rather (since such a change is what is required) how the nature of the inverse problem is changed. The role of extra constraints can often be addressed by regarding such constraints as part of the prior information used in the process of Bayesian estimation.

Deconvolution
The relation between inverse problems and differentiation can be seen by considering the convolution equation:

$$m(t) = \int \rho(t - t')s(t')\,dt' \tag{8.2.1a}$$

which, as described in Section 6.6, has a formal inverse

$$s(t) = \frac{\dot{m}(t)}{\rho(0)} + \frac{m(t)\dot{\rho}(0)}{\rho(0)^2} + \int_0^t \rho^*(t - t')m(t')\,dt' \tag{8.2.1b}$$

where $\rho^*(.)$ is a smoothly varying function. This inversion relation requires differentiation of m and so the ill-conditioning is that of first-order numerical differentation, so long as $\rho(0)$ is non-zero. If $\rho(x)$ vanishes as x^n, then the inverse problem becomes equivalent to differentiation of order $n + 1$, as discussed (including the extension to non-integer n) in Box 8.3. In inverting integral equations such as (8.2.1a), i.e. Volterra equations, the singularity in the kernel, $\rho(.)$, determines the equivalent order of fractional differentiation.

Box 8.3. Fractional-order differentiation

Additional insight into the degree of difficulty of various inverse problems can be obtained by considering the mathematical identity

$$\int_c^{x_0} dx_1 \int_c^{x_1} dx_2 \cdots \int_c^{x_{n-2}} dx_{n-1} \int_c^{x_{n-1}} h(x_n)\,dx_n$$

$$= \frac{1}{\Gamma(n)} \int_c^x (x - x_n)^{n-1} h(x_n)\,dx_n \tag{1}$$

This can be derived in terms of Laplace transforms, for which, if the transform of $F(t)$ is $f(p)$, then the transform of the integral of $F(t)$ is $f(p)/p$. A special case is that p^{-n} is the Laplace transform of $t^{n-1}/(n - 1)!$. Convolutions of functions transform into the product of their Laplace transforms. Therefore multiplying a Laplace transform by p^{-n} is equivalent both to n-fold integration and to convolution with $t^{n-1}/(n-1)!$, i.e. the equivalence indicated by (1).

- n-fold integration is equivalent to convolution with kernel $(x - x')^{n-1}$.
- Therefore deconvolution of an integral equation with kernel $(x - x')^{n-1}$ is equivalent to n-fold differentiation.

More generally, the deconvolution of a relation whose kernel goes to zero as x^{n-1} will have a degree of ill-conditioning comparable to that of nth-order differentiation. Beyond its role in indicating the degree of difficulty of deconvolutions, equation (1) forms the basis of a formalism for integration and differentiation of fractional order, since the right-hand side of (1) is not restricted to integers [415]. This in turn means that inverse problems involving kernels with fractional powers can be regarded as equivalent to fractional-order differentiation. Enting [122] gave an example related to trapping of bubbles in polar ice but this is of limited real-world applicability because it neglects the role of diffusion. However, many realistic examples of inverse problems equivalent to fractional-order differentiation occur in inversion problems in stereology [10].

Alternative presentations of flux estimates

The comparison of inverse problems with differentiation implies that the problem of estimating surface sources of trace gases is equivalent in difficulty to a numerical differentiation (in space) of the concentration data. This correspondence suggests that it may be useful to consider spatially integrated fluxes. As an illustration, we consider non-fossil sources of CO_2 for five zones bounded by 72° S, 48° S, 16° S, 16° N, 48° N and 72° N, with high-latitude fluxes taken as zero. The time-averaged sources of CO_2 were estimated using the spatial discretisation of [145], as described in Box 10.1. After performing the inversion, the estimates were cumulated onto the five latitudinal zones. (This is an approximation for the terrestrial components, which are based on biomes – the grasslands were assigned to the 16° N to 48° N zone and the 'other', mainly tundra, assigned to the 48° N to 72° N zone). Equation (8.2.2a) gives the prior covariance matrix for these five components and (8.2.2b) gives the covariance matrices for the estimated sources using CO_2 data from a network of 20 sites, with each annual mean assigned a standard deviation of 0.3 ppm.

For the ranges 72° S to 48° S, 48° S to 16° S, 16° S to 16° N, 16° N to 48° N and 48° N to 72° N, the covariance matrices are

$$\mathbf{W} = \begin{bmatrix} 1.00 & 0.00 & 0.00 & 0.00 & 0.00 \\ 0.00 & 4.25 & 0.00 & 0.00 & 0.00 \\ 0.00 & 0.00 & 11.50 & 0.00 & 0.00 \\ 0.00 & 0.00 & 0.00 & 9.00 & 0.00 \\ 0.00 & 0.00 & 0.00 & 0.00 & 5.00 \end{bmatrix} \quad \text{prior covariance} \quad (8.2.2a)$$

and

$$C = \begin{bmatrix} 0.21 & -0.37 & 0.22 & -0.06 & 0.01 \\ -0.37 & 1.04 & -0.53 & -0.04 & -0.07 \\ 0.22 & -0.53 & 1.75 & -1.37 & 0.00 \\ -0.06 & -0.04 & -1.37 & 2.43 & -0.86 \\ 0.01 & -0.07 & 0.00 & -0.86 & 0.94 \end{bmatrix} \quad \text{covariance of estimates}$$

(8.2.2b)

It will be seen that, first, the uncertainties are quite large; and secondly, there is a large negative correlation between the estimates for neighbouring zones.

If we consider the spatially integrated fluxes, i.e. the fluxes for the regions south of 48° S, 16° S, 16° N, 48° N and 72° N, then the covariance matrix for these combinations is

$$C = \begin{bmatrix} 0.21 & -0.16 & 0.06 & 0.00 & 0.01 \\ -0.16 & 0.50 & 0.18 & 0.09 & 0.03 \\ 0.06 & 0.18 & 1.64 & 0.16 & 0.10 \\ 0.00 & 0.09 & 0.16 & 1.12 & 0.20 \\ 0.01 & 0.03 & 0.10 & 0.20 & 0.23 \end{bmatrix} \quad \text{covariance of cumulated estimates}$$

(8.2.2c)

The uncertainties are somewhat reduced and the large correlations have been greatly reduced. Figure 8.2 plots the estimates of the latitudinally integrated fluxes and the ± 1- and ± 2-SD ranges, using the annual mean concentrations for 1986–7. These bands show how the -2.9 Gt C yr^{-1} net non-fossil flux can be partitioned between latitudes. Of course, the straight-line segments connecting the latitudes are a guide to the eye, rather than being actual results from the inversion. However, since the boundary latitudes are only roughly related to oceanic and biotic regimes, it is to be expected

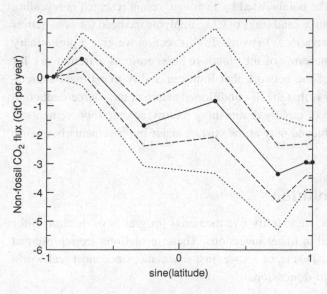

Figure 8.2 Uncertainties in latitudinally integrated flux estimates derived from synthesis inversion using 24 regions and 20 measurement sites. For each latitude, the curve represents the estimate (with ± 1- and ± 2-SD ranges) of the integrated flux south of that latitude.

that estimates of integrated fluxes will be only slightly correlated for differences in latitude exceeding 30°, regardless of where the boundaries are chosen.

This way of presenting the results of CO_2 inversions was suggested by the nature of the inverse problem as being equivalent to a numerical differentiation. Presenting the results in this 'integrated' form removes the strong negative autocorrelations associated with differentiation (or equivalent transformation) of noisy data. The integrated form also has the advantage of facilitating comparisons between results reported at different discretisations. These properties will be used in Chapter 14.

8.3   Resolution and discretisation

Discussions of tracer inversion have emphasised the issues of spatial resolution – what spatial features can be resolved and what improvements can be made with additional data. In assessing global networks, Pearman [353] estimated the requirements for interpreting CO_2 as

- ■ *about three stations,* to determine global inventories and trends;
- ■ *an additional 5–10 stations*, to determine meridional transport of CO_2;
- ■ *an additional 10–20 stations*, to determine air–surface exchange between large-scale areas; and
- ■ *over 100 stations*, to determine air–surface exchange within regions with significant anthropogenic influence.

It is an extremely difficult matter to quantify these estimates more precisely in order to determine the resolving capability of the existing network or to suggest general principles for improvements in coverage for a network. Even the results on network design presented in Section 13.3 must be regarded as preliminary 'proof-of-concept' calculations. In addition to the points listed by Pearman, recent research is revealing the complexity of *natural* sources and sinks of CO_2, implying the need for even denser spatial coverage by the observational network. In this section we give a preliminary discussion of the problem in terms of the *signal-to-noise ratio* as a framework for focusing on those aspects of the network that limit our ability to estimate source–sink strengths. What is clear is that the ill-conditioned nature of the source-deduction problem implies a condition of rapidly diminishing returns, so that improvements in spatial resolution can be achieved only at the cost of major improvements in quality and coverage of data.

Aspects of resolution

In order to clarify the issues, we identify five meanings (or groups of meanings) for the term 'resolution' as used in tracer inversions. These resolutions are represented in terms of wavenumbers, k. This is, of course, just schematic, since most real-world problems of interest are multi-dimensional.

The discussion is based on an abstract problem:

$$c_k = G_k y_k + e_k \tag{8.3.1a}$$

which is a representation in the frequency domain (spatial or temporal) of continuous functions, $y(x)$ and $c(x)$ in terms of a set of basis components $f_k(x)$ that resolve successively finer detail as k increases:

$$y(x) = \sum_k y_k f_k(x) \tag{8.3.1b}$$

The various aspects of resolution are the following.

(i) **Computational resolution.** In atmospheric-transport modelling, this means the (spatial) resolution of the transport model and is denoted k_{comp}. (The resolution of prescribed source distributions in synthesis inversion is treated as a distinct aspect – point (iv) below.) The requirement for computational resolution is that the computational basis spans the space of those c_k and y_k that appear.

(ii) **The resolution of data.** This is denoted k_{data} and represents the sampling scale. This is important both as a limit on the range of wavenumbers that can be represented in estimates based on these data and for the possibility of aliasing high-frequency variations.

(iii) **Signal resolution.** This is a question of what degree of variability is present in the signal. In practice what matters is the resolution that is imposed by the ratio of the (attenuated) signal to the observational noise, i.e. for what resolution do we have $G_k y_k > \sqrt{\text{var } e_k}$? This point is denoted k_{sn}. Note, from the example in Figure 8.1, that, in the frequency domain, the noise level will generally depend on the data-spacing.

(iv) **Component resolution.** This is the discretisation scale in synthesis inversions and is denoted k_{synth}. For computational reasons this is often significantly less than k_{comp}. An exception is when Green-function matrix elements are determined using an adjoint model, as discussed in Section 10.6. If we are estimating the y_k as $\hat{y}_k = H_k c_k$, then we can notionally define k_{synth} by $H_k G_k < 0.5$ for $k > k_{synth}$. For the optimal (minimum MSE) choice of H_k, this corresponds to $k_{synth} = k_{sn}$.

(v) **Target resolution.** This is denoted k_{target}. It applies in the situation when, rather than trying to estimate $y(x)$, we *choose* to estimate a 'smoothed' function defined by the expansion $\tilde{y}(x) = \sum_{k \leq k_{target}} x_k f_k(x)$. This form is analogous to SVD truncation. More generally, we may require k_{target} to specify a characteristic wavenumber around which a gradual cut-off is applied.

The primary requirement is

$$k_{target} \leq \min(k_{comp}, k_{data}, k_{sn}, k_{synth}) \tag{8.3.2a}$$

so that the best achievable resolution is

$$k_{\text{target}} = \min(k_{\text{comp}}, k_{\text{data}}, k_{\text{sn}}, k_{\text{synth}}) \tag{8.3.2b}$$

Other relations between these different types of resolution are

- the obvious need to have k_{comp} exceed k_{synth}; and
- that adding data is effective only if the data resolution k_{data} is less than the signal resolution k_{sn}, otherwise improvements in observational precision (i.e. in signal-to-noise ratio) are required if the resolution is to be improved.

This schematic analysis applies quite directly for the case of stationary time series expanded in terms of Fourier components. In the spatial domain, the results of the purely diffusive model [138; see also Box 2.1] suggest that the analysis can apply for a spherical-harmonic expansion.

This analysis can be applied still more generally. Following the analysis of Section 3.3, we use a basis in which both the observational covariance and the prior source covariance are given by identity matrices. We can then use the singular-value decomposition to obtain a transformed basis in which the covariance matrices $\mathbf{W}$ and $\mathbf{X}$ remain identity matrices and $\mathbf{G}$ becomes diagonal with elements G_k as in the schematic example of (8.3.1a) and (8.3.1b). Such SVD analysis was undertaken by Brown [57], who found that her methane data set allowed estimation of zonally averaged sources with a resolution of about $30°$ in latitude.

In terms of the requirement (8.3.2a), we have the ability to control k_{comp} and k_{synth} at the expense of computer time. In principle, we can expand the coverage of data to increase k_{data}. It is k_{sn} that provides the limit that is much harder to avoid. In these terms, the criterion for achievable resolution is clear: once $\tilde{G}_k$ drops significantly below unity (the standard deviation of ϵ_k), the data are not providing information. In Bayesian formalisms, the only information is coming from the prior estimates. This is reflected by having the signal-to-noise-ratio criterion involve both the precision of measurement (the noise level) and the *a priori* source strength (the expected signal) as well as the degree of attenuation in the forward problem.

An important point to note is that, if k_{sn} is to be determined as part of the estimation procedure, then the calculations need to be performed so that k_{sn} turns out to be significantly smaller both than k_{data} and than k_{comp}.

Many of these considerations apply widely in signal-processing problems. However, the issue of basis truncation (choice of k_{synth}) is particularly important in synthesis-inversion studies and became rather contentious with the study by Fan *et al.* [149].

Kaminski *et al.* [241] have quantified the importance of this truncation error in CO_2 inversions. The effects of restricted component resolution in synthesis inversion were investigated by using basis functions derived from adjoint modelling. Inversions with the full spatial resolution of 836 regions were compared with inversions constrained to 18 and 54 regions. For 18 regions, root-mean-square errors in the fluxes were found to be larger than the mean fluxes, whereas for 54 regions the errors were comparable to the fluxes. Similar aggregation studies (going from low resolution to very low resolution) were presented by Peylin *et al.* [356].

An illustrative discussion of the effect of basis truncation is given in Figure 8.3, in terms of reducing a two-parameter case to one parameter.

(i)   A schematic representation of an inversion problem involving two parameters, a and b, with unknown true values shown by *, where the quantity of interest is $a + b$. The dotted lines correspond to constant $a + b$. The filled circle is the projection of the true values onto the line $a = b$.

(ii) A data constraint that samples the distribution inhomogeneously produces a biased estimate of $a + b$ if it is analysed in terms of a uniform basis. This estimate is the heavy segment on the line $a = b$.

(iii) If sampling of data is 'homogeneous', biases from truncation cancel out to give the unbiased estimate as the heavy segment on the line $a = b$, with a smaller range than that in (ii) because of the additional data. Residuals are larger than the specified uncertainty in the data because of the neglect of the truncation error.

(iv) Wunsch and Minster [522] recommend working in a high-dimensional space, so that the modes that are resolved are determined by the data rather than imposed by arbitrary decisions of the modeller. The illustration includes a notional Bayesian constraint so that the uncertainty in the a–b space is represented by an ellipse: the uncertainty is well constrained in one direction and poorly constrained in the other. The uncertainty in $a + b$ is determined by the projection of the ellipse onto the line $a = b$. Wunsch [521: p. 128] noted that 'parameter reduction can lead to model error or biases that lead to wholly illusory results'.

(v)   Using a basis that matches the parameter distribution produces unbiased estimates with an uncertainty shown as the heavy segment on the 'shaped-basis' line. The uncertainty in $a + b$ is obtained by projecting this onto the line $a = b$, where it is also shown as a heavy-line segment.

(vi) The 'recipe' of Trampert and Snieder [484] (described in more detail below) works with a truncated basis ($a = b$ in this example) treats the missing $a - b$ component in statistical terms as an additional error to be applied to the data constraint. This reduces the bias and produces an uncertainty range that is more realistic, being comparable to that obtained from the higher-dimensional calculation (case (iv)). The $a - b$ component is characterised by an uncertainty range, shown here as a band around the line $a = b$ representing $|a - b| \leq 1$. Constraining $a - b$ constrains the amount of bias that can come from inhomogeneous constraints on data. This bias range is obtained by projecting the maximum-bias points (i.e. where the bounds of this band meet the data constraint) onto the $a - b$ line as shown by the arrows. This gives a measure of the truncation error (the inner bound around the data constraint), which is added quadratically to the original data error to give an effective data error (the outer band around the data constraint).

The need for the Trampert and Snieder correction can be understood qualitatively in terms of 'aliasing' of the source distribution. If the source distribution is truncated as k_{synth} then the neglected modes with higher wavenumbers have to be regarded as an error contribution. This term can be safely neglected only if it is much smaller than the 'measurement-error' term. Thus we have an additional requirement:

$$G_k y_k \ll \sqrt{\text{var } e_k} \qquad \text{at } k = k_{synth} \tag{8.3.2c}$$

if the Trampert and Snieder correction is neglected. One consequence of this is that adding additional data without taking account of truncation error can actually be harmful to the estimation. As illustrated in Figure 8.1, adding data leads to a reduction in var e_k. Thus, for a fixed choice of k_{synth}, adding data can transform a problem from one in which truncation error is small relative to measurement error into a problem dominated by measurement error. This needs to be considered when one is interpreting the results of 'perfect-model' experiments that explore the utility of adding additional data.

A correction for discretisation error

The mathematical analysis of the statistical treatment of truncation error has been given by Trampert and Snieder [484]. They expressed the space of parameters as $[\mathbf{x}, \mathbf{y}]^\mathsf{T}$, where the components $\mathbf{x}$ are those that are resolved and the components $\mathbf{y}$ are those that are neglected in the truncation. The prior-variance matrix is approximated by treating it as block diagonal, with prior covariances of $\mathbf{x}$ and $\mathbf{y}$ denoted $\mathbf{W}$ and $\mathbf{V}$. The data, with measurement covariance $\mathbf{X}$, are related to the sources by $\mathbf{c} \approx \mathbf{m} = \mathbf{Gx} + \mathbf{Hy}$. If both $\mathbf{x}$ and $\mathbf{y}$ are included in the estimation process, then

$$\begin{bmatrix} \hat{\mathbf{x}} \\ \hat{\mathbf{y}} \end{bmatrix} = \begin{bmatrix} \mathbf{G}^\mathsf{T}\mathbf{XG} + \mathbf{W} & \mathbf{G}^\mathsf{T}\mathbf{XH} \\ \mathbf{H}^\mathsf{T}\mathbf{XG} & \mathbf{H}^\mathsf{T}\mathbf{XH} + \mathbf{V} \end{bmatrix}^{-1} \begin{bmatrix} \mathbf{G}^\mathsf{T}\mathbf{Xc} \\ \mathbf{H}^\mathsf{T}\mathbf{Xc} \end{bmatrix} \tag{8.3.4a}$$

If we wish to estimate $\mathbf{x}$ only, then we can use the form of the block-matrix inverse (Box 13.1) that has the top-left element as a factor of the top row and put

$$\hat{\mathbf{x}} = [\mathbf{G}^\mathsf{T}\mathbf{XG} + \mathbf{W} - \mathbf{G}^\mathsf{T}\mathbf{XH}[\mathbf{H}^\mathsf{T}\mathbf{XH} + \mathbf{V}]^{-1}\mathbf{H}^\mathsf{T}\mathbf{XG}]^{-1}$$
$$\times [\mathbf{G}^\mathsf{T}\mathbf{Xc} - \mathbf{G}^\mathsf{T}\mathbf{XH}[\mathbf{H}^\mathsf{T}\mathbf{XH} + \mathbf{V}]^{-1}\mathbf{H}^\mathsf{T}\mathbf{Xc}] \tag{8.3.4b}$$

which reduces to

$$\hat{\mathbf{x}} = [\mathbf{W} + \mathbf{G}^\mathsf{T}\mathbf{Z}^{-1}\mathbf{G}]^{-1}\mathbf{G}^\mathsf{T}\mathbf{Z}^{-1}\mathbf{c} \tag{8.3.4c}$$

where

$$\mathbf{Z}^{-1} = [\mathbf{X}^{-1} + \mathbf{HV}^{-1}\mathbf{H}^\mathsf{T}]^{-1} = \mathbf{X} - \mathbf{XH}[\mathbf{H}^\mathsf{T}\mathbf{XH} + \mathbf{V}]^{-1}\mathbf{H}^\mathsf{T}\mathbf{X} \tag{8.3.4d}$$

by virtue of the matrix-inversion lemma (see problem 1 of Chapter 3).

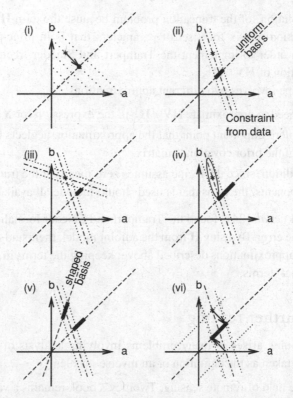

Figure 8.3 A schematic outline of basis truncation, reducing two parameters to one. (i) The inversion problem involving a and b with unknown true values shown by *, where the quantity of interest is $a + b$. (ii) Data that sample the distribution inhomogeneously produce a biased estimate of $a + b$ if they are analysed in terms of a uniform basis. (iii) If sampling of data is 'homogeneous', truncation biases cancel out. (iv) Wunsch and Minster [522] recommend working in a high-dimensional space, so that the data determine which subspace is resolved. (v) Using a 'shaped' basis that matches the parameter distribution produces unbiased estimates. (vi) The 'recipe' of Trampert and Snieder [484] that treats the missing $a - b$ component as an additional error in the data constraint.

Therefore, in an estimation that considers only the subspace of $\mathbf{x}$, the measurement-covariance matrix in the Bayesian least-squares estimate needs to be increased from $\mathbf{X}^{-1}$ to

$$[\mathbf{X}_{\text{eff}}]^{-1} = \mathbf{X}^{-1} + \mathbf{H}\mathbf{V}^{-1}\mathbf{H}^{\mathsf{T}} \tag{8.3.4e}$$

to allow for the neglected components. This allows us to obtain a more precise specification of the qualitative requirement (8.3.2c): the effect of truncation error can be safely ignored only when $\mathbf{X}^{-1} \gg \mathbf{H}\mathbf{V}^{-1}\mathbf{H}^{\mathsf{T}}$.

In terms of Bayesian estimation, relation (8.3.4c) is exact in that it fully accounts for the modes spanned by the vectors $\mathbf{y}$, subject to the assumption of independent priors for $\mathbf{x}$ and $\mathbf{y}$ and zero mean for $\mathbf{y}$. However, even under these assumptions, it is not a 'solution' of the truncation problem because the term $\mathbf{HV}^{-1}\mathbf{H}^\mathsf{T}$ involves the Green-function matrix for a set of parameters that is in principle infinite. What is required in order to implement the Trampert and Snieder 'recipe' is some form of approximation of $\mathbf{HV}^{-1}\mathbf{H}^\mathsf{T}$.

The recipe involves two different approximations.

■ The need to approximate $\mathbf{HV}^{-1}\mathbf{H}^\mathsf{T}$ in the expression for $\mathbf{X}_{\text{eff}}$.

■ The more important point that the approximation neglects the 'off-diagonal' part of the prior covariance matrix.

■ In addition, since the recipe assumes zero mean for the truncated basis components, the basis that is used should include all available information.

Kaminski *et al.* [241] found the Trampert and Snieder formalism to be effective in reducing the error. By using $\mathbf{H}$ from the adjoint model, their study assesses the second of the two approximations described above: keeping the terms in $\mathbf{V}$ but neglecting the 'off-diagonal' terms.

Further reading

Ill-conditioning arises in many problems involving analysis of indirect data and is commonly taken as the definition of an inverse problem.

■ In the field of remote sensing, Twomey's book remains a valuable introduction, as does Rodgers' paper [411] and his recently published book.

■ In seismology, classic papers include those of Parker [348], Backus and Gilbert [19] and Jackson [224].

■ From astronomy, the book by Craig and Brown [81] presents inverse problems related to remote sensing in a way that emphasises the more general applicability of the techniques.

■ The Institute of Physics (UK) publishes a journal, *Inverse Problems*, for current research.

Exercises for Chapter 8

1. Use the SVD representation to prove that the pseudo-inverse (see Box 8.2) gives the solution that is of the minimum length (for underdetermined problems) and has the minimum sum of squares of residuals (for overdetermined problems).

2. Analyse the meridional tracer transport under the assumption that, within the troposphere, the vertical profile $m(p)$ is a quadratic function of p determined by the three conditions

   $\partial m/\partial p = 0$ at the tropopause;
   $\partial m/\partial p = \kappa_p s$ at the surface; and

the vertical average concentration is described by the one-dimensional diffusion equation obtained by averaging over longitude and pressure.

Determine the relation between surface sources and concentrations in the steady state.

3. Numerically calculate the 'resolution matrix', **HG**, for a truncated expansion in terms of Legendre polynomials, $P_n(x)$, $x \in [-1, 1]$. Note that $\int_{-1}^{1} P_n(x) P_m(x)\, dx = 2\delta_{mn}/(2n + 1)$ and that the $P_n(x)$ can be obtained from $P_0(x) = 1$, $P_1(x) = x$ and $(n + 1)P_{n+1}(x) = (2n + 1)x P_n(x) - n P_{n-1}(x)$.

Chapter 9

Analysis of model error

"Therefore you don't have a single answer to your questions?"
"Adso, if I did I would teach theology in Paris."
"In Paris, do they always have the true answer?"
"Never", William said, "but they are very sure of their errors."
    Umberto Eco: *The Name of the Rose* (Trans. W. Weaver).

9.1   Model error

The effect of the ill-conditioning described in the previous chapter is to amplify errors both in the observations and in the models. This makes consideration of such errors a particularly important part of the analysis. Observational error has been discussed in Chapter 5. This chapter addresses the issue of 'model error'.

The main contributions to model error are the following.

- Errors in the atmospheric-transport model. This is the main topic of Section 9.2. The use of tracers to calibrate transport (Chapter 18) can also help characterise transport-model error.

- Errors associated with discretisation of the source. This is discussed in Section 8.3.

- Mis-match in spatial and temporal scales between models and observations, as discussed in Section 9.3.

- Errors in the statistical model. These are discussed in Section 9.4

To discuss model error, we adopt a general formalism from Tarantola [471], using $p_{model}(\mathbf{m}|\mathbf{x})$ to represent the probability (density) of the 'true' output being $\mathbf{m}$ given an input $\mathbf{x}$. For an exact deterministic relation, $\mathbf{m} = \mathbf{G}(\mathbf{x})$, this density becomes a delta-function, i.e. it is zero unless the deterministic relation holds. The more general case with non-zero probability density for $\mathbf{m} \neq \mathbf{G}(\mathbf{x})$ allows for the possibility of error in the deterministic 'model', $\mathbf{G}(\mathbf{x})$. There may also be scope for characterising some aspects of 'model error' in terms of the prior distribution of model parameters as described by $p_{prior}(\mathbf{x})$. This can be regarded as the creation of a 'meta-model' describing a family of models by adding additional parameters that define an interpolation between various special cases. This approach is discussed further at the end of this section.

The factor $p_{theory}(\mathbf{c}|\mathbf{x})$ used in Bayesian estimation represents our information about how the observations depend on the model parameters, $\mathbf{x}$. Tarantola relates this to $p_{model}(\mathbf{m}|\mathbf{x})$ as

$$p_{theory}(\mathbf{c}|\mathbf{x}) = \int p_{err}(\mathbf{c}|\mathbf{m}) p_{model}(\mathbf{m}|\mathbf{x}) \, d\mathbf{m} \qquad (9.1.1a)$$

where $p_{err}(\mathbf{c}|\mathbf{m})$ gives the probability of measuring $\mathbf{c}$ when the true value is $\mathbf{m}$. Thus it is $p_{err}(\mathbf{c}|\mathbf{m})$ that represents the distribution of 'measurement error' as distinct from 'model error'.

Relation (9.1.1a) can be incorporated into the Bayesian formalism as

$$p(\mathbf{x}|\mathbf{c}) = \frac{p(\mathbf{x}) \int p_{err}(\mathbf{c}|\mathbf{m}) p_{model}(\mathbf{m}|\mathbf{x}) \, d\mathbf{m}}{\int p(\mathbf{x}) \int p_{err}(\mathbf{c}|\mathbf{m}) p_{model}(\mathbf{m}|\mathbf{x}) \, d\mathbf{m} \, d\mathbf{x}} \qquad (9.1.1b)$$

If we characterise the measurement uncertainties by a 'measurement error' specified by the distribution $p_{err}(\mathbf{c}|\mathbf{m})$ relative to model predictions, then the theoretical knowledge of the system (as distinct from measurements of the system) is characterised by $p_{model}(\mathbf{m}|\mathbf{x})$. We describe this 'system model' as stochastic or deterministic, depending on the form of $p_{model}(\mathbf{m}|\mathbf{x})$. For a deterministic model, we have a fixed relation, $\mathbf{m}(\mathbf{x})$, and $p_{model}(\mathbf{m}|\mathbf{x})$ is expressed in terms of delta-functions. A spread in $p_{model}(\mathbf{m}|\mathbf{x})$ represents a stochastic model and/or a statistical representation of model error.

More specifically, we distinguish among the following cases.

- The fully deterministic case in which $p_{model}(\mathbf{m}|\mathbf{x})$ has the form of a delta-function centred on the deterministic prediction, $\mathbf{m}(\mathbf{x})$ – this has been the usual case in atmospheric tracer inversions.

- The case in which the system is in principle deterministic but we regard the model as subject to error so that $p_{model}(\mathbf{m}|\mathbf{x})$ represents the distribution of the error in the model relation $\mathbf{m}(\mathbf{x})$. An alternative way of describing error in a deterministic model is to incorporate it into the parameter uncertainty as described at the end of this section.

- The case in which the model is inherently stochastic so that $p_{model}(\mathbf{m}|\mathbf{x})$ represents the distribution of predictions that can be obtained for a specific set of parameters, $\mathbf{x}$ – state-space models are of this form.

It must be emphasised that, in the last case, the description of the system model as stochastic or deterministic is distinct from the model of our knowledge about the system. The latter always needs to be expressed in statistical terms, since our knowledge, especially of biogeochemical systems, is always incomplete. This specification of our knowledge is done through the prior and posterior distributions, $p(\mathbf{c}, \mathbf{x})$ and $p_{\text{inferred}}(\mathbf{x}|\mathbf{c})$, respectively.

Tarantola [471] notes two interesting aspects of $p_{\text{theory}}(\mathbf{c}|\mathbf{x})$. He expresses the spread in $p_{\text{theory}}(\mathbf{c}|\mathbf{x})$ in terms of a covariance matrix $\mathbf{C}_{\text{data}}$ and notes that the form (9.1.1a) implies the relation

$$\mathbf{C}_{\text{data}} = \mathbf{C}_{\text{measurement}} + \mathbf{C}_{\text{model}} \tag{9.1.2}$$

(The proof is set as problem 1 at the end of this chapter). Tarantola further notes that we can expect $\mathbf{C}_{\text{measurement}}$ and $\mathbf{C}_{\text{model}}$ to be of comparable magnitude. The basis for this suggestion is that, if $\mathbf{C}_{\text{measurement}} \ll \mathbf{C}_{\text{model}}$, then the observations can be used to refine the model. In general models progressively improve so that their precision is as good as (but generally not better than) can be supported by the available data. However, this is only an order-of-magnitude argument and gives no insight into the structure of the model error. Often the nature of the measurement process will give an indication as to which types of measurement error will be correlated across many measurements (e.g. a calibration error) and which will be independent. Identifying potential correlations between model errors is far more difficult. However, a particularly important case in which correlation in model error can be expected is that of multiple measurements. Many aspects of the measurement procedure will lead to independent errors and so the variance of the average of N repetitions will be reduced by a factor of $1/N$, calibration errors being an important exception. However, repeated measurements will usually lead to no reduction in model error. An early example of this problem was reported by Enting *et al.* [143], where seasonal CO_2 fluxes were represented by a single global pattern. They used only one seasonal cycle in their data set because attempts to use more seasonal data severely distorted the estimates of other fluxes. Given the discussion above, it seems clear that the problem was due to trying to treat multiple seasonal records as independent when in fact there were correlated model and/or discretisation errors.

As a formalism, incorporating model error into $p(\mathbf{c}|\mathbf{m})$ appears to be adequate for our needs in tracer inversions. Expression (9.1.1a), giving data error as a combination of measurement error and model error, emphasises that the choice of 'label' for ambiguous cases does not affect the outcome of the estimation. However, as with the other statistical modelling problems that have arisen with source and observations, the question is 'what type of distribution should one use for $p(\mathbf{m}|\mathbf{x})$?'. The characterisation of model error is a notoriously difficult part of virtually all real-world inverse problems.

Identifying model error

Enting *et al.* [145] noted that using a specified $p_{\text{theory}}(\mathbf{c}|\mathbf{x})$, as in the discussion above, was one of three approaches to estimation of error that might be applicable to such tracer inversions.

■ The first case, specifying $p_{\text{theory}}(\mathbf{c}|\mathbf{x})$, assigns uncertainties to all the inputs and propagates these uncertainties through the whole of the inversion calculation. This is an approach that is commonly used in applications and is explicit in the formalism of Sections 3.3 and 3.4. This corresponds to modelling the observations as

$$\mathbf{c} = \mathbf{m}(\mathbf{x}) + \epsilon_{\text{measurement}} + \epsilon_{\text{model}}$$

with a known $p_{\text{theory}}(\mathbf{c}|\mathbf{x})$ giving the distribution of $\epsilon_{\text{measurement}} + \epsilon_{\text{model}}$.

■ The second is to assign relative uncertainties to the inputs, propagate these through the calculations and determine an overall scaling factor from the distribution of residuals. This approach is commonly used in regression fits in cases for which it can be assumed that all data points come from the same probability distribution with an unknown parameter (or parameters). Any such unknown parameters have to be included in the parameter set $\mathbf{x}$. When one is fitting a range of disparate data, the assumption of identical distributions is hard to justify. This approach is unlikely to be readily applicable to assessing model error.

■ The third is to determine the uncertainty from the sensitivity to the input data set. This can be formalised as the *bootstrap* method and has been used in a number of mass-balance-inversion studies [78, 66].

Model error expressed through parameters

As noted above, we may work with densities $p_{\text{model}}(\mathbf{m}|\mathbf{x})$ that differ from delta-functions because we have been able to characterise the uncertainties in the model in this way. An alternative possibility is to express the model uncertainty through additional model parameters so that the model uncertainty is captured in the distribution $\mathbf{x}$. Such additional parameters are often referred to as 'nuisance variables', things that must be considered as part of the estimation process even though they are not of direct interest. This formulation avoids the problem noted above, namely that repeated measurements cannot be treated as independent if the uncertainty ranges include a contribution representing model error.

This approach has been adopted in Bayesian calibrations of global models, e.g. [140, 242], and is particularly appropriate when the parameters are lumped-average effective parameters, defined in terms that lack a direct meaning beyond the context of the model. However, although the 'model-error' formalism can be formally accommodated within the definition of $p(\mathbf{m}|\mathbf{x})$, with an expression in terms of $p(\mathbf{x})$ as a possible alternative in some cases, actually obtaining explicit expressions for either form remains a distant goal rather than a present reality.

There is no complete *a priori* answer to the requirement for statistical modelling. Each case must be considered on its merits. Chapter 12 on non-linear estimation discusses additional formulations of some of these estimation problems.

9.2   Large-scale transport error

The analysis of transport-model error largely follows that of Enting and Pearman [141].
We consider a Green-function form with a 'true model' representing concentration as

$$c_j = \sum_{\mu} G_{j\mu} s_\mu + \epsilon_j \qquad\qquad (9.2.1)$$

where ϵ_j represents measurement error. Notionally the sum over μ is over arbitrarily
fine detail.

In practice we have to work with a model representation (expressed as $G'_{j\mu}$) that
truncates the source distribution by using a finite basis and only approximates the
contributions of those components that are resolved:

$$
\begin{aligned}
c_j = &\sum_{\mu\in\text{basis}} G'_{j\mu} s_\mu && \text{assumed model}\\
&+ \sum_{\mu\in\text{basis}} (G_{j\mu} - G'_{j\mu}) s_\mu && \text{transport-model error}\\
&+ \sum_{\mu\notin\text{basis}} G_{j\mu} s_\mu && \text{truncation error}\\
&+ \epsilon_j && \text{measurement error} && (9.2.2)
\end{aligned}
$$

The truncation error has to include contributions both from sub-grid-scale fluxes
and from any small-scale detail that is neglected in restricting the basis to include
only fluxes that are aggregated in either space or time. The transport-model error has
to include both errors in the transport processes and representation errors where the
grid-cell values of the model give an erroneous representation of point concentrations.

In this last case, the distinction between 'model error' and 'measurement error'
may become blurred. The discrepancy can be regarded as an 'error' in the model since
it fails to represent small spatial scales. It could also be regarded as the measurements
'failing' to produce spatially representative data. This issue is discussed in Section
9.3. Mathematically, the ambiguity is unimportant because the estimation (9.2.1b)
depends on $p_{\text{theory}}(\mathbf{c}|\mathbf{x})$ but not on how it is partitioned into p_{err} and p_{model}. Similarly,
the source-discretisation error could, in principle, be regarded as model error (due
to truncation of the space of $\mathbf{x}$ onto which $\mathbf{G}$ acts) or as observational error (with
the observations including 'unwanted' contributions). Regardless of how they are
labelled, these errors have the characteristic that, for repeats (or near repeats) of the
measurements, such errors will be very highly correlated. This is a potential weakness
in the 'recipe' of Trampert and Snieder [484] (described in Section 8.3) as a way
of reducing the bias from 'model-truncation' error. In the early 1990s the increasing
number of transport-modelling studies of CO_2 made it apparent that a range of quite
different conclusions was being drawn. One possible reason for the difference was
the different transport models involved. A collaborative exercise making systematic
comparisons between models termed TRANSCOM (see Box 9.1) was developed. The
term 'inter-comparison' is used to distinguish comparisons within a set of models from

studies in which the models are compared with observations. There have been similar inter-comparisons of models in various areas of biogeochemical science. Examples are the inter-comparisons for atmospheric GCMs [170], ocean-carbon models [421], terrestrial-carbon models [82] and land-surface schemes in GCMs [208].

The TRANSCOM project confirmed the extent of the differences between the transport models that were being used to study the carbon cycle. Phase 1 compared the responses to the two largest components: the emission of carbon from fossil sources and the seasonal cycle from terrestrial ecosystems. One important issue raised by phase 1

BOX 9.1. TRANSCOM

The TRANSCOM transport-model inter-comparison compares a number of global atmospheric-transport models through a standard set of calculations. The TRANS-COM study is oriented towards transport modelling of atmospheric CO_2, particularly inverse modelling. It is an attempt to address the issue of model error in atmospheric-transport modelling. Section 9.2 reviews some of the key outcomes from TRANSCOM in terms of how we can characterise model error.

TRANSCOM has operated in a series of 'phases'. In each phase a set of calculations is specified. These calculations are then performed by the various participants. Phase 1 of TRANSCOM was less formal and was coordinated from the CRCSHM. For phases 2 and 3 the project has become an official activity of the IGBP GAIM programme, coordinated from the University of Colorado.

Phase 1 The initial TRANSCOM study considered the two main CO_2 fluxes: that from fossil sources and the seasonal cycle from the terrestrial biota. The results from phase 1 have been published in a technical report [398] and in a journal [277]. As emphasised above, the comparison was among models, with 12 models participating. Comparisons with observational data were not possible because the data cannot be disaggregated into the contributions from the different components (except by inversions using a model!). Even for the seasonal cycle, the fluxes had been matched to one particular model.

Phase 2 This phase used sources of SF_6. This was because data both on concentrations and on sources were available. Therefore comparisons with observations were possible. Phase 2 aimed to improve on phase 1 by having the participants produce a number of 'diagnostics' from their models, showing the transport disaggregated into different model processes [106].

Phase 3 This is, at the time of writing, on-going. The study is aimed at a more direct investigation of how differences among models (and, by implication, model error) contribute to differences (and errors) in inversions of CO_2 data.

As well as the overview TRANSCOM reports [398, 277, 106], some participants have independently published details of their own TRANSCOM contributions or related calculations [459, 105].

Box 9.2. Failure of linearity

The basis assumption of the synthesis method is that the transport behaves linearly, i.e.

$$\mathcal{G} \sum_{\mu} s_{\mu} s(\mathbf{r}, t) = \sum_{\mu} s_{\mu} [\mathcal{G}s(\mathbf{r}, t)] \tag{1}$$

Non-linearity occurs when the transport processes are modelled in a way that depends on the concentration. In fact, many atmospheric-transport schemes deliberately violate the condition of linearity. The two main reasons are

- ■ to ensure that mass is conserved in the model and
- ■ to ensure that concentrations of tracer do not become negative.

If synthesis inversions are found to be affected by non-linearity, then an iterative correction may improve the solution. However, repeating this type of correction will not provide convergence to the true Bayesian least-squares fit if the Jacobian (i.e. the matrix of derivatives of concentrations with respect to sources) is incorrect (see problem 2 at the end of this chapter). What is required is that the responses should be calculated at (or, in practice, near) the solution. This means that the responses $G_{j\mu}$ should be calculated from perturbations to a good initial guess at the solution. In this case, the Bayesian estimation needs to be similarly expressed in terms of perturbations since, if non-linearity is occurring, then the Jacobian evaluated at a first-guess solution will not describe the source–concentration relation for finite differences from a zero-flux state.

of TRANSCOM was the so-called 'rectifier effect'. This is the annual mean horizontal gradient in concentration of CO_2 established because of the seasonal (and to a lesser extent diurnal) covariance between CO_2 fluxes and transport through the boundary layer [105]. The TRANSCOM project revealed significant differences between models in terms of the strength of the rectifier effect.

Law and Rayner [275] have investigated the extent to which the rectifier effect influences inversions of CO_2, using a variation of the 'perfect-model' approach (Box 13.2). They created two versions of their transport model, one with a large rectifier effect and one with a smaller rectifier effect, by using two different parameterisations of sub-grid-scale transport. Each version of the model was used to produce a forward run. The data values at 27 locations were then extrapolated to give a surface-concentration field. This was used as input to mass-balance inversions with each of the two versions of the model, giving a total of four combinations. These represented the case of analysing the atmosphere with two 'accurate' models (with high- and low-covariance effects) and analysing the atmosphere with two 'inaccurate' models: one in which the rectifier effect in the model was too large and one in which it was too small, relative to the model from which the concentration fields had been derived. They found that the estimated sources for the four cases were quite similar in the southern hemisphere. In the tropics and northern hemisphere, the differences between the two cases in which the model was inverting data from the forward run with the same version of the model (perfect-model cases) were small. However, large differences in mean sources occurred, particularly for mid-latitudes of the northern hemisphere, when the inversion used a version of the model different from that which produced the data used in the inversion. However, even in these cases, the extent to which the reconstructed concentrations varied about these different means differed relatively little among the four cases. There was no obvious pattern of anomalous residuals that might serve as a warning of model error.

The possibilities for quantifying error are the following.

Statistical techniques. In particular, the bootstrap formalism described above can be used.

Inter-comparisons among models. Inter-comparisons can establish a spread of uncertainty. One particular advantage is the ability to investigate the relative importance of individual transport processes. This can provide suggestions of additional observational tests.

Validation of models. The majority of atmospheric-transport models have made some use of trace-constituent data for validation of the model. The main tracers are radon for small-scale effects, CFCs for global-scale transport and various radioactive species for transport from the stratosphere.

Calibration of models. In cases in which validation studies have shown the performance to be unsatisfactory, tracer data have sometimes been used to calibrate the model by determining 'corrections' to the transport. This type of calculation is discussed in Chapter 18. In principle the process of estimating such corrections could include an estimate of the degree of uncertainty, thus characterising an aspect of uncertainty in the calibrated model. A more serious difficulty is that such calibration generally applies only to a small number of 'modes' of transport, so a characterisation of uncertainty is lacking for other modes.

A common practice has been to express the results of validation or calibration in terms of indices such as interhemispheric-transport time. Chapter 10 describes an analysis by Plumb and MacConalogue [359], which points out the degree of ambiguity in such indices. This ambiguity needs to be kept in mind when one is using such indices for quantitative analysis of the performance of a model or inter-model differences.

9.3   Selection of data in models

Chapter 5 described how global observational networks aim to obtain 'baseline' data that are (conceptually) representative of large-scale air-masses. The objective is to have data that reflect the effects of large-scale biogeochemical processes of global significance. This would suggest that such baseline data are also the appropriate data to interpret when one is using global models that have (relatively) coarse resolution. There have, however, been few tests of this assumption. In practice measurement programmes replace the conceptual definition of 'baseline' by operational definitions deemed applicable to the various sites. The problem of choosing the closest correspondence to model values and observational data still requires further study.

To put this problem in perspective, one of the selection criteria considered for Cape Grim, Tasmania, is winds in excess of 5 m s^{-1} (18 km h^{-1}). An additional 'baseline' criterion requires stable concentrations (indicative of a large-scale airmass) for a period of 5 h, corresponding to a distance of 90 km at 18 km h^{-1}. This needs to be compared with model resolutions, which, for the models used in the TRANSCOM study, range from 8° × 10° (about 890 km × 840 km) down to 2.5° × 2.5°(about 300 km × 200 km).

Spatial mis-match. Trace-gas measurements generally represent the concentration at a single point, remotely sensed data being the main exception. In contrast, resolutions of global models are many hundreds of kilometres. This mis-match has been addressed by selection of baseline data and comparison of time-averages. Smoothness in time is taken as a proxy for smoothness in space. The limitations of this approach have not been well-quantified. Further insight could be gained from high-resolution limited-area modelling. Observational studies of spatial variability can be undertaken using aircraft-based sampling. To date, these approaches have not contributed significantly to addressing issues of model error, but they are likely to become more important in developing inversion techniques on smaller scales. As well as mis-matches in the horizontal directions, vertical representativeness can be a problem. The extent to which the model accurately represents the boundary layer will influence which data can be accurately fitted.

Definition. While baseline selection is designed to select concentrations representative of large spatial scales, the selection process may mean that the large regions do not correspond to the grid cells of the model. This is particularly true when the baseline criteria specify wind direction. One *ad hoc* approach has been to match such data to modelled concentrations at a grid cell upwind of the measurement location [165]. Law [273] investigated this issue using a model with resolution approximately 3.3° × 5.6°, to study the spatial variability of CO_2 concentrations. Considering the grid cell nearest a site and the eight surrounding cells, the annual mean standard deviations were found to range from about 0.2 ppm in the southern-hemisphere to about 1.0 ppm at northern-hemisphere sites. She found that using modelled radon concentrations as a selection criterion significantly reduced this variability. Ramonet and Monfray [393] used a model with 2.5° × 2.5° resolution to compare four ways of selecting model data. These were termed *autoselection*, based on wind direction in the model; *synchro-selection*, selecting model concentrations at the time of actual baseline measurements; *back-trajectories*, requiring no contact with land over the previous 5 days; and *modelled radon*, also as an indicator of contact with land. The synchro-selection approach was meaningful because transport in the model used analysed wind-fields. They concluded that the techniques based on contact with land were the most appropriate way of selecting model data. They found that failure to select model data could lead to errors in the phase and shape of the seasonal cycle at Cape Grim, due to errors of up to 0.3 ppm in the monthly mean. Haas-Laursen *et al.* [188] performed additional model studies of NOAA flask sites, quantifying the extent to which each site was affected by wind-direction sampling and filtering out of outliers.

Temporal mis-match. Mis-matches in time-scales can occur in various ways. One of the most serious problems is that GCM winds don't reproduce actual synoptic events. The most common way to address this has been to restrict matching of data to monthly means or longer-term averages when one is using GCM winds, leading to a loss of information. Matching models and data by taking one or more subsamples from each, selected by using synoptic conditions, may provide improved fits. However, it is necessary to quantify the degree of error inherent in using GCM wind-fields both in order to understand the limitations of much current practice and as a first step towards developing better approaches.

Sampling. The majority of long-term records of concentrations of CO_2 and most other long-lived trace gases come from flask sampling, leading to a potential sampling error. This has been investigated by Tans *et al.* [468] by analysis of the NOAA continuous CO_2-analyser records. The continuous records were subsampled at random in a manner that matched the flask-sampling protocols and then subjected to the normal analysis for flask data. The distribution of results from the subsamples, relative to corresponding results from the continuous record, indicated the degree of sampling error that can occur. In particular, it was estimated that the annual mean baseline CO_2 concentrations from flask samples were subject to sampling errors of up to 0.5 ppm. Haas-Laursen *et al.* [188] used their model to examine effects of sampling and quantified the mean bias. They also noted that comparing infrequent flask data with infrequently sampled model data would double the sampling bias from that involved in comparing flask data with model means.

In order to compare the temporal representativeness of models, CO_2 responses for Cape Grim are shown in Figure 9.1. The upper pane represents the response to a 1-month pulse source over south-east Australia, calculated with the MATCH model at $3.3° \times 5.6°$ resolution. The lower pane shows responses from the DARLAM model [262] run at 75-km resolution with winds nudged to follow meteorological analyses. It will be seen that the main difference arising from the greater resolution is the representation of more small-scale features.

Phase 2 of TRANSCOM included studies of small-scale variability [106]. The data sets for comparison were obtained from tall towers in North Carolina (WITN) and Wisconsin (WLEF). The spread of autocorrelation in the model was quite large and similar for each site, whereas, for the observations, the WLEF autocorrelations were at the lower end of the model spread and WITN autocorrelations dropped much

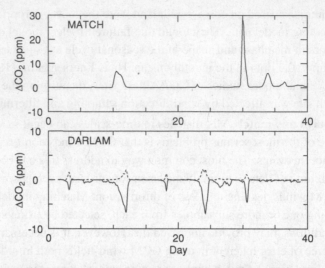

Figure 9.1 A comparison of temporal representativenesses of models for CO_2 concentrations at Cape Grim from particular flux components. The upper pane shows the response from MATCH at $3.3° \times 5.6°$ resolution, with GCM winds and a fixed source from a single grid cell in SE Australia. The lower pane shows response from DARLAM at 75-km resolution with winds nudged to follow analyses and fossil source (broken line) and terrestrial uptake for SE Australia (solid line). Since different wind-fields are used, correspondences between the models are coincidental.

faster than was found in the models. The interquartile variabilities in the models were comparable to, or less than, those observed at WITN. The interquartile range of the observations was in the middle of the model range at WLEF.

9.4   Errors in the statistical model

Throughout this book, I have emphasised that the process of modelling generally involves both deterministic and statistical modelling, with varying degrees of integration of the deterministic and statistical aspects. Consequently, any consideration of 'model error' has to address the issue of possible errors in the statistical model. This issue has been given relatively little consideration in tracer-inversion studies and clearly requires additional study. The present section simply gives a summary of some of the possible causes of error in the statistical model and a number of techniques that can either avoid some of the problems or at least identify them.

Some of the most likely causes of error in the statistical model are the following.

■ A failure to account for errors in the deterministic model. In transport
   modelling, this problem is so widespread as to be almost universal. A number
   of inversion studies make *ad hoc* adjustments to observational error in order to
   characterise the combination of model error and measurement error. This has
   the problem that the statistical characteristics of model error are quite different
   from those of measurement error.

■ A neglect of correlation between measurements, e.g. the common assumption that errors can be treated as independent white noise. This will fail if there are systematic errors in the measurement procedure. Another likely reason for non-independence of data values is when the data-covariance matrix includes contributions both from measurement error and from model error, as described by equation (9.1.2). The correlation in the model-error component for repeated measurements is likely to be close to 100%. As noted above, Enting *et al.* [143] found that attempts to fit large amounts of data on the seasonal cycle of CO_2 in their synthesis-inversion calculation led to instabilities in the solution when these data were treated as independent.

Techniques for addressing some of these problems are the following.

■ The starting point is, of course, to test the statistical assumptions. In particular, do the residuals $c - m(\hat{x})$ have the expected distribution?

■ The use of 'robust' estimation techniques that are less sensitive to the statistical assumptions. This can range from minimising sums of magnitudes of deviations (rather than sums of squares) through to techniques such as bootstrap analysis that make fewer assumptions about the statistical characteristics of the model.

■ Expanding the parameterisation of the statistical model to incorporate extra degrees of uncertainty. A simple example is estimation that treats the observational variance as an unknown (see the analysis in Section 12.1).

■ Use techniques that adapt to the statistics. An example is the work of Haas-Laursen *et al.* [187] on estimating time-varying fluxes with an underlying model, $F = I$ in state-space terminology, that specified constant sources. Although literal application of this model would produce flux estimates that grew at half the actual rate of growth (problem 1 of Chapter 11), adaptive filtering can track the fluxes more closely. Various forms of generalised cross-validation [83] provide ways of adapting to the empirical statistics of the problem.

Further reading

The issue of model error has long been a troubling aspect of any form of interpretation of observations.

■ In a number of areas of earth-system science, inter-comparisons of models have been undertaken in order to investigate inter-model differences. Of particular relevance to the inversion calculations discussed in this book are the TRANSCOM transport-model-inter-comparisons (see Box 9.1) [398, 277, 106]. Other inter-comparisons have been described [170, 421, 82].

■ In particular, Boer's review [35] of climate-model inter-comparisons covers both specifics of such inter-comparisons (especially the Atmospheric Model Intercomparison (AMIP)) and the philosophy that underlies comparing models with each other as well as with observations.

■ As summarised in Box 5.3, Morgan *et al.* [324: Chapter 4] review the origins of uncertainty and give a broad perspective on generic causes of error. Their presentation of the history of experimental determinations of the speed of light, with the estimated uncertainties that often fail to include later (presumably better) values, is a sobering reminder of the difficulties of quantifying systematic error.

■ In contrast, the issue of errors in statistical models is a well-established field of study. It is generally addressed in terms of 'robustness', which is a measure of the extent to which techniques of analysis (especially the construction of estimates) are insensitive to departures from the assumed statistical model. The issue is covered in most comprehensive textbooks on statistics. Huber's book [219] is devoted to the topic.

Exercises for Chapter 9

1. Show that
$$p(\mathbf{c}|\mathbf{x}) = \int p(\mathbf{c}|\mathbf{m}) p_{\mathrm{model}}(\mathbf{m}|\mathbf{x}) \, d\mathbf{m}$$
implies
$$\mathbf{C}_{\mathrm{data}} = \mathbf{C}_{\mathrm{measurement}} + \mathbf{C}_{\mathrm{model}}$$
for multivariate Gaussian distributions.

2. Show that, if non-linearity in the transport occurs (as described in Box 9.2), then applying an iterative correction leads (if it converges) to a source given by $(\mathbf{G}')^{\mathsf{T}}(\mathbf{X}\mathbf{G} + \mathbf{W})\mathbf{s} = (\mathbf{G}')^{\mathsf{T}}\mathbf{X}\mathbf{c}$, where $\mathbf{G}\mathbf{s}$ represents the effect of the transport model applied to the full source solution and $\mathbf{G}'$ represents the approximate Jacobian calculated away from the solution $\mathbf{s}$.

Chapter 10

Green's functions and synthesis inversion

Such schemes have much to recommend them, particularly when they work
(usually meaning that the modeler likes the results)...
Carl Wunsch: *The Ocean Circulation Inverse Problem* [521: p. 393].

10.1 Green's functions

Much of the discussion in this book has simply assumed the validity of a linear relation
between sources and concentrations, at least in a discretised form. This section aims
to clarify the origin and significance of this 'Green-function' representation.

The approach follows the ideas introduced in Chapter 2: the process of modelling
is regarded as going from the real world to a mathematical model and then to a
computer implementation of the model. A sequence of mathematical representations
of the transport equations is presented in this chapter, working backwards from those
closest to numerical implementations of transport. This is followed by analyses of the
representation of transport as sets of ordinary differential equations (ODEs) and partial
differential equations (PDEs). This explicit consideration of the mathematical model
is useful in establishing general characteristics of the solutions that are less apparent
in the computer implementation.

The Green functions used in atmospheric-transport modelling represent specific
cases of the general principle that an inhomogeneous linear differential equation (DE)
with specified boundary conditions can be 'solved' by constructing an integral operator
that is the inverse of the differential operator. The Green function is the kernel of
this integral operator. An important property of this relation is that any particular

Green-function solution is equivalent to the solution of the DE only for a specific set of boundary conditions.

The case that concerns us here is the representation of tracer transport as

$$m(\mathbf{r}, t) = m_0(\mathbf{r}, t) + \int d^3r' \int G(\mathbf{r}, t, \mathbf{r}', t') s(\mathbf{r}', t') \, dt' \tag{10.1.1}$$

where $m_0(\mathbf{r}, t)$ is the solution of the unforced tracer evolution, i.e. $s(.) = 0$. The solution (10.1.1) is required to be equivalent to the solution of integrating

$$\frac{\partial}{\partial t} m(\mathbf{r}, t) = \mathcal{T}[m(\mathbf{r}, t)] + s(\mathbf{r}, t)$$

subject to specified boundary conditions. Green-function methods of tracer inversion are based on this integral representation of atmospheric transport.

The general formalism

The specific results above are special cases of a more general mathematical formalism with wide applicability in applied mathematics [80]. In general, the Green-function formalism defines solutions, $f(\mathbf{y})$, of a general linear differential equation defined by an operator, $\mathcal{L}$, acting on functions, $f(.)$, defined in terms of a general coordinate, $\mathbf{y}$,

$$\mathcal{L} f(\mathbf{y}) = s(\mathbf{y}) \tag{10.1.2a}$$

subject to homogeneous boundary conditions expressed as

$$\mathcal{D} f = 0 \tag{10.1.2b}$$

where $\mathcal{L}$ is generally a differential operator and $\mathcal{D} f$ involves the values of f and/or its derivatives, on the boundary of the domain of $\mathbf{y}$.

The solution to (10.1.2a) and (10.1.2b) is written as

$$f(\mathbf{y}) = \int G(\mathbf{y}, \mathbf{y}') s(\mathbf{y}') \, d\mathbf{y}' \tag{10.1.3}$$

and $G(\mathbf{y}, \mathbf{y}')$, regarded as a function of $\mathbf{y}$ (with $\mathbf{y}'$ arbitrary but fixed), is a solution of the equation

$$\mathcal{L}_{\mathbf{y}} G(\mathbf{y}, \mathbf{y}') = \delta(\mathbf{y} - \mathbf{y}') \tag{10.1.4a}$$

subject to

$$\mathcal{D}_{\mathbf{y}} G(\mathbf{y}, \mathbf{y}') = 0 \tag{10.1.4b}$$

The notation $\mathcal{L}_{\mathbf{y}}$, $\mathcal{D}_{\mathbf{y}}$ means that the operators act on the $\mathbf{y}$ coordinate rather than on the $\mathbf{y}'$ coordinate of $G(\mathbf{y}, \mathbf{y}')$.

In applying this formalism to atmospheric transport we are generally interested in coordinates $\mathbf{y} = (\mathbf{r}, t)$ and the operator

$$\mathcal{L} = \frac{\partial}{\partial t} - \mathcal{T} \tag{10.1.5}$$

where $\mathcal{T}$ is an operator defining tracer transport.

The Green-function formalism can be extended to apply to inhomogeneous boundary conditions by a simple transformation. If the boundary conditions are

$$\mathcal{D}f = a \tag{10.1.6a}$$

then one chooses any function $h(\mathbf{y})$ that satisfies

$$\mathcal{D}h = a \tag{10.1.6b}$$

and writes the solution to $\mathcal{L}f = 0$ as

$$f(\mathbf{y}) = h(\mathbf{y}) + f^*(\mathbf{y}) \tag{10.1.6c}$$

where $f^*(.)$ is the solution (with $\mathcal{D}f^* = 0$) of

$$\mathcal{L}f^* = s - \mathcal{L}h \tag{10.1.6d}$$

whence

$$f(\mathbf{y}) = h(\mathbf{y}) + \int G(\mathbf{y}, \mathbf{y}')[s(\mathbf{y}') - \mathcal{L}_{\mathbf{y}'} h(\mathbf{y}')]\,d\mathbf{y}' \tag{10.1.6e}$$

An exceptional case occurs when the homogeneous DE, $\mathcal{L}f = 0$, has a non-zero solution, $f_0(\mathbf{y})$, for the specified boundary conditions. Any multiple of $f_0(\mathbf{y})$ can be added to the Green-function solution and an additional condition is needed in order to specify this multiple. This situation applies for atmospheric transport with periodic boundary conditions: an arbitrary constant mixing ratio satisfies the transport equation in the absence of sources. Therefore the mean mixing ratio needs to be specified in addition to the requirement for periodicity and the integrated source is required to be zero. For self-adjoint operators, the requirement for the Green-function formalism is that the 'forcing', $s(\mathbf{y})$, be orthogonal to $f_0(\mathbf{y})$. For the general case, the requirement is that the forcing be orthogonal to $f_0^*(\mathbf{y})$, which is the solution of the adjoint equation $\mathcal{L}^\dagger f_0^*(\mathbf{y}) = 0$ (see Box 10.3).

Most of the problems considered in this book correspond to boundary conditions specified in terms of fluxes. One exception is the case of input of constituents into the stratosphere from the troposphere via the tropical tropopause. In this case, the boundary condition can be specifed as the concentration at the tropical tropopause. This forms the basis for defining age spectra for stratospheric air as described in Chapter 18.

Green's functions for atmospheric transport

To study the implications of this general Green-function formalism for atmospheric-transport modelling, we consider a sequence of representations of atmospheric transport, following the discussion of mathematical modelling in Chapter 2.

(i) **Transport expressed as difference equations.** The expression of the model as a set of difference equations is the form that is closest to the numerical implementation. In these terms, the problem is fully discretised with a set of source values, $s_{\mu t}$, mapping onto a set of concentrations c_{xt}. The model that implements the difference equations is essentially calculating a discretised form of (10.1.1):

$$m_{xt} = m_0 + \sum_{\mu t'} G_{xt,\mu t'} s_{\mu t'} \tag{10.1.7}$$

Numerical models of transport can, in principle, accept any set of input fluxes, $s_{\mu t}$. In these terms, a numerical model of transport is simply a way of calculating any desired linear combination of discretised Green-function elements. In practice the discretisation used in the inversion calculation will be much coarser in space and time than the model discretisation, but the transformation from model output to equation (10.1.7) at a usable resolution is simply one of aggregation. This property can lead to the pragmatic view that, for the purposes of modelling, the $G_{xt,\mu t'}$ are what the model calculates when the inputs are source distributions, $\sigma_\mu(\mathbf{r}, t)$. The relation to the integral form (10.1.1) is often regarded as of little importance. However, consideration of the underlying Green-function formalism can help identify an appropriate computational representation, particularly in cases in which it is necessary to apply the extra requirement of orthogonality to solutions of the homogeneous equation.

(ii) **Transport expressed as a finite set of ODEs.** In terms of ordinary differential equations (ODEs), the mathematical model is

$$\frac{\mathrm{d}}{\mathrm{d}t} m_j(t) = \sum_k T_{jk}(t) m_k(t) + s_k(t) \qquad \text{for } j = 1\text{–}N \tag{10.1.8a}$$

where the T_{jk} are coefficients that represent the atmospheric transport as a finite-difference form of $\mathcal{T}$.

A set of N first-order ODEs requires N boundary conditions to define the solution.

Enting [124] quotes, from Coddington and Levinson [75], the theoretical result that there is a Green-function solution of the form

$$m_j(t) = \sum_k \int_{t_0}^{t_1} G_{jk}(t, t') s_k(t') \, \mathrm{d}t' \tag{10.1.8b}$$

for homogeneous boundary conditions at t_0 and t_1. Furthermore, if the boundary conditions take the form of specified initial conditions at t_0, then

$$G_{jk}(t, t') = 0 \qquad \text{for } t < t' \tag{10.1.8c}$$

as expected from causality. This generic solution is a vector generalisation of the 'integrating-factor' solution introduced in Box 1.1.

These 'initial-condition' boundary conditions are not the only possibility. An alternative is periodic boundary conditions, specified as

$$m_j(T_{\text{cycle}}) - m_j(0) = 0 \tag{10.1.9a}$$

If the transport coefficients and sources are periodic with period T_{cycle}, then these boundary conditions will ensure that periodicity of the concentrations, i.e. $m_j(t + T_{cycle}) = m_j(t)$ is achieved for all t. Since these boundary conditions also have the solution $m_j = \alpha$ for any constant α, the requirement for the Green-function solution is that the source is orthogonal to the solution of the corresponding adjoint equation. This constraint is

$$\sum_j \int_0^{T_{cycle}} s_j(t)\,dt = 0 \qquad\qquad (10.1.9b)$$

These boundary conditions can be generalised to the cyclo-stationary case corresponding to the boundary conditions

$$m_j(T_{cycle}) - m_j(0) = \beta T_{cycle} \qquad \text{for all } j \qquad\qquad (10.1.10a)$$

This case can be treated by the formalism of (10.1.6a)–(10.1.6e), leading to the requirement that

$$\sum_j \int_0^{T_{cycle}} s_j(t)\,dt = N\beta T_{cycle} \qquad\qquad (10.1.10b)$$

for consistency. Equation (10.1.10b) corresponds to the requirement of conservation of mass. (In a computational representation in which cells j differ in terms of their air-mass, the sums over j above have to be replaced by appropriately weighted sums.)

(iii) **Transport expressed as PDEs.** Atmospheric transport is generally represented in PDE form by equations that are first-order in time and second-order in space. This requires one boundary condition for 'time', defined for all spatial coordinates. The most usual cases are the 'initial-condition' and 'periodic' or 'cyclo-stationary' cases, which are essentially as described for the ODE representation above.

Representing transport using second derivatives with respect to spatial coordinates implies the need for two boundary conditions for each dimension. Considering general functions, $f(.)$, we have the following requirements on solutions of the transport equation. For continuity of the function and derivatives in the longitudinal direction:

$$f(p, \theta, \phi = 0) - f(p, \theta, \phi = 2\pi) = 0 \qquad\qquad (10.1.11a)$$

and

$$\frac{\partial}{\partial \phi} f(p, \theta, \phi = 0) - \frac{\partial}{\partial \phi} f(p, \theta, \phi = 2\pi) = 0 \qquad\qquad (10.1.11b)$$

To avoid cusps at the poles

$$\frac{\partial}{\partial \theta} f(p, \theta = 0, \phi) = 0 \qquad\qquad (10.1.12a)$$

and

$$\frac{\partial}{\partial \theta} f(p, \theta = \pi, \phi) = 0 \qquad\qquad (10.1.12b)$$

These boundary conditions can be achieved by expanding horizontal variations in terms of spherical harmonics.

In the vertical direction, a zero-gradient condition is applied at the upper boundary:

$$\frac{\partial}{\partial p} f(p = p_{min}, \theta, \phi) = 0 \qquad\qquad (10.1.13a)$$

The natural boundary conditions for defining Green's functions also apply zero gradient at the earth's surface:

$$\frac{\partial}{\partial p} f(p = p_{max}, \theta, \phi) = 0 \qquad\qquad (10.1.13b)$$

The alternative

$$\kappa_p \frac{\partial}{\partial p} f(p = p_{max}, \theta, \phi) = s(\theta, \phi) \qquad\qquad (10.1.13c)$$

can be mapped onto the zero-gradient boundary condition (10.1.13b) using the type of transformation described by (10.1.6a)–(10.1.6e).

Before leaving the discussion of Green's functions as kernels of operators, one technical point is worth noting. If the kernel $G(\mathbf{y}, \mathbf{y}')$ has a bounded 'double' integral over the spaces of $\mathbf{y}$ and $\mathbf{y}'$ then it has a property known as compactness: a continuity condition for operators. The significance of this is that no compact operator has a bounded inverse, which explains the ubiquity of ill-conditioning in this type of inverse problem.

10.2   A limiting Green function for atmospheric transport

Plumb and MacConalogue [359] have described how, for long-lived tracers, the Green function for atmospheric transport can be characterised as a time for transport from the source regions. This leads to a 'universal' spatial structure for the distributions of all long-lived tracers with slowly growing concentrations and similar source distributions. This is particularly applicable to constituents with predominantly industrial sources.

The starting point for their derivation is the transport equation:

$$\frac{\partial}{\partial t} m(\mathbf{r}, t) = \mathcal{T}[m] - \lambda(\mathbf{r}, t) m(\mathbf{r}, t) + s(\mathbf{r}, t) \qquad\qquad (10.2.1)$$

The analysis applies to the case in which the growth in concentration can be expressed as

$$\frac{\partial}{\partial t} m = \beta m(\mathbf{r}, t) \qquad\qquad (10.2.2)$$

and λ and β (which need not be constant in time) are small enough for m to be treated as approximately uniform in space for the purposes of calculating the sink term and the rate of change. Using $\langle . \rangle$ to denote averages over the whole atmosphere, (10.2.1) becomes

$$\beta \langle m \rangle = \langle s \rangle - \langle \lambda \rangle \langle m \rangle \qquad (10.2.3)$$

since the atmospheric average over transport vanishes. Substituting the approximation (10.2.3) back into (10.2.1) gives

$$\mathcal{T}[m] = (\beta + \lambda)\langle m \rangle - s(\mathbf{r}, t) = \frac{\lambda(\mathbf{r}, t) + \beta}{\langle \lambda \rangle + \beta} \langle s \rangle - s(\mathbf{r}, t) \qquad (10.2.4)$$

In the case of spatially uniform λ or $\lambda \ll \beta$, this reduces to

$$\mathcal{T}[m] = \langle s \rangle - s(\mathbf{r}, t) \qquad (10.2.5)$$

Note that equations (10.2.4) and (10.2.5) describe particular solutions $m(\mathbf{r}, t)$ and are not general statements about the operator $\mathcal{T}[.]$. The consequence of (10.2.5) is that, subject to the conditions, the spatial distribution of the tracer does not depend on either the lifetime or the growth rate. Plumb and MacConalogue confirmed that these results applied in a two-dimensional model, for a range of growth rates and sink strengths.

Plumb and MacConalogue went on to explain this result by interpreting the Green function for sources as a mean transport time. They defined a transport time, τ, for a parcel of air travelling from the source region to a location $(\mathbf{r}, t)$. For a mixture of parcels, the probability density function $p_t(\tau)$ characterises the range of transport times, τ. The concentration is thus

$$m(\mathbf{r}, t) = m(\mathbf{r}_0, t) \int p_t(\tau) e^{-(\beta + \lambda)\tau} \, d\tau \qquad (10.2.6)$$

For long-lived, tracers with slowly growing concentrations, $(\beta + \lambda)\tau \ll 1$, except for a small range of trajectories with low probability density, $p_t(\tau)$. Therefore

$$\frac{m(\mathbf{r}_0, t) - m(\mathbf{r}, t)}{m(\mathbf{r}_0, t)} = \beta_{\text{dyn}} \bar{\tau} \qquad (10.2.7a)$$

where $\bar{\tau}$ is the average transport time

$$\bar{\tau} = \int \tau p_t(\tau) \, d\tau \qquad (10.2.7b)$$

and β_{dyn} is a dynamic growth rate (the inverse of the dynamic lifetime defined by Plumb and MacConalogue) given by

$$\beta_{\text{dyn}} = \frac{\langle s \rangle}{\langle m \rangle} = \langle \lambda \rangle + \beta \qquad (10.2.7c)$$

Expressing these results as

$$\frac{m(\mathbf{r}_0, t) - m(\mathbf{r}, t)}{m(\mathbf{r}_0, t)} = (\langle \lambda \rangle + \beta)\bar{\tau} \qquad (10.2.7d)$$

shows the information in normalised gradients as the product of a universal spatial distribution from transport scaled by a factor involving rates of growth and loss. This form clarifies some of the disputes, reviewed in Section 16.3, about the use of gradients of ALE/GAGE CH_3CCl_3 data to estimate loss rates.

Plumb and MacConalogue extended their analysis to show that, for compounds for which $\lambda \ll \beta$ was violated in the stratosphere, the 'universal' spatial distribution could still occur in the troposphere so long as $\langle \lambda \rangle \ll \beta$. They also noted that, in the approach to the universal spatial distribution, the concentration isopleths would equilibrate faster than the concentration values.

For quantitative comparisons, it is useful to note how the formalism applies to the two-reservoir atmosphere of toy model A.2 (Box 2.1). For a constant source, s, in reservoir 1, the solution is

$$M = m_1 + m_2 = (1 - e^{-\lambda t})s/\lambda \qquad (10.2.8a)$$

$$\Delta m = m_1 - m_2 = (1 - e^{-(\lambda + 2\kappa)t})s/(\lambda + 2\kappa) \qquad (10.2.8b)$$

This gives a dynamic growth rate

$$\beta_{\text{dyn}} = \frac{\langle s \rangle}{\langle M \rangle} = \frac{\lambda}{1 - e^{-\lambda t}} \to \lambda \qquad (10.2.9)$$

Plumb and MacConalogue defined an interhemispheric exchange time based on hemispheric averages as

$$\Delta \Gamma_{\text{I}} = \frac{\Delta m}{s} \qquad (10.2.10a)$$

for which the two-reservoir model gives

$$\Delta \Gamma_{\text{I}} = \frac{1}{\lambda + 2\kappa} \qquad (10.2.10b)$$

They noted that, in their two-dimensional model, $\Delta \Gamma_{\text{I}}$ varies from 0.73 to 0.96 years as the source moves from mid-northern latitudes to high northern latitudes.

When looking at the more detailed spatial distribution in terms of latitude, θ_L, and pressure, p, they defined a gradient, $\Delta(\theta_L, p)$, and a transport-time difference, $\Delta \Gamma(\theta_L, p)$, as

$$\Delta(\theta_L, p) = 2\frac{m(\theta_L, p) - m(-\theta_L, p)}{m(\theta_L, p) + m(-\theta_L, p)} = \Delta \Gamma(\theta_L, p)/\tau \qquad (10.2.11)$$

They noted that, for p at the surface and a mid-latitude source, this time-scale varies from 1.92 to 2.27 years as θ_L, the latitude range across which the gradient is specified, goes from 30° to 60°. For a high-latitude source the change is from 2.01 to 3.96 years.

In the two-reservoir model, the corresponding quantity would be

$$\Delta = 2\frac{\Delta m}{M} = \frac{2\,\Delta m}{\tau s} \qquad (10.2.12a)$$

whence

$$\Gamma_{2\text{-box}} = \frac{2}{\lambda + 2\kappa} \tag{10.2.12b}$$

These comparisons imply first, a need to define a consistent normalisation before comparing time-scales; and secondly, a need to appreciate that interhemispheric time-scales will vary with the location of the source. This becomes important when such time-scales are used to report the results of using tracer distributions to calibrate model transport, as described in Chapter 18.

10.3   Synthesis inversion

For trace atmospheric constituents, the solution of atmospheric transport in terms of Green's functions takes the form (10.1.1). If the time-integration starts sufficiently far back in the past, then $m_0(\mathbf{r}, t)$ can be taken as constant.

Synthesis inversion discretises the sources in terms of processes (indexed by μ) as an unknown scale factor, s_μ, times a specified source distribution, $\sigma_\mu(\mathbf{r}, t)$, termed a basis function,

$$s_\mu(\mathbf{r}, t) = s_\mu \sigma_\mu(\mathbf{r}, t) \tag{10.3.1a}$$

so that

$$m(\mathbf{r}, t) = \sum_\mu s_\mu G_\mu(\mathbf{r}, t) \tag{10.3.1b}$$

with

$$G_\mu(\mathbf{r}, t) = \int d^3r' \int dt' \, G(\mathbf{r}, t, \mathbf{r}', t') \sigma_\mu(\mathbf{r}', t') \tag{10.3.1c}$$

At this point, we need to clarify the terminology. Commonly, the terms 'Green-function (inversion) methods' and 'synthesis inversions' have been treated as synonymous, reflecting the common practice. However, it seems preferable to use the term Green-function inversions to refer to techniques based on the integral representation (10.1.1) (as opposed to the mass-balance inversions based on the differential equation describing mass transport). The term synthesis inversions will be applied to techniques for which the discretisation of sources specified by (10.3.1a) involves synthesising the result as a linear combination of pre-specified source patterns. This chapter mainly discusses methods that have both characteristics, i.e. Green-function synthesis inversions. However, the association is not inevitable. The adjoint methods (described in Section 10.6) make it possible to perform inversions based on (10.1.1) using the full spatial (and potentially temporal) resolution of the transport model and so lack the 'synthesis' aspect of combining pre-specified components. Conversely, some of the techniques described in Chapter 11 work with a small number

of source components with pre-specified spatial distributions. These should be regarded as synthesis inversions even though the time-evolution is developed in a manner that is closer to the differential approach than it is to the integral approach derived
from (10.1.1).

Box 10.1. An example of synthesis inversion

This box lists the discretisation used by Enting *et al.* [145] as an example of early
synthesis inversions and as documentation of some of the examples presented in
this book.

The ocean regions are split into tropical (16° S to 16° N), mid-latitude (16° to
48°) in each hemisphere and high latitude (>48°) and are each characterised by a
prior estimate of the flux.

Region	Prior flux
High North Atlantic	-0.2 ± 0.5
High North Pacific	0.2 ± 0.5
North Atlantic	-0.3 ± 1.0
North Pacific	-0.3 ± 1.0
Equatorial Atlantic	0.3 ± 1.0
Western Equatorial Pacific	0.3 ± 1.0
Eastern Equatorial Pacific	0.6 ± 1.0
Equatorial Indian	0.3 ± 1.0
South Atlantic	-0.5 ± 1.0
South Pacific	-1.0 ± 1.5
South Indian	-0.7 ± 1.0
Southern	-0.2 ± 1.0

Terrestrial regions have similar latitudinal boundaries.

Region	Prior	Seasonality
Tropical forest in the Americas	0.0 ± 1.5	8.5 ± 2.0
Tropical forest in Africa	0.0 ± 1.5	2.8 ± 2.0
Tropical forest in Asia	0.0 ± 1.5	6.7 ± 2.0
Temperate evergreen forest	0.0 ± 1.5	1.4 ± 2.0
Temperate deciduous forest	0.0 ± 1.5	7.3 ± 2.0
Boreal forest	0.0 ± 1.5	3.5 ± 2.0
Grassland	0.0 ± 1.5	9.4 ± 2.0
Other	0.0 ± 1.5	6.0 ± 2.0

Additional components are the fossil CO_2 emission, the free-atmosphere source
of CO_2 from CO (see Box 14.1) and fluxes due to changes in use of land and forestry
(LUCF).

Fossil	5.3 ± 0.3
CO	0.9 ± 0.2
Tropical America LUCF	0.6 ± 0.5
Tropical Africa LUCF	0.4 ± 0.5
Tropical Asia LUCF	0.6 ± 0.5
Other LUCF	0.2 ± 0.5

Each component is further characterised by specified isotopic ratios and the non-anthropogenic components are also assigned prior estimates of isofluxes.

The formal analyses of the Green-function method are expressed in terms of the generic discretised relation

$$c_j = \sum_\mu G_{j\mu} s_\mu + \epsilon_j = m_j + \epsilon_j \qquad (10.3.2)$$

where c_j is an item of observational data, ϵ_j is the error in c_j, s_μ is a source strength and $G_{j\mu}$ is a discretisation of the Green function relating concentrations to source strengths. For simplicity of notation, it is often convenient to express the homogeneous solution (taken as a constant) as a pseudo-source defining an additive background contribution for each species.

For each specified 'basis-function' source distribution, $\sigma_\mu(r, t)$, integration of a numerical model of atmospheric transport will produce a response, $G_\mu(r, t)$. From these integrations, the specific time and space values that correspond to each observation j can be extracted to produce the matrix $G_{j\mu}$ of (10.3.2). Thus each numerical integration produces one row (a particular μ value) of the matrix $G_{j\mu}$. In contrast, adjoint methods (see Section 10.6) produce a column (a particular j value) of $G_{j\mu}$ for each integration. This is suitable for problems in which there are many possible source components and relatively few observations.

It is common to normalise the 'basis' source distributions in (10.3.1) by integration over the whole source–sink region and a full annual cycle, $[0, T_{\text{cycle}}]$, as

$$\int d^3 r' \int_0^{T_{\text{cycle}}} \sigma_\mu(\mathbf{r}, t, \mathbf{r}', t')\, dt' = 1 \qquad (10.3.3)$$

so that the scale factors, s_μ, have a direct interpretation in budget studies as mass exchange rates. Of course, for purely seasonal fluxes with zero annual mean, some other normalisation must be chosen.

The representation of synthesis inversion as a linear-estimation problem using the formalism described in Section 3.3 [143, 145] allows the systematic propagation of uncertainty in data to give estimates of uncertainty for the sources. The use of prior estimates in a Bayesian formalism has been found to be necessary. Without such constraints, fitting (10.3.2) was found to be unstable [104]. In the early, less formal, synthesis studies, it appears that instability was suppressed through a 'reality-check' deriving from the judgement of the investigators.

The components of the Bayesian synthesis-inversion estimation procedure for a set of parameters, $\mathbf{x}$, are

INPUTS

The matrix of responses	$\mathbf{G}$
Prior estimates of the parameters $\mathbf{x}$	$\mathbf{z}$
The inverse covariance matrix for $\mathbf{z}$	$\mathbf{W}$
The observational data set	$\mathbf{c}$
The inverse covariance matrix of $\mathbf{c}$	$\mathbf{X}$

OUTPUTS

Estimates of the parameters	$\hat{\mathbf{x}} = [\mathbf{G}^{\mathrm{T}}\mathbf{X}\mathbf{G} + \mathbf{W}]^{-1}[\mathbf{G}^{\mathrm{T}}\mathbf{X}\mathbf{c} + \mathbf{W}\mathbf{z}]$
The covariance matrix of $\hat{\mathbf{x}}$	$[\mathbf{G}^{\mathrm{T}}\mathbf{X}\mathbf{G} + \mathbf{W}]^{-1}$

Given this general formalism, it is necessary to choose the way in which the continuum relation (10.1.1) is discretised. Usually, the spatial index, $\mathbf{r}$, is discretised by the observational records: there is only a small number of sites and generally as many as possible are used. The main aspect regarding which a choice needs to be made is how to discretise the observations and sources in the time domain. At most sites, the records are time series and we would not usually want to use all values. We can discretise time as a sequence of points, n, with

$$c_{xn} = \sum_{\mu} G_{xn,\mu} s_{\mu} + \epsilon_{xn} \tag{10.3.4a}$$

but these 'points' would often represent time-averages of the actual measurements and the Green-function elements should represent corresponding averages.

Actually this takes the form

$$c_{xn} = \sum_{\mu} \sum_{\tau \leq n} G_{xn,\mu\tau} s_{\mu\tau} + \epsilon_{xn} \tag{10.3.4b}$$

There is considerable freedom of choice in the degree of temporal discretisation of the sources, ranging from a pulse decomposition to cases in which each process has a single source, $s_{\mu}\sigma_{\mu}(r, t)$, with a fully specified time-dependence.

The Green-function approach to time-dependent synthesis inversion is computationally demanding. For example, 25 source components for 12 months of 20 years implies 6000 source components, i.e. the need to invert a 6000×6000 matrix. Even assuming that interannual variability in transport can be neglected, $20 \times 12 = 300$ tracer-response calculations are needed in order to determine the Green-function matrix.

Because of these computational demands, the first formal synthesis inversions for CO_2 considered the special case in which the source components had an annual periodicity. In this case, (10.3.2) can be written using a frequency index ω as

$$c_{x\omega} = \sum_{\mu} G_{x\omega,\mu} s_{\mu} + \epsilon_{x\omega} \qquad\qquad (10.3.4c)$$

This corresponds to the special 'cyclo-stationary' boundary conditions (10.1.10a) of periodicity plus a globally applicable linear increase for each species. The equations describing the global trends, $\beta_{\eta} t$, of species η took the form

$$\beta_{\eta} = \sum_{\mu} \kappa_{\eta\mu} s_{\mu} \bar{\sigma}_{\mu} \qquad\qquad (10.3.5)$$

where $\bar{\sigma}$ denotes the annual mean source strength of the basis function $\sigma_{\mu}(\mathbf{r}, t)$, the sum is over those sources affecting species η and the $\kappa_{\eta\mu}$ are known scale factors relating source strengths, s_{μ}, to rates of change in concentration of species η. This is derived from the consistency constraint (10.1.10b), which becomes an extra item of data to be fitted.

Even for steady-state inversions, there are still several alternative choices for the form of the time-dependence:

(i) Enting *et al.* [143, 145] used pre-specified seasonal variations for each region and estimated only the overall amplitudes;

(ii) Taguchi [460] has used both monthly discretisation and discretisation in terms of Fourier components with 12- and 6-month periods; and

(iii) Peylin *et al.* [356] compared inversions using monthly discretisation of seasonal cycles with inversions in which the seasonal cycle was the difference between pre-specified uptake and release, constrained to be approximately equal.

A special case of time-dependent inversion is that used by Brown [56] for CH_4 based on

$$\sigma_{jm}(t) = \cos[(m-1)\pi t / T] \qquad \text{for } m = 1\text{--}N \qquad (10.3.6)$$

for latitude bands $j = 1$–18 in a two-dimensional model to fit N time-points over an interval of $t = 5$ years. This Fourier-series representation of a multi-year period was chosen in order to avoid the discontinuities associated with discretising sources in the time domain.

10.4   A history of development

Most of the development of synthesis inversions has taken place in the context of applications. This section reviews the history of applications in terms of what has been learnt about techniques, mainly in the context of long-lived greenhouse gases. Much of this development can be regarded as a process of introducing and investigating the

Table 10.1. *Some global steady-state synthesis inversions with a number of source components. The first three cases lacked a formal parameter-fitting procedure. The CH_4 analyses also involve sinks in the free atmosphere. Two studies [143, 145] included a source in the free atmosphere from oxidation of CO (c.f. Box 14.1) and also components for purely isotopic effects. The four groupings of source components are LU (terrestrial fluxes, changes in use of land etc.), seasonal cycles, ocean and industrial. The ^{18}O inversion (*) used a composite data set from various time periods and used three terrestrial sources with prescribed seasonal variation. The last case is the* TRANSCOM-3 *level-1 specification.*

Species	Period	Model	Winds	LU	Cycles	Ocean	Industrial	Reference
CO_2	1962, 1968, 1980, 1984	TM1	FGGE	2	1	3	1	[249]
CO_2	1980–87	GISS	GCM	4	1	5	1	[467]
$CH_4,^{13}C,^{14}C$	1984–87	GISS	GCM	9	–	2	4	[165]
$CO_2,^{13}C$	1986–87	GISS	GCM	12	1	9	1	[143]
$CO_2,^{13}C$	1986–87	GISS	GCM	12	8	12	1	[145]
$CH_4, ^{13}C$	1983–89	TM2	ECMWF	6	–	–	5	[206]
^{18}O	c. 1990 (*)	TM2	ECMWF	–	3(*)	1	1	[68,69]
CO_2	1992–96	16	Various	11	Fixed	11	Fixed	

types of error models used for the measurements and for the prior estimates of sources. In some cases, the analysis was in terms of choice of basis functions. Of course, with the assumption of independent errors, the choice of basis components is simply a special aspect of the general issue of choice of error model. The contribution to the understanding of biogeochemical cycles is described in later chapters. Table 10.1 lists some of the early steady-state synthesis-inversion calculations.

Ad hoc fitting procedures. Several early inversion studies used a synthesis approach, without any formal estimation procedure, generally including additional information. Keeling *et al.* [249] used three ocean components, two net land-flux components, a seasonal biota and an industrial component. The global budget was constrained to agree with a box-diffusion-model calculation. The results were checked against ^{13}C data. Tans *et al.* [467] used p_{CO_2} data to constrain the fluxes from the northern and tropical oceans. Fung *et al.* [165] studied the distribution of CH_4 using 16 source components, including a sink in the free atmosphere (and apparently coined the term 'synthesis' for this process).

Steady-state Bayesian synthesis. Bayesian synthesis inversion was developed to provide a technique that included a systematic analysis of uncertainty and also incorporated additional information [143, 145]. The problems of attempting this type of fit for CO_2 without the use of additional information are illustrated by Denning [104]. Note, however, that Prahm *et al.* [364] were able to perform a successful inversion for emissions of sulfur from Europe by least-squares fitting without any additional constraint. As described in the previous section, the Bayesian synthesis takes the form of a statistical (multi-component-regression) fit to the observational data, solving (10.3.2) using the formalism of Section 3.2. As part of the statistical approach, Enting *et al.* [145: Figure 5] plotted the cumulative distribution of normalised residuals to show that their results were consistent with the initial assumption that the observational errors were normally distributed.

In order to have a computationally manageable calculation, these initial synthesis inversions were restricted to the 'steady-state' or 'cyclo-stationary' case of sources whose only time-variation was a prescribed annual cycle.

Time-dependent inversions. Time-dependent inversions based on (10.3.4a) were used by Bloomfield *et al.* [34] for CCl_3F and CH_3CCl_3. This study introduced the feature of having the synthesis inversion embedded within an outer loop that adjusted the stratospheric lifetime to implement a non-linear fitting procedure.

Around the time that the steady-state synthesis inversions of CO_2 were being developed, it was becoming apparent that the assumption that the carbon cycle was in a steady state (with a slowly varying anthropogenic perturbation) was not valid on the short (2–5-year) time-scales that were being considered. In particular studies based on joint CO_2–$^{13}CO_2$ budgets suggested that the interannual variability could be comparable in size to the anthropogenic perturbation [159, 250]. Because CO_2 has a strong seasonal cycle, the annual resolution used for CCl_3F and CH_3CCl_3 is insufficient for CO_2. Time-dependent inversions of CO_2 have been developed by Rayner *et al.* [402], who found [P. J. Rayner, personal communication, 1996] that a time-resolution of 3 months was insufficient and that monthly resolution was required.

Time-series statistics of sources. Mulquiney and co-workers [330, 332] used a random-walk model as their model for the prior distribution of CFC sources.

Cyclo-stationary inversions involve a different type of choice. Enting *et al.* [143, 145] used terrestrial components with prescribed seasonality, based on simple terrestrial modelling from Fung *et al.* [163]. Taguchi [460] used a Fourier expansion and estimated the mean, annual cycle and (in the case of land regions) the semi-annual cycle; he found that the estimated cycles were plausible for most regions.

Peylin *et al.* [356] compared calculations with prescribed seasonality (based on the SiB-2 terrestrial model) with those allowing adjustment of the fluxes for each month, subject to a constraint on month-to-month changes. They found that the adjustable case gave a better fit to the observations. P. J. Rayner [personal communication, 1999] has investigated issues of the appropriate statistics for the priors and suggested a representation as mean plus anomaly. The issue is that, if a flux is specified as a set of N monthly values with prior variance W, then the prior variance of the long-term mean is W/N.

Time-series statistics of measurements. An autoregressive error model for the observations was used in the synthesis inversion for CFCs reported by Bloomfield et al. [34]. In the steady-state inversions by Enting et al. [143, 145], the measurements were fitted as Fourier coefficients, $c_{j\omega}$, with errors taken as independent. For a particular site, this independence of frequency components corresponds to assuming that the errors are a stationary time series. More recent studies by P. J. Rayner [personal communication, 2000] have identified the need for consistency in the time-series modelling of observations and priors. In particular, representing a length-N time series as if all the errors were independent with variance X would imply that the error in the long-term mean was X/N. In cases in which X is intended to include a contribution from model error (i.e. all cases in which model error is not included in some other way), the assumption of independent data errors is clearly incorrect.

Spatial statistics of sources. Spatial statistics have the potential for great complexity and only a few special cases have been studied. One issue is that of specifying the spatial distribution as uniform, compared with the alternative of using a 'pre-shaped' distribution defined by a prior estimate.

Fan et al. [151] compared inversions for which fluxes within regions were uniformly distributed with inversions for which terrestrial fluxes were distributed in proportion to modelled NPP. They found terrestrial sinks of 2.4 and 1.1 Gt C yr^{-1} with their two models when using uniformly distributed fluxes and 2.1 and 0.9 Gt C yr^{-1} when they distributed fluxes in proportion to NPP. The differences were mainly due to southern-hemisphere fluxes. Gloor et al. [181] found that this 'bias' problem was reduced when networks were optimised to reduce the variance due to propagation of measurement error.

Bousquet et al. [42] compared inversions with alternative forms of the spatial distributions within the terrestrial and oceanic regions and found that the largest change was for the North-American sink. Bousquet et al. also compared inversions with 66, 54, 45, 27 and 14 regions and found the greatest changes for estimates of tropical fluxes as regions were aggregated. The study by Kaminski et al. [241], described in Section 8.3, addresses the issue of the truncation involved in low-resolution synthesis inversions.

Model-dependence. The third stage of the TRANSCOM inter-comparison (see Box 9.1) aims to address the extent to which inversion results are sensitive to model error. At the time of writing, this exercise is still in progress. Several groups have made smaller-scale comparisons, usually comparing only two models. Fan et al. [151] compared inversions with the GCTM and SKYHI models and found many differences over 0.5 Gt C yr^{-1}, mainly for northern-hemisphere flux estimates. Bousquet et al. [42] also found the main differences in the northern hemisphere on comparing results from TM2 and TM3.

Regularisation. Fan et al. [151] compared the techniques of SVD truncation, use of a tunable quadratic constraint on the source components and use of Bayesian priors. They also considered tuning the strength of the Bayesian constraint and found that the northern-hemisphere estimates were relatively insensitive over an eight-fold range of weighting of the Bayesian constraint. Baker [20] has analysed the statistics of time-dependent inversions of CO_2. He compared, as regularisation techniques, combining regions (from 19 to seven), smoothing the time series (going from monthly data to estimates with cycles less than 4 months removed) and SVD truncation. Various approaches led to errors of order 1 Gt C yr^{-1} with a data error set at 1 ppm. Calculations with 'perfect data' and an 'imperfect prior' suggested that SVD truncation at about 1000 components (for about 2000 source components and 5000 data values) minimised the error.

Resolution. Bruhwiler *et al.* [58] performed a set of 'perfect model' experiments. They found that monthly means from 89 sites from the GlobalView CO_2 dataset (see Box 5.2) could determine long-term fluxes from each of 14 regions to within about 0.25 Gt C yr^{-1}, assuming that the data error was 0.3 ppm for land sites and 0.15 ppm for marine sites. Increasing the number of stations gave only slow improvement – their data suggest that the reduction in error variance behaves as approximately $1/N$. Gloor *et al.* [180] have investigated the uncertainties in flux estimates as a function of source discretisation and data coverage. They considered the errors from uncertainty in data, the bias from source discretisation and errors from atmospheric transport, which they judged to be the most important.

Further applications of steady-state inversions. Although time-dependent inversions have assumed the greater importance, steady-state inversions can still play useful roles because of their lesser computational demands, particularly in studies of sensitivity [42], in which multiple inversions are required. Other applications are network-design studies [400, 181; see also Chapter 13]; studies with high spatial resolution of fluxes (such as those from adjoint model calculations) [239] and studies of the time-resolution e.g. [460].

Regional-scale analyses. The development of regional-scale inversions appears somewhat fragmented. Several applications of synthesis techniques on regional scales are described in Chapter 17. In addition there have been several studies exploring the theoretical capabilities of the techniques, prior to analysing (or in some cases prior to obtaining) observational data, e.g. [491, 529, 530].

10.5   Multi-species analysis

The generic discretised Green-function relation (10.3.2) forms the basis for linear estimation of source/sink parameters, s_μ. The estimation procedure depends on the nature of the data 'only' through the inverse covariance matrices, **X** and **W**, but this dependence can play a dominant role in determining how particular inverse problems are addressed. The previous section considered issues that arise in the common situation in which we wish to estimate sources that evolve in time and the data are a relatively small number of time series. This section considers special issues that arise when the data are measurements of concentrations of several different chemical or isotopic species whose fluxes are related through being caused by a small number of processes.

Expanding the notation to indicate species, η, (10.3.2) becomes

$$c_{j\eta} = \sum_\mu G_{j\eta\mu} s_\mu + \epsilon_{j\eta} \tag{10.5.1a}$$

In content, this is equivalent to (10.3.2), being simply a partial recognition of the fact that the data index, j, can in principle range over arbitrarily disparate data. Useful applications of multi-species inversions come when there are known relations between the fluxes so that, for example, (10.5.1a) becomes

$$c_{j\eta} = \sum_\mu G_{j\mu} \alpha_{\eta\mu} s_\mu + \epsilon_{j\eta} \tag{10.5.1b}$$

where the $\alpha_{\eta\mu}$ are known coefficients that give a minimal 'process model' relating fluxes of the different species. One interesting point to note is that, if the data sets indexed by j for each species represent the same time and place, then the 'model-error' contribution to the $\epsilon_{j\eta}$ is likely to be very similar. The extent to which CO_2 and ^{13}C data are co-located in inversions does appear to influence inversions [P. J. Rayner, personal communication, 2000].

Global budgeting

A starting point for understanding multi-species analysis is global budgeting, in which one has, for each species η, a set of relations connecting global fluxes, Φ_μ, for various processes:

$$\dot{M}_\eta = \sum_\mu \alpha_{\eta\mu} \Phi_\mu \tag{10.5.2a}$$

When they are used in pairs, such relations can be represented in a vector form:

$$[\dot{M}_1, \dot{M}_2] = \sum_\mu [\alpha_{1\mu}, \alpha_{2\mu}] \Phi_\mu \tag{10.5.2b}$$

and plotted graphically as in Figures 10.1, 10.2, 14.3 and 14.4. The observable changes, $\dot{M}_\eta$, are expressed as a sum of vectors whose directions are known (from the $\alpha_{\eta\mu}$) but whose lengths involve Φ_μ, some of which are unknown. These unknown values are the objectives of the budget analysis which can be visualised as vectors on a two-dimensional graph.

One of the simpler examples is the joint carbon–oxygen budget of the atmosphere, written (either in molar units or in molar equivalents to CO_2) as

$$[\dot{M}_{CO_2}, \dot{M}_{O_2}] = \dot{M}_{CO_2}[1, 0] + \dot{M}_{O_2}[0, 1]$$
$$= [1, -1.3]\Phi_F + [1, -1.1]\Phi_B + [1, 0]\Phi_O \tag{10.5.3}$$

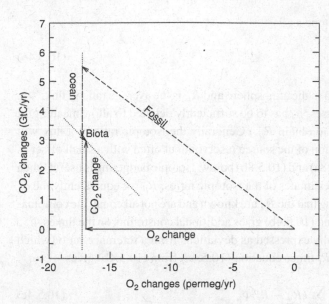

Figure 10.1 A combined CO_2–O_2 budget. The rates of changes $\dot{M}_C[1, 0]$ and $\dot{M}_O[0, 1]$ in concentrations of CO_2 and O_2 represented as the vertical and horizontal solid arrows are equated to the fluxes from fossil material (dashed arrow) plus the unknowns: the ocean (chain arrow) plus terrestrial biota (short arrow), which are determined by the intersection of the two dotted lines. One permeg is approximately 4.8 ppm [252].

The factor 1.3 reflects the average hydrogen content of fossil fuel, resulting in more molecules of oxygen being lost than CO_2 molecules produced by the fossil flux, Φ_F. The fluxes Φ_O and Φ_B represent the net fluxes of oxidised carbon into the oceans and of reduced carbon into terrestrial biota. The net flux Φ_B includes a component that returns to the atmosphere via rivers followed by outgassing from the ocean (see Chapter 14). Figure 10.1 shows an example of such CO_2–O_2 budgeting whereby measured changes and the known stoichiometric factors for each term in (10.5.3) are combined to give estimates of Φ_B and Φ_O [269]. Knowing $\dot{M}_{CO_2}$, $\dot{M}_{O_2}$ and Φ_F constrains $\Phi_O[1, 0]$ to lie along the vertical dotted line and $\Phi_B[1, -1]$ to lie on the diagonal dotted line. The intersection gives the solution vectors as shown. The appropriate stoichiometric $O_2 : CO_2$ ratio for Φ_B depends on the space- and time-scales involved. At the plant level, a small but significant amount of the carbon taken up as CO_2 returns to the atmosphere in other compounds, but, on the global scale, most of these are oxidised to CO_2, in many cases via CO.

Isotopic budgeting

If the two species involved are isotopes, then the principles remain the same, but the convention for notation changes. Isotopic budgeting has played an important role in reducing uncertainties in greenhouse-gas budgets. Criss [85] has described the underlying principles of physical chemistry that result in the occurrence of differences in isotopic composition.

The budget for the sum over isotopes is written

$$\dot{M} = \sum_\mu \Phi_\mu \tag{10.5.4a}$$

and that for the minority isotope as

$$\frac{d}{dt}R_A M = \sum_{\mu} R_{\mu}\Phi_{\mu}$$

(10.5.4b)

where R_A is the isotopic ratio in the atmosphere and R_{μ} is the average ratio in flux, Φ_{μ}. (The subdivision into processes, μ, has to be sufficiently detailed to allow meaningful averaging of isotopic ratios to obtain R_{μ}.) Generally the isotopic ratio of a flux will reflect the isotopic composition of the source reservoir, but often with a small offset as indicated by equations (10.5.8a) and (10.5.8b) below. Isotopic budgeting is useful when it is possible to obtain good estimates of the isotopic ratios, R_{μ} (or equivalently the δ_{μ} defined below), and, assuming that the R_{μ} are known and are not all equal, the combination of equations (10.5.4a) and (10.5.4b) gives additional constraints on the fluxes Φ_{μ}.

Isotopic ratios are generally expressed as deviations from a reference ratio (which we denote R_r). Subtracting R_r times equation (10.5.4a) from (10.5.4b) gives

$$\frac{d}{dt}(R_A - R_r)M_A = \sum_{\mu}(R_{\mu} - R_r)\Phi_{\mu}$$

(10.5.4c)

Box 10.2. The δ notation for isotopic differences

Most elements occur as a number of different isotopes: varieties of atoms that differ in the number of neutrons. In general, chemical reactions and transport processes on the molecular scale discriminate against the heavier isotopes. This discrimination has led to small differences in isotopic ratios in natural systems and these ratio differences can be used to diagnose particular groups of biogeochemical processes.

The main isotopes of interest are those of hydrogen, carbon and oxygen. For hydrogen the isotopes are ^1H, the majority; ^2H (usually denoted D); and ^3H, a radioactive isotope (generally denoted T). The carbon isotopes are ^{12}C (about 99% of carbon), ^{13}C (about 1%) and ^{14}C (known as radiocarbon), a radioactive isotope with a half-life of 5730 years. The oxygen isotopes are ^{16}O, 99.63%; ^{17}O, 0.037%; and ^{18}O, 0.204%.

The measured isotopic ratios are generally expressed in δ notation, which is a measure of the proportion by which the ratio differs from a standard. Virtually all measurements of isotopic composition are measurements of the isotopic ratio, compared with the ratio measured in some reference material. Thus $\delta^{13}C$ is defined as

$$\delta^{13}C = \frac{[^{13}C]/[^{12}C]_{obs}}{[^{13}C]/[^{12}C]_{standard}} - 1$$

Because the differences are so small, $\delta^{13}C$ and other isotopic differences are generally expressed in units of ‰, i.e. parts per thousand:

$$\delta^{13}C = \left[\frac{[^{13}C]/[^{12}C]_{obs}}{[^{13}C]/[^{12}C]_{standard}} - 1\right] \times 1000 \qquad \text{in ‰}$$

as a mass budget for the isotopic anomaly. Normalising by a factor α gives

$$\frac{d}{dt}\delta_A M_A = \sum_\mu \delta_\mu \Phi_\mu \qquad (10.5.4d)$$

with δ_μ defined as

$$\delta_\mu = \alpha(R_\mu/R_r - 1) \qquad (10.5.4e)$$

Combining (10.5.4a) and (10.5.4d) as a vector equation gives

$$[1, \delta_A]\frac{d}{dt}M + \left[0, \frac{d}{dt}\delta_A\right]M = \sum_\mu [1, \delta_\mu]\Phi_\mu \qquad (10.5.4\,f)$$

The notation δ_μ (as used by Tans [463]) is meant to be suggestive of the $\delta^{13}C$ used to communicate measurements of isotopic ratios (see Box 10.2). Any such correspondence cannot be exact and will depend on the choice of the constants R_r and α used in the definitions. If we denote the 'standard' definition of $\delta^{13}C$ (in units of $\permil$) by δ' and $^{13}C : ^{12}C$ ratios by r, we have

$$r = (1 + \delta'/1000)r_s \qquad (10.5.5a)$$

where r_s is the standard $^{13}C : ^{12}C$ ratio of a material with $\delta^{13}C = 0$. The anomaly δ values defined here are

$$\delta = \alpha\left(\frac{r}{1+r}\Big/R_r - 1\right)$$

Putting $R_r = r_s/(1 + r_s)$ gives

$$\delta = \alpha\frac{r - r_s}{(1 + r)r_s} \qquad (10.5.5b)$$

Putting $\alpha = 1000(1 + r)$ would make δ exactly equivalent to the standard definition of δ' (in $\permil$). However, we need to use the same α value for all terms in equation (10.5.3) and so the factor $1 + r$ needs to be replaced by a fixed value. For CO_2 fluxes, all the δ values lie between -10 and -30 and so using the central value and replacing $1 + r$ by $1 + 0.98r_s$ will lead to errors of only 1%. Therefore, in all the calculations listed below, we substitute δ' values (i.e. observed $\delta^{13}C$) in place of the corresponding δ values.

 - Figure 10.2 shows an example of the vector representation of the joint atmospheric budgets of CH_4 and $^{13}CH_4$. The rates of change and fluxes are expressed in the form of a sum of two-component vectors $[a, b]$ with b in Tg CH_4 yr^{-1} and a converted to $\permil$Pg CH_4 yr^{-1} in order to express it in terms of smaller numerical values. The left-hand half of Figure 10.2 shows the budget with δ chosen to correspond to the normal definition of $\delta^{13}C$. The right-hand side shows the budget if the reference ratio, R_r, is chosen to be R_A, the atmospheric $\delta^{13}C$ of CH_4. In geometric terms, the transformation is an affine transformation: a skewing of the axes. Normalising isotopic ratios relative to the atmosphere can be convenient for inversion calculations because the errors

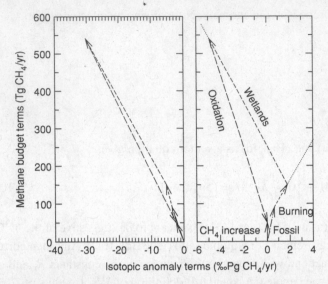

Figure 10.2 A joint atmospheric budget for CH_4–$^{13}CH_4$. Values are based on Quay *et al.* [386]. The version on the right-hand side has isotopic anomalies expressed as differences from the current atmosphere rather than differences from the PDB standard. The change in atmospheric concentration (solid arrow), with no change in $\delta^{13}C$, is the sum of fluxes from burning of biomass (dotted arrow) plus fossil CH_4 (chain arrow, estimated from $^{14}CH_4$), other sources such as wetlands (shorter dashes) and a loss, mainly due to oxidation by OH (longer dashes), that is depleted in ^{13}C, relative to the atmospheric ratio. The dotted lines show how this relation can determine unknown fluxes for wetlands and burning of biomass if their isotopic ratios are known and all other budget terms are known.

in the measurements of the two components of atmospheric change will tend to be uncorrelated [145]. The right-hand pane represents the process of solving for unknown fluxes from burning and wetlands, assuming that their isotopic ratios are known and all other fluxes and ratios are known. The known ratios mean that the unknown vectors lie along the two dotted lines and, as with Figure 10.1, the intersection of these lines determines the magnitudes of the unknown fluxes.

Tans [463] has noted that the budget equations (10.5.4a) and (10.5.4d) imply that δ equilibrates on time-scales that reflect the extent of atmospheric dilution and notes potential problems for inversions that assume a steady state. In this context, 'steady state' does not have to imply equilibrium, but rather that the approach to equilibrium is globally uniform. The test for adequacy of the steady-state assumption is whether the isotopic trends are, to within the accuracy of measurement, globally uniform when they are averaged over the time period represented by the inversion.

Isofluxes

In considering the atmospheric budget of CO_2, we need to recall that the anthropogenic perturbations represent small residuals from pairs of opposing gross fluxes Φ_O^+ and Φ_O^- and Φ_B^+ and Φ_B^-.

The net isotopic change due to two opposing gross fluxes from reservoir X can be re-expressed as

$$\Phi_X^+ \delta_X^+ - \Phi_X^- \delta_X^- = (\Phi_X^+ - \Phi_X^-)\delta_X^+ + \Phi_X^-(\delta_X^+ - \delta_X^-) \qquad (10.5.6a)$$

or

$$\Phi_X^+ \delta_X^+ - \Phi_X^- \delta_X^- = (\Phi_X^+ - \Phi_X^-)\delta_X^- + \Phi_X^+(\delta_X^+ - \delta_X^-) \qquad (10.5.6b)$$

Notionally, we could express the anomaly flux in terms of a mean $\bar{\delta}_X$ as

$$\Phi_X^+ \delta_X^+ - \Phi_X^- \delta_X^- = (\Phi_X^+ - \Phi_X^-)\bar{\delta}_X = \Phi_X \bar{\delta}_X \qquad (10.5.7a)$$

with $\bar{\delta}_X$ defined by

$$\Phi_X \bar{\delta}_X = \frac{\Phi_X^+ \delta_X^+ - \Phi_X^- \delta_X^-}{\Phi_X^+ - \Phi_X^-} \qquad (10.5.7b)$$

In practice, this definition is inappropriate since the denominator can go to zero and change sign. The anomaly flux needs to be expressed, as in (10.5.6a) and (10.5.6b), as a term proportional to the net flux and a purely isotopic term, which has become known as the isoflux, given by the product of the strength of the gross fluxes and the isotopic difference between them. The two parts of Figure 10.3 show the net flux and normalised anomaly fluxes as the resultants of large gross fluxes (left-hand side) and (on the right-hand side, enlarged by a factor of ten) as the resummations (10.5.6a) (dashed arrows) and (10.5.6b) (solid arrows).

The isotopic ratios $R_X^{\pm}$ refer to the isotopic composition of the fluxes, not the source reservoirs. Frequently, we have relations between isotopic compositions of sources and fluxes of the form

$$R_X^+ = R_X f_X^+ \qquad (10.5.8a)$$

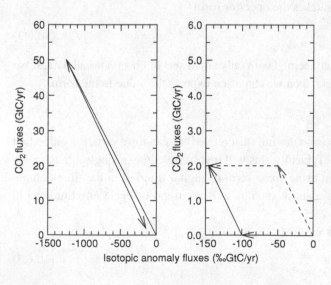

Figure 10.3 The relation between carbon fluxes and anomaly fluxes for gross fluxes described by (10.5.6a)–(10.5.6b). The plot on the left-hand side illustrates the combination of gross fluxes. The plot on the right-hand side (on a scale ten times larger) shows the two ways in which the net effect of the gross fluxes can be expressed as a net flux (affecting carbon content and isotopic composition) and an isoflux that affects isotopic composition only.

and

$$R_X^- = R_A f_X^-$$ (10.5.8b)

where $f_X^\pm$ are fractionation factors, generally close to unity.

The significance for inversions

Experience with inversion calculations for CO_2 has shown that the global mean budget is mainly determined by the global mean concentration data. Therefore the global budgeting relations will appear as a subproblem of a general inversion calculation. This remains true when more than one species is involved (see, for example, Figure 13.1). For the spatially varying component of the fluxes, the budgeting relations are equivalent to the global relations, with the addition of transport terms. Therefore it is necessary to determine the spatial variabilities of the relative weighting factors in the budgets (the stoichiometric factors and/or isotopic ratios) as well as 'offset' terms such as the isofluxes.

10.6   Adjoint methods

As described in Section 10.1, the generic Green-function formalism for tracer inversion is developed for the operator form

$$\mathcal{L}\mathbf{m} = \mathbf{s}$$ (10.6.1)

where, for tracer transport, we are concerned with the specific case

$$\mathcal{L} = \frac{\partial}{\partial t} - \mathcal{T}$$ (10.6.2)

The Green-function relation takes the operator form

$$\mathbf{m} = \mathcal{G}\mathbf{s}$$ (10.6.3)

In tracer studies, we are concerned with only small sets of observational data. If we consider a single item of data, then we can often express the value in the form

$$a = \mathbf{e} \cdot \mathbf{m}$$ (10.6.4)

where the vector $\mathbf{e}$ projects from the full (discretised) space–time variation embodied in $\mathbf{m}$ onto a single value a. (Typically, this will involve selecting a single site from the spatial distribution and then using some form of digital filtering in the time domain to extract a time-averaged quantity such as a mean, trend or cycle amplitude as in Section 4.2.)

We can write (10.6.4) as

$$a = \mathbf{e} \cdot \mathcal{G}\mathbf{s} = \mathbf{s} \cdot \mathcal{G}^\dagger \mathbf{e}$$ (10.6.5)

where the operator $\mathcal{G}^\dagger$ is the adjoint of $\mathcal{G}$. (The adjoint is defined as the operator for which equation (10.6.5) holds for arbitrary $\mathbf{e}$ and $\mathbf{s}$.)

Adjoint models are programs that take inputs $\mathbf{e}$ and calculate $\mathcal{G}^\dagger\mathbf{e}$. The form of the adjoint model can best be understood by considering the non-linear case. We define a set of ODEs as

$$\frac{\mathrm{d}}{\mathrm{d}t}\mathbf{g} = \mathbf{f}(\mathbf{g}, \mathbf{s}, t) \tag{10.6.6a}$$

where, in analogy with the state-space representation, the vector $\mathbf{g}$ is a discrete set of values evolving in time. We put the specific values to be matched to observations as

$$m_j = g_{k(j)}(t_j) \tag{10.6.6b}$$

and ask for the sensitivity of the m_j to one of the $\mathbf{s}_\mu$.

A forward model takes a vector $\mathbf{s}$ with components s_μ and produces the output

$$\sum_\mu \frac{\partial m_j}{\partial s_\mu} s_\mu \qquad \text{for all } j \tag{10.6.7a}$$

while the adjoint model takes a vector $\mathbf{e}$ with components e_j specifying projections of concentration and calculates

$$\sum_j \frac{\partial m_j}{\partial s_\mu} e_j \qquad \text{for all } \mu \tag{10.6.7b}$$

The formalism to do this is to differentiate (10.6.6a) to obtain

$$\frac{\partial}{\partial t}\frac{\partial}{\partial s_\mu} g_n = \frac{\partial}{\partial s_\mu} f_n(\mathbf{g}, \mathbf{s}, t) + \sum_k \frac{\partial}{\partial s_\mu} g_k \frac{\partial}{\partial g_k} f_n(\mathbf{g}, \mathbf{s}, t) \tag{10.6.8}$$

In computational terms, evaluating $\partial m_j/\partial s_\mu$ (for all j) involves setting up computer routines to implement and integrate (10.6.8). This integration will need to be done in parallel with the integration of (10.6.6a) since, in the non-linear case, the terms on the right-hand side will depend on $\mathbf{g}$. Equation (10.6.8) defines the tangent linear model.

Subject to relatively mild restrictions, there are computer programs that can take code for equations of the form (10.6.6a) and automatically generate code to implement (10.6.8) through a process of symbolic differentiation. The TAMC (tangent linear and adjoint model compiler) described by Giering [176] takes Fortran code as input and can generate the tangent linear model defined by (10.6.8) or the adjoint model.

The adjoint model represents essentially the same process as the tangent linear model, except that (10.6.6a) is (notionally) integrated backwards in time and the sensitivity equation is for the sensitivity to an initial condition of this notional integration.

Box 10.3. Adjoint operators

The description of the adjoint requires careful specification of the vector spaces involved. For any space $\mathcal{V}$ of vectors, $\mathbf{x}$, we can consider linear functionals, y, of these vectors. We use the notation $[x, y]$ to denote the value of functional y acting on vector $\mathbf{x}$. The set of all linear functionals on $\mathbf{x}$ is itself a vector space, denoted $\mathcal{V}'$, and termed the dual space for $\mathcal{V}$. The composition of linear functionals is defined as

$$[x, ay_1 + by_2] = a[x, y_1] + b[x, y_2] \qquad \text{for all } x$$

In the tracer-transport problem, the Green-function operator, $\mathcal{G}$, maps source vectors (from space $\mathcal{V}$) onto sets of concentrations, in a vector space that we denote $\mathcal{U}$. Comparisons of data are made by projecting vectors in the space $\mathcal{U}$ to obtain a specific value. This may be a delta-function, selecting a particular component, or it may be a more general average. So long as the selection/projection process is linear, it corresponds to a vector in $\mathcal{U}'$, the dual space for $\mathcal{U}$, and a particular selection/projection functional, $\mathbf{e}$, produces a value $[\mathcal{G}\mathbf{x}, \mathbf{e}]$.

The adjoint, $\mathcal{G}^\dagger$, of $\mathcal{G}$ is defined by

$$[\mathcal{G}\mathbf{x}, \mathbf{e}] = [\mathbf{x}, \mathcal{G}^\dagger\mathbf{e}] \qquad \text{for all } \mathbf{x}, \mathbf{e}$$

This means that $\mathcal{G}^\dagger$ is an operator that acts on the vectors in the space $\mathcal{U}'$ and produces vectors in the space $\mathcal{V}'$.

In other words, $\mathcal{G}$ acts on a particular source vector to produce a vector in the concentration space, to which we can apply any number of chosen projections to fit to observations. In contrast $\mathcal{G}^\dagger$ acts on the space of selection/projections and, for a single fixed selection, produces a vector of linear functionals that can be applied to vectors in the source space $\mathcal{V}$.

Generally, adjoint-model compilers produce code that evaluates $\mathcal{G}^\dagger\mathbf{e}$ for a fixed choice of $\mathbf{e}$. Except to the extent that $\mathbf{e}$ can be made to depend on model parameters, changing the selection of data, $\mathbf{e}$, usually requires not only re-running the calculation, but also a re-compilation.

Some of the principles can be illustrated with the mass-balance equation

$$\frac{d}{dt}m = \sigma_s\delta(t - s) - \lambda m \tag{10.6.9a}$$

for which we define a sensitivity

$$f(s, t) = \frac{\partial m(t)}{\partial \sigma_s} \tag{10.6.9b}$$

In this case we know the answer – the sensitivity is given by the response function

$$f(s, t) = \rho(t - s) = \Theta(t - s)\exp(-\lambda(t - s)) \tag{10.6.9c}$$

where $\Theta(.)$ is the unit step function.

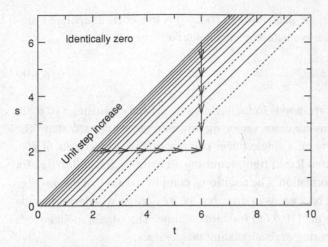

Figure 10.4 A contour plot of the sensitivity, $f(s, t)$, of concentration at time t to a source at time s. The tangent linear model provides a DE for $f(s, t)$ that can be integrated forwards in time t for fixed s (following the horizontal arrows). The adjoint model provides a DE that can be integrated backwards in 'time', s, for fixed t (following the vertical arrows).

The tangent linear equations for $f(s, t)$ can be obtained by differentiating (10.6.9a) with respect to σ_s as

$$\frac{\mathrm{d}}{\mathrm{d}t} f(s, t) = \delta(t - s) - \lambda f(s, t) \tag{10.6.9d}$$

For fixed s, these equations can be integrated forwards in time to give $f(s, t)$ for a succession of t values. In terms of Figure 10.4, the integration path in the t–s plane is the horizontal path. The adjoint equation is an expression for $(\mathrm{d}/\mathrm{d}s)f(s, t)$ for fixed t and is integrated along the vertical path in Figure 10.4, to give $f(s, t)$ for a succession of values of s (in decreasing order), for fixed t. These adjoint equations do not arise by simple manipulation of (10.6.9a) because the boundary conditions are involved.

The adjoint equations can be formally derived by taking term-by-term adjoints of the tangent linear equation, with adjoint, $\mathcal{A}^\dagger$, of an operator, $\mathcal{A}$, satisfying

$$\int g(t)\mathcal{A}f(t)\,\mathrm{d}t = \int f(t)\mathcal{A}^\dagger g(t)\,\mathrm{d}t \tag{10.6.10a}$$

This relation has to apply for the 'causal' boundary conditions with $f(t)$ going to zero for sufficiently small t and $g(t)$ going to zero for sufficiently large t. Constants and multiplicative factors are self-adjoint. Using the boundary conditions with integration by parts shows that the adjoint of $\mathrm{d}/\mathrm{d}t$ is $-\mathrm{d}/\mathrm{d}t$, so that the adjoint of (10.6.9d) is

$$-\frac{\mathrm{d}}{\mathrm{d}s} f(s, t) = \delta(t - s) - \lambda f(s, t) \tag{10.6.10b}$$

which may be seen to agree with the solution (10.6.9c).

Computationally, the tangent linear and adjoint compilers work with the model regarded as a composition of functions:

$$\mathbf{g}^{(n)} = \mathbf{F}^{(n)}(\mathbf{F}^{(n-1)}(\mathbf{F}^{(n-2)}(\cdots \mathbf{F}^1 \mathbf{g}^0)\cdots)) \tag{10.6.11a}$$

For ease of notation, the parameters s_μ are included in the state-vector $\mathbf{g}$. Initially (10.6.10a) would represent the succession of transformations from one time-step to the next, but ultimately this needs to be further subdivided so that conceptually each

arithmetic operation in the implementation of (10.6.6a) is treated as a separate transformation. The derivatives are evaluated by the chain rule:

$$\frac{\partial g_k}{\partial s_\mu} = \frac{\partial F_k^{(n)}}{\partial g_{k'}} + \frac{\partial F_{k'}^{(n-1)}}{\partial g_{k''}} \cdots \frac{\partial g_q^0}{\partial s_\mu} \tag{10.6.11b}$$

The tangent linear model corresponds to taking a single s_μ and evaluating (10.6.8b)
from right to left so that only an extra vector $\partial \mathbf{g}/\partial s_\mu$ is required at each step. The
adjoint model involves taking m, a single linear combination of components of $\mathbf{g}^{(n)}$,
and evaluating (10.6.11b) from left to right, requiring an extra vector of $\partial m/\partial g_k$ for
the current step in the transformation. The additional complication is that the $\partial m/\partial g_k$
are being evaluated working backwards in time, but the $\partial F_k^{(n)}/\partial g_{k'}$ will depend on the
$\mathbf{g}$ that are defined by integrating (10.6.6a) forwards in time. The adjoint model needs
to adopt some strategy for storing or recalculating these values.

The tangent linear and/or adjoint models have a range of applications [176],
including

- calculation of gradient vectors (with respect to parameters) for optimisation of
  some chosen objective function;
- sensitivity analysis;
- assimilation of data when the sensitivity to initial conditions is used as a basis
  for refining the estimates of these initial conditions; and
- error analysis, using the sensitivity calculations to propagate uncertainties from
  initial conditions and/or model parameters through to uncertainties in a final
  result.

There has been a range of applications of adjoint techniques in the earth sciences,
including ocean-model studies [510, 425]. For atmospheric-tracer inversions, the importance of the adjoint models lies in their ability to generate sets of $\partial m_j/\partial s_\mu$ for a
given m_j, performing the calculation for all μ simultaneously. This makes it possible
to avoid the problems arising from the coarse discretisation of source components
in synthesis inversion. Atmospheric-tracer inversions using adjoint models have been
used in a number of inversion studies [240, 214]. Section 8.3 includes a description
of a study [241] in which adjoint calculations were used to determine the extent of
aggregation error in conventional synthesis inversion.

Further reading

- Green's functions are reviewed in many books on differential equations,
  although often the treatment concentrates on DEs defined by self-adjoint
  operators.
- Material on transport modes occurs in various papers by Plumb and
  co-workers. Transport modes are noted by Prather in his account of chemical
  modes [366, 367], but the concept of transport modes is not developed.

■ Synthesis inversion is documented in various papers, including the review from the Heraklion conference [133]. Other papers in that volume [244] cover applications.

Exercises for Chapter 10

1. Show that, for the DE

$$\frac{d}{dt}m(t) = -\lambda(t)m(t) + s(t)$$

the $m(t_0)\exp[\int_0^t \lambda(t')\,dt']$ term in Box 1.1 is a particular case of the Green-function formalism of inhomogeneous boundary conditions as described by equations (10.1.6a)–(10.1.6e).

2. For the differential equation

$$\frac{d}{dt}m(t) = -\lambda(t)m(t) + s(t)$$

of Box 1.1, consider boundary conditions $m(0) = m(T)$.

   (a) Find the condition for which there is no non-zero solution for $s(t) \equiv 0$.
   (b) Subject to that condition, use the solution for initial conditions from Box 1.1 to find $m(0)$ in terms of $s(t)$.
   (c) Use that relation to obtain the Green function, $G(t, t')$, that applies to these boundary conditions to give $m(t) = \int_0^T G(t, t')s(t')\,dt'$.

3. Use equations (3) and (4) of Box 2.1 to derive a Green-function expression for the two-box model with constant $\lambda_N = \lambda_S$ and constant κ.

Chapter 11

Time-stepping inversions

Knowledge advances by steps and not by leaps.
Thomas Babington Macauley (1812).

11.1   Mass-balance inversions

The standard numerical implementation of atmospheric-transport models is in the form
of procedures for the numerical integration of the transport equations:

$$\frac{\partial}{\partial t} m(\mathbf{r}, t) = \mathcal{T}[m(\mathbf{r}, t)] + s(\mathbf{r}, t) \tag{11.1.1}$$

calculating the time-evolution of the tracer concentration $m(\mathbf{r}, t)$ at all points in the
atmosphere. This is subject to inputs from a specified source/sink flux, $s(\mathbf{r}, t)$, acted
on by atmospheric transport, which is modelled by the transport operator, $\mathcal{T}[m(\mathbf{r}, t)]$.

The synthesis techniques described in Chapter 10 are based on an integral form
of (11.1.1) with the estimation being largely detached from the integration of the
transport-model equations. There are, however, situations in which it is desirable to
have the estimation proceed in step with the model integration. The mass-balance
techniques described in this section have the estimation occur at each time step of the
transport model (generally of order hours) whereas the techniques described in the following section perform the estimation at much larger time intervals, typically monthly.

The mass-balance-inversion technique divides the atmosphere into two domains
(i.e. two sets of grid points), for one of which (set A) it is assumed that the source/sink
term $s(\mathbf{r}, t)$ is known (with $m(\mathbf{r}, t)$ unknown) and for the other set (B) it is assumed that
$m(\mathbf{r}, t)$ is known and $s(\mathbf{r}, t)$ is unknown. In many applications, the known sources in

set A are known to be zero, but this is not an essential requirement for the mass-balance formalism.

For set A, the time-evolution of $m(\mathbf{r}_a, t)$ ($a \in A$) is obtained as usual, by numerical integration of (11.1.1). For set B, the transport equation (11.1.1) is rewritten as

$$s(\mathbf{r}_b, t) = \frac{\partial}{\partial t} m(\mathbf{r}_b, t) - \mathcal{T}[m(\mathbf{r}, t)] \qquad \mathbf{r}_b \in B \tag{11.1.2}$$

and, from the known values of $m(\mathbf{r}_b, t)$, an estimate of the sources $s(\mathbf{r}_b, t)$ is obtained. Since $\mathcal{T}[.]$ involves concentrations from both sets, A and B, equations (11.1.1) and (11.1.2) must be stepped forwards synchronously. The technique is described as a 'differential' because of the $(\partial/\partial t)m$ term. Since equation (11.1.2) represents conservation of mass, the technique is often referred to as the 'mass-balance' technique, although this name is also applied to what we call the 'boundary-integral' technique, which is described in Section 17.2.2.

The mass-balance formalism requires a specification of concentrations $m(\mathbf{r}_b, t)$ at all points, $\mathbf{r}_b$, in set B. In addition, the time derivative $(\partial/\partial t)m(\mathbf{r}_b, t)$ is needed at all points, $\mathbf{r}_b$, in set B, i.e. usually at all surface grid points. To produce such a specification from observational data requires that the concentrations $c(\mathbf{r}_j, t)$ be interpolated (and generally smoothed) both in space and in time. The most common application of mass-balance inversion is deducing surface sources from surface observations, so that B is the set of surface grid points, while A is the set of grid points in the free atmosphere.

As illustrated schematically in Chapter 7 (Figures 7.5 and 7.6), the use of (11.1.2) involves modifying the time-stepping loop of the transport model, using the transport specification to evaluate $\mathcal{T}[.]$ for all grid points but using this transport term in different ways for the sets A and B.

While equations (11.1.1) and (11.1.2) show the principle of the method and Figures 7.5 and 7.6 illustrate it graphically, the specific implementation in a numerical model needs to interface with the numerical integration scheme.

The basic Euler time-stepping takes the form

$$m(\mathbf{r}_a, t + \Delta t) = m(\mathbf{r}_a) + [\mathcal{T}[m(\mathbf{r}, t)] + s(\mathbf{r}_a, t)] \times \Delta t \qquad \text{for } \mathbf{r}_a \in A \tag{11.1.2a}$$

$$s(\mathbf{r}_b, t) = [m(\mathbf{r}_b, t + \Delta t) - m(\mathbf{r}_b)]/\Delta t - \mathcal{T}[m(\mathbf{r}, t)] \qquad \text{for } \mathbf{r}_b \in B \tag{11.1.2b}$$

but higher-order schemes in space and time are desirable.

- The Euler predictor–corrector scheme is second-order accurate in time. In this scheme equations (11.1.2a) and (11.1.2b) are used to produce a prediction, $m^*(\mathbf{r}, t + \Delta t)$, and a source estimate, $s^*(\mathbf{r}, t)$. The corrector is based on the transport calculated by $\mathcal{T}[m^*]$ and gives the numerical integral for $m(\mathbf{r}, t + \Delta t)$ and a source, $s'(\mathbf{r}, t + \Delta t)$, from which the source estimate is calculated as $\hat{s}(\mathbf{r}, t + 0.5\Delta t) = 0.5[s^*(\mathbf{r}, t) + s'(\mathbf{r}, t + \Delta t)]$
- The lowest-order discretisation in the vertical uses $m(\mathbf{r}_b, t) = c_{obs}(\mathbf{r}, t)$.

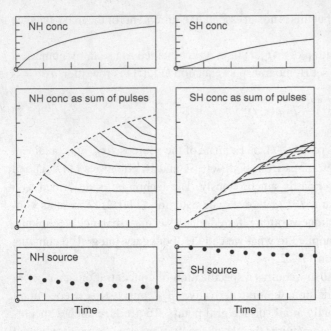

Figure 11.1 A schematic representation, using a two-reservoir atmosphere, of how mass-balance inversions deduce sources from the differences between observed concentrations and concentrations calculated as due to earlier sources. Units are arbitrary; the origin is circled on each plot.

■ Higher-order spatial schemes treat the relation (11.1.2) as schematic and actually apply (11.1.1) at all grid points. The sources are deduced from additional flux-gradient relations, relating the surface concentrations (prescribed) and the concentrations at the lowest level (calculated) to deduce a source that is taken as an input to the lowest level [304]. The same principle can be used with quadratic extrapolations based on the lowest two layers.

The way in which the mass-balance-inversion calculations evolve in time is illustrated in Figure 11.1. This extends, for the case of the two-box atmosphere, the graphical representation introduced in Figure 6.4 to represent the inversion of convolution relations. The particular case corresponds, using the notation of Box 2.1, to a fixed northern-hemisphere source, s_N, with $s_S = 0$. Given the transport (i.e. an interhemispheric flux $K(m_N - m_S)$), the inversion calculates the fluxes $s_N - \lambda m_N$ and $-\lambda m_S$ for the two hemispheres. The analytical solutions for m_N and m_S are used as the 'observed' concentrations for the illustration. The source estimates, shown in the lower two panes, are obtained as the difference between the 'observed' concentrations and the calculated concentrations resulting from earlier sources. These are represented in the left- and right-hand centre panes, as a series of pulses, showing the effect in each hemisphere of the combination of the sources from the lower panes. As with Figure 6.4, the division into pulses is an illustrative device. At any point in time the model works with the concentration distribution resulting from the combined past flux history and does not retain other information about the sequence of pulses. In the northern hemisphere, the behaviour is similar to that shown in Figure 6.4. A net source results as the 'observed' concentration grows faster than would have been expected if the source dropped to zero.

In the southern hemisphere, the growth at any time is less than would apply if fluxes were set to zero – in such a case, the southern-hemisphere concentration would increase as the two hemispheres equilibrated. Having the 'observed' concentration grow more slowly than the 'zero-current-source' calculation implies that there is a net sink in the southern hemisphere, as shown in the lower righthand pane.

The initial applications of mass-balance inversion were with two-dimensional models [136, 465]. Extending the technique to three dimensions involves extrapolating the observational data to cover the whole of the earth's surface. This is problematic, given the degree of variability in most trace-gas concentrations. One approach that has been developed is to use a perturbative approach, whereby mass-balance inversion is applied to the residuals from a forward calculation with a 'reference-source' distribution [274].

Advantages of mass-balance inversion

■ The mass-balance method is based on a single run of the model. Compared with synthesis inversions based on Green's functions (see Chapter 10), this makes it very efficient, particularly when high spatial resolution is required. The advantage, relative to Green-function inversions, becomes even greater when one is using transport fields that differ from year to year.

■ Mass-balance inversions can handle non-linear problems, such as the known source having a non-linear dependence on the concentrations.

■ The inversion calculation involves no extra complications when transport fields vary from year to year, apart from the requirements of storage and access for the larger data sets.

■ The mass-balance technique can also be easily applied to two tracers simultaneously, as in the inversions of CO_2 and $^{13}CO_2$ by Ciais et al. [66, 67].

Disadvantages of mass-balance inversions

■ The method requires the fitting of data to the whole of the earth's surface.

■ The problem of resolving spatial detail is ill-conditioned and subject to amplification of errors in the absence of appropriate smoothing.

■ There is no direct formalism for uncertainty analysis from the single-inversion integration. Using multiple integrations, uncertainty estimates have been obtained using the bootstrap approach [78]. A Monte Carlo approach based on simulating estimated data error would also be possible.

■ There are no obvious criteria for determining the appropriate degree of smoothing of data.

The presence of the time derivative in the estimation equation (11.1.2) implies a degree of ill-conditioning and a necessity for smoothing in time. The ill-conditioning associated with the spatial aspects is shown by the $1/n$ attenuation for spatial variations of wavenumber n, determined numerically by Enting and Mansbridge [136], in agreement with the results from purely diffusive models. A manifestation of this ill-conditioning was found on attempting to modify the method so that Antarctic regions were treated as having known (zero) fluxes and the equation (11.1.2) applied only outside Antarctic regions. This produced spurious oscillations – a form of Gibbs phenomenon – in the source estimates to the north of the boundary.

Enting and Mansbridge [136] used a spectral fit to the spatial variation. However, spline fits were found to be preferable [465, 137]. Spectral fits generate spurious variability in the southern hemisphere when one is fitting details of northern-hemisphere distributions. Conway *et al.* [78] applied a two-dimensional mass-balance inversion to CO_2 data from 32 sites in the NOAA/CMDL network over the period 1981–92. The results for semi-hemispheres are shown in Figure 14.8 in Chapter 14. This study was notable for the introduction of uncertainty analysis using bootstrap statistics, repeating the calculation with multiple realisations of random subsets of the 32 time series.

A series of papers by Ciais and co-workers extended two-dimensional mass-balance inversions to the simultaneous inversion of CO_2 and $^{13}CO_2$ [66, 67]. These calculations required a specification of the space–time distribution of isotopic disequilibrium, as described in Section 14.2.

As noted above, extending the mass-balance-inversion technique to three dimensions involves a considerable difficulty in constructing a concentration field for all points on the earth's surface. The applicability of mass-balance inversion in three dimensions has been the subject of a series of studies by Law and co-workers. The developments include applying mass-balance inversions to residuals from reference 'best-guess' forward calculations and the use of 'objective-analysis' techniques for defining the full spatial dependence given these residuals at a small number of sites [274].

A variation on the spatial fitting is the 'data-assimilation' approach of Dargaville [96, 97]. This is a mass-balance inversion in which the observational data notionally specify the surface concentrations only over a limited region, but with an iterative fitting until a self-consistent set of concentrations and sources is obtained.

The computational efficiency of mass-balance inversion has made it possible to perform a relatively large number of sensitivity studies. These include (as described in Section 9.2 above) the extent to which errors in modelling the 'rectifier effect' influence inversions [275] and the extent to which inversions are affected by repeated use of a single year's transport fields rather than using interannually varying transport [98], assessing the effects of selection of data [273] and sensitivity to the form of interpolation of surface data [274].

11.2   State-space representations

Time-stepping synthesis involves estimating fluxes as a time-dependent linear combination of pre-specified flux distributions. Generally, this estimation occurs with a time interval much greater than the integration time-step of the transport model. Most of the tracer-inversion studies using time-stepping synthesis have been based on a state-space representation, using the Kalman filter as the estimation procedure [e.g. 197, 58]. As discussed in Section 4.4, this formalism defines a prior statistical model in terms of a 'state-vector' describing the system and its evolution from one time-point to the next as

$$\mathbf{g}(n+1) = \mathbf{F}(n)\mathbf{g}(n) + \mathbf{u}(n) + \mathbf{w}(n) \tag{11.2.1a}$$

where $\mathbf{F}(n)$ defines the unforced evolution of the state-vector $\mathbf{g}(n)$, $\mathbf{u}(n)$ is deterministic forcing and $\mathbf{w}(n)$ is a stochastic forcing with zero mean and covariance matrix $\mathbf{Q}(n)$. The model relates the state-vector to time series of observed quantities, $\mathbf{c}(n)$, by

$$\mathbf{c}(n) = \mathbf{H}(n)\mathbf{g}(n) + \mathbf{e}(n) \tag{11.2.1b}$$

where $\mathbf{H}(n)$ defines the projection of the state onto the observations and $\mathbf{e}(n)$ represents noise with zero mean and covariance matrix $\mathbf{R}(n)$.

Given a state-space model of this form, the Kalman-filter formalism (see Section 4.4) gives the optimal one-sided estimate of the state-vector $\mathbf{g}(n)$, i.e. the best estimate of $\mathbf{g}(n)$ that can be obtained using observations $\mathbf{c}(1)$ to $\mathbf{c}(n)$. In general, truly optimal estimates, i.e. the best estimate using *all* of a data set $\mathbf{c}(1)$ to $\mathbf{c}(N)$, requires a technique known as *Kalman smoothing* [173; and Section 4.4]. In special cases, known as *non-smoothable* systems, the smoothing does not improve the estimate. The most important example of *non-smoothable* models is when the state is constant, i.e. $\mathbf{F} = \mathbf{I}$ and $\mathbf{Q} = \mathbf{0}$, so that the Kalman-filter estimate $\hat{g}(N)$, which is based on all the $\mathbf{c}(n)$, is also an estimate (and the optimal estimate) of $\hat{g}(n)$ for $n < N$, as in example (i) below.

Many different types of inversion can be accommodated within this general state-space-model/Kalman-filter formalism. The relation between state-space forms and the time-dependent synthesis is noted in example (ii) below. Mass-balance inversions can be expressed in a state space form (see [133] and example (v) below), but the estimation process does not correspond to a Kalman filter and so, even within the class of one-sided estimates, mass-balance inversions will be sub-optimal.

The Kalman-filtering formalism described in Section 4.4 is a procedure for estimating the components of a time-evolving state-vector, $\mathbf{g}(n)$. Therefore, if the Kalman filter is to be used as an estimation technique, the quantity to be estimated has to be part of the state-vector. In order to represent the estimation problem in state-space form, all the quantities in (11.2.1a) and (11.2.1b) need to be defined and, in many cases, need to be specified by giving numerical values. Specifically,

$\mathbf{g}(n)$: the state-vector needs to be defined, i.e. what quantities make up its components;

$\mathbf{F}(n)$: the deterministic component of the prior model for the estimation of the state-vector needs to be specified;

$\mathbf{u}(n)$: the deterministic forcing (if any) needs to be specified;

$\mathbf{Q}(n)$: the stochastic forcing, $\mathbf{w}(n)$, of the prior model for $\mathbf{g}(n)$ needs to be characterised by specifying its covariance matrix $\mathbf{Q}(n)$;

$\hat{\mathbf{g}}(0)$: an initial estimate of the state-vector at time zero is required;

$\hat{\mathbf{P}}(0)$: the covariance for this initial estimate needs to be specified;

$c(n)$: values of the observations need to be specified;

$H(n)$: the relation between the state-vector and the observations needs to be specified; and

$R(n)$: the covariance matrix of the observations needs to be specified.

Various applications of state-space modelling can be characterised in terms of the way in which the real-world problem is mapped onto the mathematical description given above. The following list includes some of the ways of formulating the tracer-inversion problem. The relations given in this section are conceptually exact. As described in the following section, practical applications will, in general, require approximations.

(i) **The state is a constant source.** The model for a system with a constant source is

$$c(n) = G(n)s + e(n) \tag{11.2.2a}$$

This is mapped onto the state-space formalism by putting

$$g(n) = s \tag{11.2.2b}$$

since it is necessary for the state to include the quantity to be estimated. It is taken as constant, so that, in state-space form, its evolution is defined by

$$F(n) = I \tag{11.2.2c}$$

and

$$Q(n) = 0 \tag{11.2.2d}$$

There is also no additional deterministic forcing:

$$u(n) = 0 \tag{11.2.2e}$$

The Green-function matrix, $G(n)$, plays the role of the operator that maps the state onto the observations at time n so that

$$H(n) = G(n) \tag{11.2.2f}$$

and the variance of the measurement is given by

$$R(n) = \text{cov}(e, e) \tag{11.2.2g}$$

In addition we need to specify the initial estimate of $\hat{g}(0)$ expressed as $s = \hat{s}(0)$ and its variance, W. Non-Bayesian estimation corresponds to the limit of arbitrarily large diagonal elements in W.

Equations (11.2.2c)–(11.2.2e) apply when one is using the Kalman filter as a recursive implementation of linear regression. As described in Box 4.4, they lead to significant simplifications of the Kalman-filtering formalism, giving $\tilde{P}(n) = \hat{P}(n-1)$ and $\tilde{x}(n) = \hat{x}(n-1)$. The estimation equations reduce to

$$\hat{x}(n) = \hat{x}(n-1) + K(n)[c(n) - H(n)\hat{x}(n-1)] \tag{11.2.3a}$$

with

$$\begin{aligned}
\hat{\mathbf{P}}(n) &= \hat{\mathbf{P}}(n-1) - \mathbf{K}(n)[\mathbf{H}(n)\hat{\mathbf{P}}(n-1)\mathbf{H}(n)^{\mathsf{T}} + \mathbf{R}(n)]\mathbf{K}(n)^{\mathsf{T}} \\
&= \hat{\mathbf{P}}(n-1) - \mathbf{K}(n)\mathbf{P}_{\mathrm{I}}(n)\mathbf{K}(n)^{\mathsf{T}}
\end{aligned} \tag{11.2.3b}$$

where

$$\mathbf{K}(n) = \hat{\mathbf{P}}(n-1)\mathbf{H}(n)^{\mathsf{T}}[\mathbf{H}(n)\hat{\mathbf{P}}(n-1)\mathbf{H}(n)^{\mathsf{T}} + \mathbf{R}(n)]^{-1} \tag{11.2.3c}$$

The variant on the lower line of (11.2.3b) for $\hat{\mathbf{P}}(n)$ has the computational advantage that, once $\mathbf{P}_{\mathrm{I}}(n)$ has been calculated (for use in defining $\mathbf{K}$), $\hat{\mathbf{P}}(n)$ can be obtained by overwriting $\hat{\mathbf{P}}(n-1)$.

This approach has been used on a global scale by Hartley and Prinn [197]. On a regional scale, Zhang et al. [529, 530] have undertaken studies of the feasibility of estimating emissions of CH_4 from north-east Europe.

For the 'scalar' case of a single time series of observations, $c(n)$, and a single constant source, s, the Kalman-filter formalism of Section 4.4 gives results identical to those of the direct Green-function solution to the same problem as that given in Box 3.3. (The proof of this is exercise 7.2). This equivalence occurs because this case is non-smoothable. In general, the estimates from the Green-function technique will be equivalent to those from the fixed-interval Kalman smoother rather than to those from the Kalman filter.

Note that the estimates, $\hat{\mathbf{g}}(n)$, evolve in time, but that this does not represent a time-evolution of $\mathbf{s}$. Rather it represents a convergence from the initial estimate, $\hat{\mathbf{g}}(0)$, towards the true *constant* value, $\mathbf{s}$ (compare this with problem 1 of this chapter).

(ii) **The state is a sequence of sources.** The state-space representation becomes more complex when the source evolves in time. One needs to define a prior model for the evolution of the source. We denote such models

$$\mathbf{s}(n+1) = \mathbf{B}(n)\mathbf{s}(n) + \mathbf{W}(n) \tag{11.2.4a}$$

where $\mathbf{B}$ is our generic notation for prior autoregressive modelling of the time-evolution of sources.

The concentration represents the cumulative effects of past sources and so one has

$$\mathbf{c}(n) = \sum_{m \le n} \mathbf{G}_{nm}\mathbf{s}(m) + \mathbf{e}(n) \tag{11.2.4b}$$

This introduces a serious complication: the concentration depends on $\mathbf{s}(n)$ from past times and, therefore, in order to use the formalism of (11.2.1a) and (11.2.1b), all such sources must be part of the state $\mathbf{g}(n)$.

This can be brought about by defining a state-vector in terms of $N + 1$ blocks,

$$\mathbf{g}(n) = [\mathbf{s}(n), \mathbf{s}(n-1), \mathbf{s}(n-2), \ldots]^{\mathsf{T}} \tag{11.2.4c}$$

with zero subvectors in the last $N - n$ blocks. This has the evolution matrix

$$
\mathbf{F}(n) = \begin{bmatrix} \mathbf{B} & \mathbf{0} & \mathbf{0} & \cdots \\ \mathbf{I} & \mathbf{0} & \mathbf{0} & \cdots \\ \mathbf{0} & \mathbf{I} & \mathbf{0} & \cdots \\ \mathbf{0} & \mathbf{0} & \mathbf{I} & \cdots \end{bmatrix}
$$

(11.2.4d)

The Green-function expression (11.2.3b) reduces to the state-space form (11.2.1b) if

$$
\mathbf{H} = \mathbf{H}_0[\mathbf{G}_{n,n}, \mathbf{G}_{n,n-1}, \mathbf{G}_{n,n-2}, \ldots]
$$

(11.2.4e)

As n increases, the complexity of this approach would seem comparable to that of the direct least-squares fit of (10.3.2). In practice, most applications of state-space modelling (Kalman-filter analysis) have involved additional approximations to the Green-function relation.

This mapping of the Green-function formalism onto a state-space form is introduced here to illustrate the relation – no examples of direct applications are known to the author. However, a transformation of this form (case (iv) below) was used by Mulquiney *et al.* [330, 332]. For systems with high temporal and spatial resolution, the direct inversion required in the Green-function formalism has proved to be a computationally demanding task, requiring careful consideration of the stability of the matrix inversions. In contrast, recursive estimation based on state-space formulations of the equations has appeared to offer a less computationally intensive pathway to time-dependent inversions, at the expense of less-than-optimal estimates. A fixed-lag estimation procedure is a possible compromise between computational efficiency and statistical efficiency.

The state-space forms are less relevant to steady-state inversions because of the mis-match between the boundary conditions: initial information versus periodicity.

(iii) **The state includes sources and concentrations.** Formally, an alternative way of retaining all the information in the transport modelling is to construct a state that includes both sources, **s**, and modelled concentrations, **m**:

$$
\mathbf{g}(n) = [\mathbf{m(n)}, \mathbf{s(n)}]^{\mathsf{T}}
$$

(11.2.5a)

evolving with the evolution matrix

$$
\mathbf{F}(n) = \begin{bmatrix} \mathbf{T} & \mathbf{G}_{n,n-1} \\ \mathbf{0} & \mathbf{B} \end{bmatrix}
$$

(11.2.5b)

where $\mathbf{T}$ specifies the unforced evolution of $\mathbf{c}$ and $\mathbf{G}_{n,n-1}$ is the Green function relating $\mathbf{s}(n - 1)$ to $\mathbf{c}(n)$. In order to keep all the information for the transport model, **m** needs to include all grid points of the model. This means that $\mathbf{H}$ will select the subset of **m** that corresponds to the observed data set **c**. This approach leads to very large state-vectors. The point of serious difficulty comes with the requirement for calculating $\mathbf{FPF}^\mathsf{T}$ for the propagation of the uncertainty in the state. One approach is to express $\mathbf{P}$ using the SVD as

$$\mathbf{P} = \sum_q \mathbf{u}_q \lambda_q \mathbf{u}_q^\mathsf{T} \qquad (11.2.6a)$$

and truncate the expansion so that

$$\mathbf{FPF}^\mathsf{T} \approx \sum_{q \leq q'} (\mathbf{Fu}_q) \lambda_q (\mathbf{Fu}_q)^\mathsf{T} \qquad (11.2.6b)$$

with relatively small q'. This approximation is common in assimilation of meteorological data.

As with case (ii), this case is used to illustrate the principle – no direct applications are known to the author and the computational demands make any full implementation unlikely. However, the description here can serve as a reference for the derivation of approximations and it also serves as a basis for comparison with the state-space representation of mass-balance inversions (case (v) below).

(iv) **The state includes sources and a set of response components.** Mulquiney *et al.* [330, 332] have recently devised a way of retaining the essential time-dependence of (11.2.4a)–(11.2.4e) in a more compact form. Using a representation of the Green-function elements as sums of exponentials allows them to be represented in a compact recursive form.

The Green-function matrix in the block form with indices representing time steps is expressed as

$$\mathbf{G} = \begin{bmatrix} \mathbf{G}_{11} & \mathbf{0} & \cdots & \mathbf{0} \\ \mathbf{G}_{21} & \mathbf{G}_{22} & \cdots & \mathbf{0} \\ \vdots & \vdots & \vdots & \vdots \\ \mathbf{G}_{N1} & \mathbf{G}_{N2} & \cdots & \mathbf{G}_{NN} \end{bmatrix} \qquad (11.2.7a)$$

and is represented as a sum of K components, $\mathbf{G}_{nm}^*$, that have exponential decay:

$$\mathbf{G}_{nm} = \sum_{\alpha=1}^{K} \mathbf{G}_{\alpha nm}^* \qquad (11.2.7b)$$

where the exponential decay of the components in the decomposition is expressed as

$$\mathbf{G}_{\alpha nm}^* = \mathbf{L}_\alpha \mathbf{G}_{\alpha, n-1, m}^* \qquad (11.2.7c)$$

This decomposition is obtained by fitting each response $G_{xt,\mu\tau}$ as a sum of exponential functions of t. Complex exponentials are included to represent the oscillations induced by seasonally varying transport. Mulquiney *et al.* then define a state

$$\mathbf{g}(n) = \left[\mathbf{s}(n), \sum_{n' < n} \mathbf{G}^*_{1,n,n'} \mathbf{s}(n'), \ldots, \sum_{n' < n} \mathbf{G}^*_{K,n,n'} \mathbf{s}(n') \right] \qquad (11.2.7d)$$

This state evolves with the evolution matrix

$$\mathbf{F}(n) = \begin{bmatrix} \mathbf{B} & \mathbf{0} & \mathbf{0} & \cdots & \mathbf{0} & \mathbf{0} \\ \mathbf{G}^*_{1nn} & \mathbf{L}_1 & \mathbf{0} & \cdots & \mathbf{0} & \mathbf{0} \\ \mathbf{G}^*_{2nn} & \mathbf{0} & \mathbf{L}_2 & \cdots & \mathbf{0} & \mathbf{0} \\ \cdots & \cdots & \cdots & \cdots & \mathbf{0} & \mathbf{0} \\ \mathbf{G}^*_{Knn} & \mathbf{0} & \mathbf{0} & \cdots & \mathbf{L}_{K-1} & \mathbf{0} \end{bmatrix} \qquad (11.2.7e)$$

and the observations are obtained from

$$\mathbf{H} = [\mathbf{G}_{nn}, \mathbf{I}, \ldots, \mathbf{I}] \qquad (11.2.7f)$$

The random-walk prior model for the sources, as used by Mulquiney *et al.*, corresponds to $\mathbf{B} = \mathbf{I}$. (Actually, Mulquiney *et al.* included a delay in the response – this increases the complexity of the representation slightly, but does not affect the underlying principle.)

(v) **The state-space representation of mass-balance.** We can extend the state-space description of the transport model (case (iii)) to represent a simplified mass-balance equation in a vector form comparable to that used in state-space estimation. This can illustrate the difference between the mass-balance inversions and other techniques and also provides a formalism for analysing the uncertainties in the estimates. We restrict the analysis to the case in which the known source for set A is zero.

In the simplified form of Euler time-stepping, the mass-balance-estimation equations (11.1.1) and (11.1.2) become

$$\hat{\mathbf{m}}_A(n + 1) = \hat{\mathbf{m}}_A(n) + \Delta t \, [\mathbf{T}_{AA} \hat{\mathbf{m}}_A(n) + \mathbf{T}_{AB} \hat{\mathbf{m}}_B(n)] \qquad (11.2.8a)$$

$$\hat{\mathbf{m}}_B(n + 1) = \sum_k \mathbf{D}_k \mathbf{c}(t_n - \alpha k) \qquad (11.2.8b)$$

$$\hat{\mathbf{s}} = [\mathbf{T}^*]^{-1} \, [\hat{\mathbf{m}}_B(n + 1) - \hat{\mathbf{m}}_B(n) - \Delta t \, [\mathbf{T}_{BA} \hat{\mathbf{m}}_A(n) + \mathbf{T}_{BB} \hat{\mathbf{m}}_B(n)]] \qquad (11.2.8c)$$

where the matrix $\mathbf{D}$ represents the smoothing and interpolation of the observed concentrations and $\mathbf{T}^*$ describes the mapping of sources onto concentrations over a single time-step.

This is a non-Bayesian estimation technique since any prior source model (e.g. as specified by $\mathbf{B}$ below) plays no role in the estimation. The only statistical considerations are those involved in specifying the projection of observations, possibly smoothed in time, onto the grid points of set B.

If we define the state as

$$g(n) = \begin{bmatrix} \mathbf{m}_A(n) \\ \mathbf{m}_B(n) \\ \mathbf{s}(n) \end{bmatrix} \qquad (11.2.9a)$$

then we can represent the evolution equation (11.2.8) in the form

$$g(n + 1) = \mathbf{F}(n)\mathbf{g}(n) + \mathbf{w}(n) \qquad (11.2.9b)$$

with

$$\mathbf{F}(n) = \begin{bmatrix} \mathbf{I} + \Delta t\, \mathbf{T}_{AA} & \Delta t\, \mathbf{T}_{AB} & \mathbf{0} \\ \Delta t\, \mathbf{T}_{BA} & \mathbf{I} + \Delta t\, \mathbf{T}_{BB} & \mathbf{T}^* \\ \mathbf{0} & \mathbf{0} & \mathbf{B} \end{bmatrix} \qquad (11.2.9c)$$

where the stochastic forcing $\mathbf{w}$ is introduced to complete the description of the prior model for the evolution of the source. The matrix $\mathbf{B}$ describes a 'prior' statistical model for the source components. This plays no role in mass-balance inversions, but is included to allow comparison with other methods. For consistency with the deterministic form of the transport model, the covariance matrix for $\mathbf{w}$ has the block form

$$\mathbf{Q} = \begin{bmatrix} \mathbf{0} & \mathbf{0} & \mathbf{0} \\ \mathbf{0} & \mathbf{0} & \mathbf{0} \\ \mathbf{0} & \mathbf{0} & \mathbf{Q}_{ss} \end{bmatrix} \qquad (11.2.9d)$$

This representation would allow us, in principle, to use the Kalman filter as an estimation procedure. It is, however, impractical because the calculation requires the propagation of the uncertainties in the state and in particular $\mathbf{T}\hat{\mathbf{P}}\mathbf{T}^\mathsf{T}$.

The actual mass-balance-estimation equations (11.2.8a)–(11.2.8c) can be written as

$$\hat{\mathbf{g}}(n + 1) = \mathbf{K}_1(n)\mathbf{F}(n)\hat{\mathbf{g}}(n) + \mathbf{K}_2\mathbf{c}(n) \qquad (11.2.10a)$$

with

$$\mathbf{K}_1(n) = \begin{bmatrix} \mathbf{I} & \mathbf{0} & \mathbf{0} \\ \mathbf{0} & \mathbf{0} & \mathbf{0} \\ \mathbf{0} & -\mathbf{I} & \mathbf{0} \end{bmatrix} \qquad (11.2.10b)$$

and

$$\mathbf{K}_2(n) = \begin{bmatrix} \mathbf{0} \\ \mathbf{D} \\ \mathbf{D} \end{bmatrix} \qquad (11.2.10c)$$

Consequently, the covariance matrix $\hat{\mathbf{P}}(n)$ for the estimates $\mathbf{g}(n)$ evolves as

$$\hat{\mathbf{P}}(n + 1) = \mathbf{K}_1\mathbf{F}(n)\hat{\mathbf{P}}(n)\mathbf{F}(n)^\mathsf{T}\mathbf{K}_1^\mathsf{T} + \mathbf{K}_1\mathbf{Q}(n)\mathbf{K}_1^\mathsf{T} + \mathbf{K}_2\mathbf{R}(n)\mathbf{K}_2^\mathsf{T} \qquad (11.2.11)$$

As with the Kalman filter, actual computation of this propagation of uncertainty is impractical with high-resolution numerical models of atmospheric transport.

The discussion above has used a state-space representation to define a formal error-propagation formalism for mass-balance inversions. However, while this shows what is required formally, it is essentially impractical for operational calculations. As noted in case (iii), the complexity can be reduced by approximating the covariance matrix using a truncation of the singular-value decomposition. Even with such approximations, the direct propagation of measurement uncertainty through each time-step is probably too cumbersome to be practical. This type of time-evolving error propagation is probably of greater applicability in the 'hybrid' inversions described in the following section.

11.3   Time-stepping synthesis

Chapter 10 and Section 11.1 identify two main techniques of tracer inversion, described as 'differential' or 'integral' depending on whether the transport equation was fitted in its differential or integral form. The differential techniques are often denoted 'mass-balance' because the estimation uses conservation of mass at each point to deduce the sources. There seems to be little scope for constructing non-local estimates using differential techniques because, if only one time-point is considered, then there is no time for a concentration to be influenced by sources at other locations.

The integral methods are often termed Green-function methods. Furthermore, the term 'synthesis inversion' has often been used. However, as noted in the previous chapter, synthesis methods, i.e. those that combine pre-specified patterns, are not confined to the 'pure' Green-function methods and, conversely, with the use of adjoint methods, Green-function methods can avoid the use of pre-specified patterns.

As well as those involving mass-balance and Green's functions, there are techniques that can best be described as hybrids of the differential (mass-balance) and integral (Green-function) approaches.

These hybrid methods generally use the synthesis approach in that they use pre-specified spatial distributions in the source discretisation. In other words, they interpolate in the 'source space' rather than in the concentration space as is done in mass-balance inversions. However, the hybrid techniques described here are similar to the differential inversions in that they follow a single forward time-integration of the concentrations resulting from the estimated sources.

The sources are expressed as a sequence of pulses

$$s(\mathbf{r}, t) = \sum_{\mu, n} \alpha_{\mu, n} s_{\mu, n}(\mathbf{r}, t) \tag{11.3.1a}$$

where

$$s_{\mu n} = 0 \qquad \text{for } t \notin [t_n, t_{n+1}]$$

The coefficients $\alpha_{\mu, n}$ are determined by fitting:

$$c(\mathbf{r}, t_{n+1}) \approx m_n(\mathbf{r}, t_{n+1}) + \sum_{\mu} \alpha_{\mu, n} G_{\mu, n}(\mathbf{r}, t_{n+1}) \tag{11.3.1b}$$

where $m_n(\mathbf{r}, t)$ is the solution of the transport model, integrated forwards from time t_n with zero sources, and $G_{\mu,n}(\mathbf{r}, t)$ is the response to source component $s_{\mu,n}$.

In a number of applications, the coefficients $\alpha_{\mu,n}$ for each time interval n have been treated as the components of a state-vector $\mathbf{x}(n)$ and estimated using the Kalman-filtering formalism.

There are several variations on this basic approach.

- The actual fits are usually to time-averages of concentration data, rather than instantaneous values (using appropriate averages of the responses $G_{\mu,n}(\mathbf{r}, t)$).

- The $\alpha_{\mu,n}$ can be re-expressed as changes from the sources in the previous time-step, with $m_n(., .)$ being the model integration with the previous sources.

- The method could be extended to produce continuous piecewise linear estimates of sources, by using basis functions $s_{\mu,n}$ that were proportional to $t - t_n$ and defining $m_n(., .)$ as the forward integral using fixed sources $s(r, t_n)$ over the whole period $[t_n, t_{n+1}]$.

Comparison with the Kalman-filter formalisms described in the previous section shows that the main limitation of this approach is that any error estimate based on (11.3.1a) and (11.3.1b) will omit the propagation of errors from earlier time-steps.

An early application of these hybrid methods was by Brown [56]. She used a two-dimensional model with 18 latitude zones and a chemical sink due to OH in the free troposphere. Forward integrations of CCl_3F and ^{85}Kr were used to validate the model transport and CH_3CCl_3 calculations were used to validate the OH sink. Inverse calculations with these tracers were performed as a test of the inversion technique. It was also argued that this provided an extra degree of validation of the model due to inversions being highly sensitive to model error. The sources were represented as piecewise constant with the discretisation interval being either a year or a season. The estimation procedure was essentially that of fitting equation (11.3.1b). Specifying concentration for each latitude zone (as in mass-balance inversion) meant that the matrices were square and had an exact inverse, although Brown [56] noted that this aspect of the method was not essential.

Note, however, that the main result from Brown's study was a CH_4 inversion that was essentially a Green-function inversion rather than the hybrid form discussed in this chapter. Rather than use the piecewise-constant representation of sources, a Fourier representation was used. Therefore, rather than a sequence of short integrations for each constant segment, each source component was independently integrated forwards over the whole time interval for which data were fitted.

In the time-stepping synthesis, Brown [56] calculated the uncertainty in each source component (for each time interval) due to uncertainties in the data (at the end of the time interval) that had been fitted. However, she did not calculate the effect that this uncertainty would have on source estimates for later periods.

The applicability of this type of approach to CO_2 has been studied by Bruhwiler et al. [58]. They presented a test inversion using pseudo-data for CO_2 to determine the

ability of the 89-site GlobalView data set to resolve 14 source regions with a time-step
of 1 month.

Further reading

■ Rather little has been written in general terms about mass-balance inversions.
Several methodological issues have been explored by Law [274, 276].

■ Approaches based on state-space modelling are covered in the review by Prinn
and Hartley [376] and by Prinn's overview [375] from the Heraklion
conference.

Exercises for Chapter 11

1. Show that, if a system with model $\mathbf{c}(n) = \mathbf{G}(n)\mathbf{s}(n) + \mathbf{e}(n)$ with $\mathbf{s}(n) = \alpha + \beta n$
   is analysed as if $\beta = 0$, i.e. assuming the model (11.3.2a)–(11.3.2g) with
   constant $\mathbf{G}(n)$, then the rate of increase of $\hat{\mathbf{g}}(n)$ tends to $\beta/2$.

2. Examine the role of AR(1) noise on mass-balance inversions for the N-box
   model in the absence of smoothing of data. Numerical solution for $N \geq 2$ and
   analytical, using AR(1) noise, for $N = 2$, using centred time-differencing.

Chapter 12

Non-linear inversion techniques

I have yet to see any problem, however complicated, which when you looked at it in the right way, did not become still more complicated.

Poul Anderson: *New Scientist*, p. 638, 25 September 1969.

12.1 Principles

This chapter considers a range of non-linear estimation techniques. In reality, the majority of real-world estimation problems would be expected to be non-linear, with linear problems as a minor special case. This book's emphasis on linear problems reflects current practice – most progress has been made on these more tractable problems. This reflects the simplicity of the linear problems and their suitability for illustrating principles of indirect estimation. Most of the cases for which there are results with explicit mathematical expressions are linear problems.

Non-linear problems are defined by what they lack, rather than by any common type of behaviour. Furthermore, there have been relatively few applications of non-linear techniques in tracer studies. Consequently, this chapter is rather more exploratory than most of the rest of this book. Simple cases are used to illustrate the techniques so that analytical solutions can illustrate the key features. Analysing the corresponding cases with real-world complexity will almost always require numerical techniques.

Taking the case of linear estimation as a starting point, there are a number of approaches of varying degrees of generality.

■ Cases in which linear estimation is used because it is optimal. The requisite conditions are described below.

■ Cases in which linear estimation is used for convenience, even though it is not optimal.

■ Explicit linearisation is used to approximate a non-linear problem. The example presented in Section 16.3, estimating the sink due to tropospheric OH, illustrates such a linearisation. The extended Kalman filter [173: Chapter 6] is a generic approach to estimation for non-linear models of vector time series. Applications to trace-gas inversions are discussed by Segers *et al.* [429]. The tangent linear model formalism (see Section 10.6) provides the basis for automatic generation of the linearisation of a non-linear computation.

■ Separation of the linear and non-linear aspects of the estimation. One common form of this is when the non-linear aspect appears as an iterative adjustment to a linear subproblem. An example in CFC studies [34] is the linear estimation of sources as part of an estimation that also estimates a lifetime.

■ Use of a special case for which simplifications lead to computationally tractable inversions. The following section gives examples based on calculations involving linear programming (see Section 12.2 and Box 12.3). Other special cases are noted in Section 12.3.

■ Use of numerical techniques to perform a non-linear estimation such as ML or MPD, often using an iterative application of a linear approximation. This involves an explicit form of $p(\mathbf{c}|\mathbf{x})$ or $p_{\text{inferred}}(\mathbf{x}|\mathbf{c})$. Tarantola [471] describes a range of numerical optimisation techniques, some being quite general whereas others are designed to take advantage of special properties of particular estimation problems. Athias *et al.* [16] and Mazzega [313] give examples of inversions in highly non-linear problems.

■ Use of Monte Carlo inversion as the estimation technique. Note the distinction between the use of Monte Carlo methods for obtaining estimates versus for performing sensitivity analyses on estimates obtained in other ways (see Section 7.1).

■ Use of an inversion technique that does not rely on linearity. In particular the mass-balance-inversion technique described in Section 11.1 does not assume linearity in the relations between sources and concentrations. In addition, non-linear data-smoothing techniques can be applied to the observational data that are used as inputs to inversion calculations.

The remainder of this section deals with examples of the special cases in which a linear subproblem can be identified. Other special cases are dealt with in remaining sections of this chapter.

As a starting point, we recall the conditions for Bayesian maximum-of-posterior-distribution (MPD) estimates to be linear functions of the observations.

■ The model values are linearly related to the parameters to be estimated through $\mathbf{m} = \mathbf{Gx}$.

Box 12.1. **Extending the Bayesian approach**

The Bayesian approach is based on the relation

$$p(\mathbf{x}, \mathbf{c}) = p(\mathbf{c}|\mathbf{x})p(\mathbf{x}) \tag{1}$$

giving the joint probability distribution of $\mathbf{x}$ and $\mathbf{c}$ *prior to any measurements*.

Tarantola [471: Section 1.5] proposed a more general formalism for the joint posterior probability density, given by

$$p_{\text{post}}(\mathbf{x}, \mathbf{c}) \propto p_{\text{obs}}(\mathbf{c})\, p(\mathbf{x}, \mathbf{c})\, \tilde{p}(\mathbf{c}) = p_{\text{obs}}(\mathbf{c})\, p(\mathbf{c}|\mathbf{x})\, p(\mathbf{x})\, \tilde{p}(\mathbf{c}) \tag{2}$$

where $\tilde{p}(\mathbf{c})$ is a 'null' non-informative prior density and the proportionality factor is determined by the normalisation of $p_{\text{post}}(\mathbf{x}, \mathbf{c})$.

The main application of this generalisation is when the process of measurement produces probability distributions rather than discrete values.

It puts

$$p(\mathbf{c}, \mathbf{c}') = p_{\text{err}}(\mathbf{c}'|\mathbf{c})\tilde{p}(\mathbf{c}) = p_{\text{obs}}(\mathbf{c}|\mathbf{c}')\tilde{p}(\mathbf{c}') \tag{3}$$

so that, to obtain the distribution $p_{\text{obs}}(\mathbf{c}|\mathbf{c}')$ associated with a measurement, $\mathbf{c}'$, we put

$$p_{\text{obs}}(\mathbf{c}|\mathbf{c}') = \frac{p_{\text{err}}(\mathbf{c}'|\mathbf{c})\tilde{p}(\mathbf{c})}{\int p_{\text{err}}(\mathbf{c}'|\mathbf{c})\tilde{p}(\mathbf{c})\,d\mathbf{c}} \tag{4}$$

Substituting this expression of p_{obs} into (2) gives

$$p_{\text{post}}(\mathbf{x}, \mathbf{c}) = \frac{p(\mathbf{x}) \int p_{\text{err}}(\mathbf{c}'|\mathbf{c})p(\mathbf{c}|\mathbf{x})\,d\mathbf{c}}{\int p(\mathbf{x}) \int p_{\text{err}}(\mathbf{c}'|\mathbf{c})p(\mathbf{c}|\mathbf{x})\,d\mathbf{c}\,d\mathbf{x}} \tag{5}$$

■ The distribution of observations is multivariate normal with mean $\mathbf{m}$ and inverse covariance matrix $\mathbf{X}$.

■ The prior distribution of the parameter values is multivariate normal with mean $\mathbf{z}$ and inverse covariance matrix $\mathbf{W}$.

Under these conditions, the MPD estimate is

$$\hat{\mathbf{x}} = [\mathbf{G}^{\mathsf{T}}\mathbf{X}\mathbf{G} + \mathbf{W}]^{-1}[\mathbf{G}^{\mathsf{T}}\mathbf{X}\mathbf{c} + \mathbf{W}\mathbf{z}] \tag{12.1.1}$$

Estimating noise in data

The next level of complication is when the measurement errors are assumed to have identical independent normal distributions as before, but with the variance unknown. We put $\mathbf{X} = \tilde{\mathbf{X}}/\sigma^2$ with $\tilde{\mathbf{X}}$ known but σ unknown. Formally, this means that σ needs to be estimated. To emphasise this, we write the parameter vector as $(\mathbf{x}, \sigma)$.

For maximum-likelihood estimates, the relevant probability distribution is

$$p(\mathbf{c}|\mathbf{x}, \sigma) = \sqrt{\frac{|\mathbf{X}|}{(2\pi)^M}} \exp\left[-\tfrac{1}{2}(\mathbf{c} - \mathbf{Gx})^{\mathsf{T}}\mathbf{X}(\mathbf{c} - \mathbf{Gx})\right] \tag{12.1.2a}$$

or

$$\ln[p(\mathbf{c}|\mathbf{x}, \sigma)] = \tfrac{1}{2}\ln(|\mathbf{X}|) - \tfrac{1}{2}\ln[(2\pi)^M] - \tfrac{1}{2}(\mathbf{c} - \mathbf{Gx})^{\mathsf{T}}\mathbf{X}(\mathbf{c} - \mathbf{Gx}) \tag{12.1.2b}$$

The maximum-likelihood solution is given by

$$\frac{\partial}{\partial \alpha_\mu} \ln[p(\mathbf{c}|\mathbf{x})] = 0 \tag{12.1.2c}$$

which still leads to equation (12.1.1), but with $\mathbf{X}$ defined in terms of the unknown σ. There is also the additional equation

$$\frac{\partial}{\partial \sigma} \ln[p(\mathbf{c}|\mathbf{x}, \sigma)] = 0 \tag{12.1.2d}$$

which becomes

$$\sigma^{-1} = (\mathbf{c} - \mathbf{Gx})^{\mathsf{T}}\sigma^{-3}\tilde{\mathbf{X}}(\mathbf{c} - \mathbf{Gx}) \tag{12.1.2e}$$

whence

$$\sigma^2 = (\mathbf{c} - \mathbf{Gx})^{\mathsf{T}}\tilde{\mathbf{X}}(\mathbf{c} - \mathbf{Gx}) \tag{12.1.2f}$$

so that σ is estimated from the residuals.

For Bayesian estimation, we also need to define a prior distribution for $p_{\text{prior}}(\sigma)$ for σ. The relevant probability distributions are

$$p(\mathbf{x}, \sigma|\mathbf{c}) = p_{\text{prior}}(\sigma)\sqrt{\frac{|\mathbf{G}^{\mathsf{T}}\mathbf{XG} + \mathbf{W}|}{(2\pi)^M}}$$
$$\times \exp\left[-\tfrac{1}{2}(\mathbf{c} - \mathbf{Gx})^{\mathsf{T}}\mathbf{X}(\mathbf{c} - \mathbf{Gx}) - \tfrac{1}{2}(\mathbf{x} - \mathbf{z})^{\mathsf{T}}\mathbf{W}(\mathbf{x} - \mathbf{z})\right] \tag{12.1.3a}$$

or

$$\ln[p(\mathbf{x}, \sigma|\mathbf{c})] = \ln[p_{\text{prior}}(\sigma)] + \tfrac{1}{2}\ln(|\mathbf{G}^{\mathsf{T}}\mathbf{XG} + \mathbf{W}|)$$
$$- \tfrac{1}{2}\ln[(2\pi)^M] - \tfrac{1}{2}(\mathbf{c} - \mathbf{Gx})^{\mathsf{T}}\mathbf{X}(\mathbf{c} - \mathbf{Gx}) - \tfrac{1}{2}(\mathbf{x} - \mathbf{z})^{\mathsf{T}}\mathbf{W}(\mathbf{x} - \mathbf{z})$$
$$\tag{12.1.3b}$$

In non-Bayesian regression (i.e. $\mathbf{W} = \mathbf{0}$) the estimates of $\mathbf{x}$ are independent of σ because σ cancels out between the numerator and denominator of (12.1.1). In Bayesian problems with an unknown measurement-error covariance, the MPD equations need to be solved iteratively.

Non-linear models

The general case of a non-linear model will usually require general numerical techniques such as numerical optimisation or Monte Carlo searches. However, if only a few parameters appear non-linearly, it is likely to be worth using techniques that take advantage of this. We write the case with a single parameter appearing non-linearly as

$$\mathbf{m} = \mathbf{G}(\lambda)\mathbf{x} \tag{12.1.4a}$$

An example of such a problem is the simultaneous estimation of lifetimes and sources for halocarbons [34].

The maximum-likelihood solution is obtained by minimising

$$J = [\mathbf{c} - \mathbf{G}(\lambda)\mathbf{x}]^{\mathsf{T}}\mathbf{X}[\mathbf{c} - \mathbf{G}(\lambda)\mathbf{x}] \tag{12.1.4b}$$

This can be performed iteratively, using (12.1.1) (with $\mathbf{W} = \mathbf{0}$) to find the local minimum for a fixed λ and then adjusting λ to find the global minimum. The extension to the Bayesian case does not bring any additional computational difficulty in this case.

12.2   Linear programming

Enting [130] used toy model B.2 to give two illustrative examples of non-linear estimation of CO_2 sources based on the technique of linear programming (see Box 12.3). These used objective functions based on the ℓ_1 and ℓ_∞ norms.

In each case, zonal mean sources were estimated from concentrations that were assumed to be zonally representative. Transport was represented by the purely diffusive 'toy model' described in Box 2.2. In terms of the coordinate $x = \cos \theta$, the modelled concentrations were represented as

$$m(x) = m_0 + \beta t + \sum_{n=1}^{\infty} m_n P_n(x) \tag{12.2.1a}$$

where the $P_n(.)$ are Legendre polynomials. (Note that the present usage of m for modelled concentrations and c for observed concentrations reverses the usage of [130].) Sources are given by

$$s(x) = \beta + \sum_{n=1}^{n_{\max}} \gamma_n m_n P_n(x) \tag{12.2.1b}$$

where (see Box 2.1)

$$\gamma_n = \frac{n(n+1)\kappa_\theta (p_0 - p_1)^2}{\kappa_p R_{\mathrm{e}}^2} \tag{12.2.1c}$$

The observations were fitted by adjusting the coefficients m_n and β. Fitting the growth rate introduces the additional constraint

$$\beta_{\min} \leq \beta \leq \beta_{\max} \tag{12.2.1d}$$

where $\beta_{\min}$ and $\beta_{\max}$ specify the range of growth rates consistent with observations.

Box 12.2. Norms

Norms are numbers that are used to characterise the sizes of vectors on a single numerical scale. By extension, the 'closeness' of two vectors can be characterised by the norm of the difference between the two vectors. In estimation, objective functions, which measure how close the estimates are to the initial values and how close the reconstructed data are to the observations, are often expressed in terms of norms.

In the mathematical theory of vector spaces, the norm of a vector $\mathbf{x}$ is defined as any real function, denoted $||\mathbf{x}||$, such that

(i) $||\mathbf{x}|| = 0$ if and only if $\mathbf{x} = \mathbf{0}$;

(ii) $||a\mathbf{x}|| = |a| \, ||\mathbf{x}||$ for all scalars, a; and

(iii) $||\mathbf{x} + \mathbf{y}|| \leq ||\mathbf{x}|| + ||\mathbf{y}||$ for all vectors $\mathbf{x}$ and $\mathbf{y}$.

A general class of norms, denoted ℓ_p for $1 \leq p < \infty$, is defined in terms of components x_j of a vector $\mathbf{x}$ as

$$||\mathbf{x}||_p = \left(\sum_j |x_j|^p \right)^{1/p}$$

with the ℓ_∞ defined by the limit as

$$||\mathbf{x}||_\infty = \max_j |x_j|$$

Tarantola defines norms with additional normalising factors to ensure that the norms are defined as sums over dimensionless quantities. This is not required for the mathematical definition of norms, but such a definition (or the use of a basis that makes the x_j non-dimensional) is required for analyses using norms to be physically meaningful.

The most widely used norm is the ℓ_2 or least-squares norm, but the ℓ_1 norm is useful in robust estimation since it gives a much lower weight to outliers.

Robust estimation

The first case from [130] used the ℓ_1 norm and illustrates a form of robust regression. Robust estimation techniques are those that are designed to be insensitive to errors in the statistical model. One of the most important cases is a requirement for estimates to be insensitive to occasional large outliers.

A class of robust estimators based on the ℓ_1 norm can be obtained by replacing the assumption of normally distributed errors with the assumption that the error distribution is a double exponential.

Thus

$$p(\mathbf{c}|\mathbf{x}) = \prod_j \frac{\alpha}{2} \exp(-\alpha|c_j - m(x_j)|) = \left(\frac{\alpha}{2} \right)^N \exp(-\alpha J) \qquad (12.2.2a)$$

with

$$J_{\ell_1} = \sum_j \left| m_0 + \sum_n m_n P_n(x_j) - c_j \right| \qquad (12.2.2b)$$

The estimate is obtained from the coefficients that minimise J_{ℓ_1}. It is the maximum-likelihood estimate, given the assumed error distribution (12.2.2a).

To express this formally as a linear programming (LP) problem, we define additional variables f_j and g_j representing positive and negative residuals, respectively. In the LP formalism these are automatically constrained to be non-negative. We define a set of constraints

$$c_j = m_0 + \sum_n m_n P_n(x_j) + f_j - g_j \qquad (12.2.2c)$$

and seek to maximise

$$J_{\ell_1} = -\sum_j (f_j + g_j) \qquad (12.2.2d)$$

The calculations were actually performed in terms of $m_n^* = m_n + K$ with K chosen to be sufficiently large to ensure that all the m_n^* were non-negative.

Using toy model B.2, it was found [130] that the ℓ_1 estimates led to very poorly defined source estimates and that the extremes of source distributions generally occurred when data from Samoa were poorly fitted. In other words, if the estimation technique allows Samoa to be treated as an outlier, then the fluxes were very poorly constrained. Since the time of that study, NOAA sampling has been expanded, now including regular ship-based sampling, thus reducing the importance of Samoa.

Figure 12.1 shows a fit of CO_2 data as a function of latitude in terms of Legendre polynomials P_0 to P_6. The solid line is obtained by minimising J_{ℓ_1}. This had $J_{\ell_1} = 4.18$.

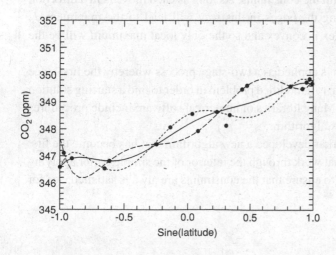

Figure 12.1 An example of robust estimation, fitting annual mean CO_2 data as a function of latitude.

Box 12.3. Linear programming

The problem of linear programming (expressed using the sign conventions of Press *et al.* [373]) is that of finding a set of variables, x_j, $j = 1$–N to maximise a linear objective function,

$$J = \sum_j d_j x_j \tag{1}$$

subject to the constraints

$$\sum_j a_{j,m} x_j \leq \alpha_m \geq 0 . \qquad \text{for } m = 1, m_1 \tag{2}$$

$$\sum_j b_{j,m} x_j \geq \beta_m \geq 0 \qquad \text{for } m = 1, m_2 \tag{3}$$

$$\sum_j c_{j,m} x_j = \gamma_m \geq 0 \qquad \text{for } m = 1, m_3 \tag{4}$$

$$x_j \geq 0 \qquad \text{for } j = 1\text{–}N \tag{5}$$

There are three possible outcomes to this problem:

■ a set of x_j that maximise J;
■ the result that there are sets of x_j for which J can be arbitrarily large; and
■ the result that there is no set of x_j that satisfies all the constraints (2)–(5).

The classic algorithm for solving the problem of linear programming is known as the simplex method. It operates by performing a succession of updates to a possible solution, following the 'edges' of the multi-dimensional polyhedron (the simplex) of sets of x_j consistent with the constraints. As long as each move is in a direction that increases J, following the edges in this way will lead to the maximum J. This is because the simplex is convex and so the only local maximum will be the absolute maximum.

The simplex algorithm is applied as a two-stage process whereby the first stage applies the same algorithm to a modified problem in order to find a starting solution for the 'real' calculation. Many libraries of numerical software include procedures implementing the simplex algorithm.

More recently, Karmarkar developed a new algorithm that has become the first of a class of algorithms that work through the interior of the simplex after applying a sequence of distortions to ensure that the constraints are always satisfied, see for example [341].

To consider the sensitivity of the fit, additional calculations were performed, minimising

$$J(\Lambda) = J_{\ell_1} + \Lambda \sum_n n m_n P_n(0) \tag{12.2.2e}$$

The two cases shown as the dotted and dashed lines each had $J(\ell_1) \approx 4.89$ (using $\Lambda = -1.0$ and $\Lambda = 1.35$). This choice of $J(\Lambda)$ corresponds, in the framework of the toy model, to looking for extremes of the source near the equator (using the approximation $\gamma_n \approx n$). However, (12.2.2e) is introduced here mainly to illustrate the sensitivities of the ℓ_1 fits.

It will be seen that this approach tends to produce an exact fit of as many of the data points as possible. As a function of Λ, this leads to discontinuous changes in $J(\Lambda)$ each time there is a change in this subset of exactly fitted points.

The box-car distribution

The other case considered by Enting [130] using the linear-programming formalism can be related to the ℓ_∞ norm. This arises from using the uniform, or 'box-car', distribution which can be regarded as the 'opposite' of the long-tailed distribution:

$$p(c_j) = \begin{cases} \dfrac{1}{2\delta_j} & \text{for } |c_j - m(x_j)| < \delta_j \\ 0 & \text{otherwise} \end{cases} \tag{12.2.3a}$$

This statistical model is of potential relevance when one is fitting data records that have been pre-processed to remove outliers. Since the LP formalism requires non-negative variables, the spectral coefficients in the model were expressed as

$$m_n = a_n - b_n \tag{12.2.3b}$$

with $a_n, b_n \geq 0$.

The equations for the ℓ_∞ estimates were expressed as the constraints

$$c_j - \delta_j \leq m_0 + \sum_{n=1}^{N} (a_n - b_n) P_n(x_j) \leq c_j + \delta_j \tag{12.2.3c}$$

The solutions for specific sets of observations, c_j, were obtained by looking for the extremes of objective functions, J_s, representing integrated sources over specified bands of latitudes. The extremes of J_s were sought, by varying β, the a_n, the b_n and m_0 subject to (12.2.3b) and (12.2.3c) with the requirement that each of the adjustable variables be non-negative. Other linear-programming problems associated with (12.2.3c) include the case of finding the minimum range δ consistent with the data.

12.3   Other special solutions

Several other statistical models are of potential interest in tracer inversions. In order
to illustrate the techniques, we consider a single-component estimation problem with
the model

$$c_n = G_n s + \epsilon_n \tag{12.3.1}$$

with $X = \mathrm{var}\,\epsilon$.

For precisely known G_n, we have our standard linear problem. For a prior of 0, the
MPD solution is

$$\hat{s} = \sum_{j=1}^{N} (G_j X c_j) \bigg/ \left(W + \sum_{j=1}^{N} G_j X G_j \right) \tag{12.3.2}$$

as derived directly in Box 3.3. Problem 2 of Chapter 7 shows that the recursive (Kalman-
filter) solution gives the same answer for this problem. The ML estimate is the $\mathbf{W} = \mathbf{0}$
case of (12.3.2). The examples following consider ML estimates for simplicity; the
extension to Bayesian estimation is usually straightforward, although in some cases it
will be necessary to use numerical solutions. In any case, it is likely that expanding
these illustrative cases to deal with real-world problems will require numerical solution,
even for ML estimation.

Random transport

This case generalises (12.3.1) to the case in which the Green-function elements, G_n,
are taken as random variables with mean zero and variance g,

$$p(\mathbf{c}|s) = [2\pi(g^2 s^2 + X)]^{-N/2} \prod_{n=1}^{N} \exp\left(-\frac{c_n^2}{2(g^2 s^2 + X)} \right) \tag{12.3.3a}$$

or

$$\ln[p(\mathbf{c}|s)] = -\tfrac{1}{2} N \ln[2\pi(g^2 s^2 + X)] - \sum_{n=1}^{N} \left(\frac{c_n^2}{2(g^2 s^2 + X)} \right) \tag{12.3.3b}$$

so the maximum-likelihood solution is

$$g^2 s^2 + X = \frac{1}{N} \sum_{j=1}^{N} c_n^2 \tag{12.3.3c}$$

The assumption of unknown transport but known variability has a potential application
when one is interpreting synoptic-scale variability using a transport model based on
wind-fields derived from a GCM. However, in this case, it is likely that a more realistic
statistical representation will be obtained by considering autocorrelations in the Green-
function elements, G_n.

Changes in the degree of variability have been interpreted (mainly in qualitative terms) in terms of changes in regional emissions where halocarbon emissions have been reduced under the Montreal Protocol. This behaviour has been noted on towers in the USA [221] and at Cape Grim [P. J. Fraser, personal communication, 1999]. One possible extension [Y. P. Wang, personal communication, 1999] is to fit the power spectrum of the observations. The TRANSCOM-2 study [106] shows the extent to which different models manage to represent such smaller-scale correlations.

Multi-tracer analysis: the tracer-ratio method

There is a class of techniques that deal with the issue of uncertain transport by using multiple tracers. The requirement is for the various tracers to have a common space–time distribution of sources that creates a characteristic atmospheric pattern. The ratios of observations correspond to the ratios of source strengths. This approach is commonly used in urban-pollution studies. One of the earliest applications of the technique [162] used the variant of estimating unknown atmospheric destruction rates from measured ratios of two tracers with known sources. Another variant is the 'fingerprint' analysis whereby the relative contributions of several sources are estimated by fitting a suite of measured ratios of gas concentrations to a linear combination of measured source ratios, e.g. [258, 5].

For a set of tracers, indexed by x, and a sequence of multi-tracer observations, indexed by n, we represent the relation as

$$c_{nx} = m_{nx} + \epsilon_{nx} = G_n s_x + \epsilon_{nx} \qquad (12.3.4a)$$

The transport coefficients G_n, notionally Green-function elements, are regarded as unknown.

For two tracers, we can rewrite the relation as

$$m_{n2} = \alpha m_{n1} \qquad (12.3.4b)$$

with

$$\alpha = s_2/s_1 \qquad (12.3.4c)$$

Observations of the two tracers can be used to deduce s_2 if s_1 is known. This problem is often analysed by regression, estimating α from regression of c_2 against c_1. This has two limitations: first, a different estimate is obtained if c_1 has been expressed as a regression on c_2; and secondly, each regression captures the uncertainty due to errors in only one of the sets of observations.

The ambiguity has been discussed by Kendall and Stuart [254: Chapter 29] in the more general form of a relation with a possible offset α_0:

$$m_{n2} = \alpha m_{n1} + \alpha_0 \qquad (12.3.5a)$$

They point out that, without some additional knowledge, or assumptions, about the error distributions, it is not possible to estimate α. The standard case that they consider is when there is knowledge of the ratio of the error variances:

$$\xi = \operatorname{var} \epsilon_2 / \operatorname{var} \epsilon_1 \tag{12.3.5b}$$

If the sample variances and covariances are denoted

$$C_{\eta\nu} = \sum_n (c_{n\eta} - \bar{c}_\eta)(c_{n\nu} - \bar{c}_\nu) \tag{12.3.5c}$$

then the maximum-likelihood estimate of α is given by the solution of

$$\hat{\alpha}^2 C_{12} + \hat{\alpha}(\xi C_{11} - C_{22}) - \xi C_{12} = 0 \tag{12.3.5d}$$

If the variances of the two error distributions are both known, then (12.3.5d) still gives the maximum-likelihood estimate of α. For the case in which only one of the variances is known, Kendall and Stuart give alternative expressions for the ML estimate of α. It will be seen that the two limits $\xi \to 0$ and $1/\xi \to 0$ reduce the solutions of (12.3.5d) to the two regression estimates that arise from ignoring errors in one or other of the records. Kendall and Stuart [254: Section 29.21] show how to calculate confidence regions for the estimates $\hat{\alpha}$ obtained from (12.3.5d).

Given the estimate of α, the estimate of the intercept, α_0, is

$$\hat{\alpha}_0 = \bar{c}_2 - \hat{\alpha}\bar{c}_1 \tag{12.3.5e}$$

Figure 12.2 shows an example of such a fit using values of 1000, 1 and 1/1000 for the assumed variance ratio. This plot uses synthetic data (with a variance ratio of 100) and is solely for the purpose of illustrating the way in which the fit depends on the assumptions about ξ. It will be seen that, for moderate degrees of variability, $\hat{\alpha}$ depends only weakly on the assumed ξ. However, using regression analysis (the $\xi \to \infty$ limit) will underestimate the uncertainties in the estimates of α.

Source estimation based on ratio analysis of such data often requires more comprehensive selection of data. The analysis by Dunse et al. [115] used integrated peak area as the basis of comparison in order to reduce the effect of inexact matching between the locations of sources of CO and CH_4.

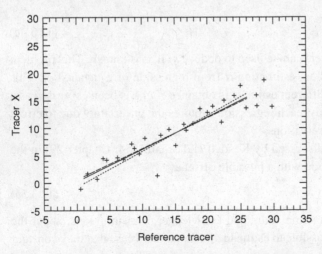

Figure 12.2 An example of ratio analysis of synthetic data showing how the linear fit varies with the assumptions that the ratio of variance in the reference versus that in the unknown is 1000 (dotted line), 1 (solid line) and 1/1000 (broken line).

More complicated cases occur when more than two tracers are present. In principle, measurements of n tracers will make it possible to estimate a partitioning into $n - 1$ source components if these have known (and distinct) ratios for the emissions [258, 5]. Pollution studies have classified inferences of source characteristics under the terminology of 'receptor modelling'. The review by Henry *et al.* [209] covers several different fitting techniques appropriate to different circumstances. Wiens *et al.* [505] also discuss a range of techniques, emphasising the desirability of robust procedures.

12.4   **Process inversion**

The introduction to this book identified one of the main aims of tracer inversion as using the spatial distributions of trace-gas concentrations to determine the spatial distributions of sources and sinks in order to help understand the processes involved. An obvious extension to this research programme is the attempt to construct techniques that estimate characteristics of flux processes more directly.

Several types of studies have foreshadowed this type of analysis.

■ Multiple-tracer analysis (see Sections 12.3 and 17.3) provides a starting point for separating the combined effects of multiple processes – isotopic budgeting (see Sections 10.5 and 14.2) is a special case.

■ Fung *et al.* [164] used their transport model to analyse the amplitude of the seasonal cycle of exchange of CO_2 with terrestrial ecosystems, basing the timing on remote sensing. The comparison was used to select among alternative functional forms of the fluxes. This sort of analysis has been re-visited in an inter-comparison of terrestrial models [335] and again, although the atmospheric data were not used for estimating model parameters, they were used to make comparisons among models.

■ The analysis by Tans *et al.* [467] suggested that their results favoured an air–sea carbon-exchange coefficient closer to that estimated from calibrating low-resolution models using ^{14}C than that estimated experimentally.

■ The estimation of atmospheric lifetimes for trace gases (see Section 16.2) and the extension of this formalism to the estimation of concentrations of OH (see Section 16.3).

A pioneering example of process inversion is the work of Knorr and Heimann [261]. They considered terrestrial carbon models characterised by two free parameters representing efficiency of use of light and the sensitivity of respiration to temperature. The models were run in diagnostic mode with inputs of NDVI, temperature, precipitation and solar radiation from ECMWF analyses. The free parameters were estimated by fitting the seasonal cycles of CO_2 at five northern-hemisphere sites. The minimisation involved iterative adjustment of the respiration parameter as an 'outer loop' with the 'inner' problem of minimisation with respect to the efficiency of use of light.

The analysis actually incorporated an additional non-parametric phase, in that eight different two-parameter formulations were compared. However, the CO_2 data were found to provide little basis for distinguishing among these, apart from tending to exclude cases with a low-temperature cut-off.

Among the benefits of approaching the problem of 'process inversion' in a more systematic way is the potential for a unified approach to analysis of uncertainty. An outline has been given by Rayner [397], showing how a synthesis-inversion formalism that produces estimates of fluxes can be extended to estimate parameters in process models.

The representation of modelled concentrations

$$m(r_j, t) = \sum_{\mu} s_{\mu} G_{\mu}(r, t) \qquad (12.4.1a)$$

is replaced by

$$m(r_j, t) = \int G(r, t, r', t') \sum_{k} h_k F_k(r', t', \mathbf{x}) \, dr' \, dt' \qquad (12.4.1b)$$

where the sum over k is introduced to reflect the fact that, in most applications, several models will be needed.

In these terms,

$$\frac{\partial m(r, t)}{\partial s_{\mu}} = G_{\mu}(r, t) \qquad (12.4.2a)$$

is replaced in the inversion calculations by

$$\frac{\partial m(r, t)}{\partial x_{\mu}} = \sum_{k} h_k \int G(r, t, r', t') \frac{\partial F_k}{\partial x_{\mu}} \, dr' \, dt \qquad (12.4.2b)$$

In this type of calculation, the modelling of atmospheric transport can provide information that contributes to the 'research phase' of either prognostic or diagnostic terrestrial modelling.

This type of estimation can be implemented in at least two ways. The first is to run the atmospheric model with source components

$$\sigma_{\mu}(r, t) = \sum_{k} h_k \frac{\partial F_k(r, t)}{\partial x_{\mu}} \qquad (12.4.3)$$

These calculations may need to be repeated during the fitting if the terrestrial model has significant non-linearity, in order to prevent the $\partial F_k(r, t)/\partial x_{\mu}$ from being treated as constant.

The second approach is to calculate $G(r_j, t, r', t')$ for all locations (r', t'), for a restricted data set specified at (r_j, t). If the form of dependence on t' can be restricted then this type of calculation could be achieved by using an adjoint model (see Kaminski et al. [240] and Section 10.6).

Our experience with synthesis inversion suggests one class of problem that can arise in the type of formalism outlined above. In the set of source components listed in Box 10.1 there is the component 'other' as part of the terrestrial biota. This was intended as a 'catch-all' for ecosystems outside the main groups. In the inversion calculations, the 'other' component often acted as a 'catch-all' that the estimation procedure used in order to fit anomalous aspects of the data, presumably including many features that could more realistically be interpreted as 'model error', in the most general sense of that term. The origin of the difficulty seemed to be the geographic spread of the contributions to the 'other' component. It is likely that the use of (12.4.2b) will need to be modified so that similar problems are avoided. Indeed, this problem may apply to many of the 'biome-oriented' basis components listed in Box 10.1 and could render synthesis inversions based on these functions vulnerable to the type of sampling bias described in Section 8.3.

Problems of using trace-gas data for 'process inversion' are comparable to the problems of performing assimilation of data for models of surface phenomena in weather forecasting, when one is incorporating hydrological and other surface processes. This type of assimilation is becoming increasingly important as models are refined and longer-range forecasts are sought. Daley [94] has noted that such calculations incur a 'spin-up' problem, since many of the time-scales for surface processes are much longer than the time-scale of the assimilation cycle. For global-scale inversions, for which typical analyses have used monthly data, the difficulties arising from the spin-up problem may be minor. The problems are likely to be more serious for regional inversions using data on time-scales of hours to interpret processes operating on time-scales of days to months.

Further reading

■ Bard's book [22] is devoted to non-linear estimation.

■ Huber's book [219] deals with robust estimation.

■ Inversion algorithms described by Tarantola [471] include many that are applicable to non-linear problems.

■ The specific application area of assimilating data into surface schemes of GCMs is an active area of research. The report [117] and other reports from the ECMWF re-analysis project present the current state of research.

Exercises for Chapter 12

1. Show that the maximum-likelihood estimator for the two-sided exponential distribution is the median.

2. Simulate and compare the sampling distributions of the mean and median for (a) normal distributions and (b) a two-sided exponential.

3. (a) For a multivariate normal distribution, obtain a factorisation into the form

$$p_{\text{inferred}}(\mathbf{x}) = \prod p(x_N | x_1 \cdots x_{N-1}) p(x_{N-1} | x_1 \cdots x_{N-2}) \cdots p(x_2 | x_1) p(x_1)$$

(b) Show that, if one has new information about each component and uses $p_{\text{inferred}}(.)$ above as the new prior, the new posterior distribution can still be written in the same factorised form.

4. (Numerical) Explore ℓ_1 fits to a CO_2 gradient. Linear programming for small problems can be done with the 'solver' component of MicroSoft Excel (if you have installed it). The book *Numerical Recipes* [373 (and editions for other programming languages)] gives linear-programming routines.

Chapter 13

Experimental design

How odd it is that anyone should not see that any observation must be for or against some view if it is to be of any service.

Charles Darwin (1861) in a letter to Henry Fawcett, quoted by
S. J. Gould in *Dinosaur in a Haystack*.

13.1 Concepts

Many factors can contribute to the way in which observational programmes to measure concentrations of trace gases are established. These include factors such as

- perceptions of scientific need,
- recognition of the potential of new measurement technology,
- availability of researchers with specific skills; and
- funding.

Apart from the choices of which gases are measured and by what technique, an important question is the locations at which air is sampled or measurements are made. Often initial sampling, particularly in a development phase, is based on opportunity. Samples are collected, or measurements made, as close as possible to the home base of the researchers involved. Beyond this, *ad hoc* criteria are often used: remote sites may have good signal-to-noise ratios, but may undersample important regions.

In view of the considerable cost and effort required to establish measurement programmes, it is desirable to assess the likely utility of such programmes in advance. As well as the question of where to sample, there are issues such as the extent to which new measurements might duplicate existing measurements. Particular cases are the relative role of isotopes and other multi-tracer studies.

This chapter concerns the use of modelling of atmospheric transport in this type of 'experimental-design' mode. In order to determine what we require from such studies, we can pose a sequence of successively refined questions about a proposed measurement programme:

- ▨ 'do the proposed measurements add to our knowledge?';
- ▨ more specifically: 'do the proposed measurements tell us specific things we want to know about a biogeochemical (or human) system?'; and,
- ▨ expressing this in terms of estimation: 'do the proposed measurements reduce the uncertainty in our estimates of particular quantities of interest?'.

As usual, the linear-estimation case (i.e. least-squares fits for the case of multivariate normal distributions) is the simplest. The uncertainty analysis (see Section 3.4) shows that the uncertainty in the estimates does not depend on the values of the observational data. This allows us to perform experimental-design studies using least-squares estimation to evaluate the effectiveness of hypothetical measurement programmes. The inversion calculation is performed with arbitrary values for putative data, but using realistic values for the achievable precision. The reduction in uncertainty gives a measure of the effectiveness of adding new data (or improving the precision of existing data).

Two important points stand out from the discussion above.

- ▨ First, the question 'what measurements are best?' has to be answered relative to some specific criterion of 'best'.
- ▨ Secondly, such an evaluation generally needs a Bayesian approach to define a meaningful context. The potential value of new measurements lies in what they would add to what is already known.

Problem 3 of Chapter 3 involved showing that, if an initial data set (c_1) is augmented by additional data (c_2), then the Bayesian estimate, $p_{inferred}(x|c_2, c_1)$, can be calculated either from a prior $p_{prior}(x)$ by using $p(c_1, c_2|x)$, which is assumed to be of the form $p(c_1|x) p(c_2|x)$, or by using $p(c_2|x)$ with the prior taken as $p_{inferred}(x|c_1)$. The equivalence obviously generalises to estimates based on a chain of successive measurements, $c_1, c_2, \ldots, c_N$. The calculations of the distributions $p_{inferred}(x|c_1, \ldots, c_N)$ and $p_{inferred}(x|c_1, \ldots, c_{N-1})$, which are required for assessing the value of measurements c_N, can be performed using any of the steps as a starting point. The two obvious cases are

(i) to use $p_{inferred}(x|c_1, \ldots, c_N) \propto p(c_N|x) p_{inferred}(x|c_1, \ldots, c_{N-1})$; and

(ii) $p_{inferred}(x|c_1, \ldots, c_N) \propto p_{model}(c_1, \ldots, c_N|x) p_{prior}(x)$ with c_1 to c_N representing all the data that can be calculated using the model.

Case (i) has the apparent advantage of having $p_{inferred}(x|c_1, \ldots, c_{N-1})$ available from the reference case (i.e. that without c_N) but in practice the distribution

---

Box 13.1. **The inverse of a block matrix**

The use of block-matrix notation to partition different aspects of linear estimation leads us to analyse the inversion of such block matrices. Generally these are covariance matrices and so we restrict ourselves to symmetric block matrices. Specifically, if

$$\begin{bmatrix} \mathbf{A} & \mathbf{B} \\ \mathbf{B}^\mathsf{T} & \mathbf{C} \end{bmatrix}^{-1} = \begin{bmatrix} \mathbf{X} & \mathbf{Y} \\ \mathbf{Y}^\mathsf{T} & \mathbf{Z} \end{bmatrix}$$

we require explicit representations of $\mathbf{X}$, $\mathbf{Y}$ and $\mathbf{Z}$ in terms of $\mathbf{A}$, $\mathbf{B}$ and $\mathbf{C}$.

The solution, which can be checked by back substitution, is

$$\mathbf{X} = \mathbf{A}^{-1} - \mathbf{A}^{-1}\mathbf{B}[\mathbf{B}^\mathsf{T}\mathbf{A}^{-1}\mathbf{B} - \mathbf{C}]^{-1}\mathbf{B}^\mathsf{T}\mathbf{A}^{-1} = [\mathbf{A} - \mathbf{B}\mathbf{C}^{-1}\mathbf{B}^\mathsf{T}]^{-1}$$
$$\mathbf{Z} = [\mathbf{C} - \mathbf{B}^\mathsf{T}\mathbf{A}^{-1}\mathbf{B}]^{-1}$$

and

$$\mathbf{Y}^\mathsf{T} = -\mathbf{Z}\mathbf{B}^\mathsf{T}\mathbf{A}^{-1} = -\mathbf{C}^{-1}\mathbf{B}^\mathsf{T}\mathbf{X}$$

Note that, for scalar blocks, the solution reduces to the expression from Cramer's rule in terms of co-factors and the determinant, e.g. the solution for $\mathbf{X}$ becomes $x = c/(ac - b^2) = (a - b^2c^{-1})^{-1}$.

A special case occurs when we augment a matrix by one additional row and column:

$$\begin{bmatrix} \mathbf{A} & \mathbf{b} \\ \mathbf{b}^\mathsf{T} & c \end{bmatrix}^{-1} = \begin{bmatrix} \mathbf{X} & \mathbf{y} \\ \mathbf{y}^\mathsf{T} & z \end{bmatrix}$$

for which the expressions above give

$$\mathbf{X} = \mathbf{A}^{-1} + \mathbf{A}^{-1}\mathbf{b}z\mathbf{b}^\mathsf{T}\mathbf{A}^{-1}$$

with

$$z = (c - \mathbf{b}^\mathsf{T}\mathbf{A}\mathbf{b})^{-1}$$

and

$$\mathbf{y}^\mathsf{T} = -z\mathbf{b}^\mathsf{T}\mathbf{A}^{-1} = -c^{-1}\mathbf{b}^\mathsf{T}\mathbf{X}$$

Apart from the determination of $\mathbf{A}^{-1}$, the only inverses are of scalars, so this provides an iterative algorithm for matrix inversion.

$p_{\text{inferred}}(\mathbf{x}|\mathbf{c}_1, \ldots, \mathbf{c}_{N-1})$ is likely to be awkward to handle (even in the case of a multivariate Gaussian characterised by its mean and covariance matrix). For performing both calculations starting from a prior that makes no use of atmospheric-transport modelling (and thus has a greater likelihood of independent prior estimates of the components of $\mathbf{x}$), a simpler form of $p_{\text{prior}}(\mathbf{x})$ may be applicable. For linear forms, if approach (i) is used, then the block-matrix-inversion procedure described in Box 13.1 may simplify the calculations.

13.2 Applications

There have been several examples of calculations using hypothetical data to evaluate the degree to which additional data would reduce uncertainties in inferences from model studies. In their study of the inversion of CCl_3F data from the AGAGE stations, Hartley and Prinn [197] considered the improvement that could be obtained by replacing data from Oregon (USA) with data from Kamchatka (Russia) or Hateruma (Japan).

Enting *et al.* [145] presented several calculations of this type using synthesis inversion of CO_2 data. These included

- reducing the assumed measurement error from 0.3 to 0.1 ppm at all sites;
- including additional data (of unspecified type) to constrain the uptake by the ocean; and
- removing items of data to determine how much this increased the uncertainties (or equivalently how much the presence of such data reduced the uncertainties).

Several other 'experimental-design' calculations have been undertaken. Some preliminary calculations have investigated the use of synoptic-scale data to resolve fluxes from regions near the observational site. In part this work is an exploratory study for the type of regional inversion described in Chapter 17. The results suggest [R. Law and P. Rayner, personal communication, 2001] that such data can significantly improve local source estimates. By avoiding the use of baseline selection and/or the need to take monthly (or longer) means, it is sometimes possible to avoid the k^{-1} attenuation associated with averaging that makes transport appear diffusive.

More recently, inversion calculations have been used to assess the potential utility of satellite data [399]. This suggests that column averages of concentration of CO_2 covering a significant fraction of the earth to a precision of 1 or 2 ppm could make a significant improvement to our present (2001) knowledge of the carbon budget.

Figure 13.1 shows the 'uncertainty ranges' for net land and ocean fluxes of CO_2, based on various data sets. Successively nested ellipses (covering 68% of the distribution) show estimates from priors, priors plus rate of growth in concentration of

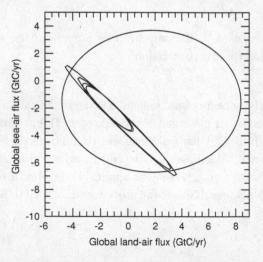

Figure 13.1 Ranges of estimates for various data sets, indicating the relative utility of various measurements. Successively nested ellipses (covering 68% of the distribution) show estimates from priors, priors plus rate of growth in concentration of CO_2, priors with growth and a 14-site distribution of CO_2, priors plus rates of growth of concentration of CO_2 and $\delta^{13}C$, priors with growth and a 14-site distribution of CO_2 and rate of growth of $\delta^{13}C$.

Box 13.2. Perfect-model experiments

In the discussion of the process of linear estimation, the estimates were described as coming from a matrix $\mathbf{H}$ that was the approximate inverse of the model matrix $\mathbf{G}$. There are a number of reasons why $\mathbf{H}$ might not be an exact inverse of $\mathbf{G}$.

■ The matrix $\mathbf{G}$ is generally not square and so an inverse will not exist. Other criteria have to be introduced, such as the minimum norms of solution and/or residuals that gives the pseudo-inverse solution.

■ Even if $\mathbf{G}$ has an inverse, this need not give the 'best' solution. This is the basis of regularisation techniques and the Bayesian formalism advocated in this book.

■ The Kalman-filter technique gives the 'best' one-sided estimate, but (except in the case of non-smoothable models) this will have a larger variance than will two-sided estimates.

■ The various state-space formalisms that give a complete description of the time-evolution of a model lead to very large state-spaces and computationally demanding estimation calculations. Practical calculations may demand the use of approximations either in the definition of the state or in the estimation procedure.

Given that the estimation procedure will often (i.e. virtually always) be equivalent to determining a matrix $\mathbf{H}$ that is not an exact inverse of $\mathbf{G}$, a common calculation is to calculate $\mathbf{HGs}$ for various specified sources $\mathbf{s}$, i.e. to apply the inversion procedure to the output of a forward model integration. A variation is to add noise to the output of the forward calculation to determine the extent to which this influences the inversion. It is a technique for designing and developing inversion techniques.

These types of calculation have become known as 'perfect-model', 'perfect-data' or 'identical-twin' experiments. Another application of such perfect-model experiments is in testing the correctness of adjoint models by comparison with sensitivities of the forward model [118].

The modelled concentrations, $\mathbf{Gs}$, produced in this type of calculation are often termed 'pseudo-data' or 'synthetic data'.

CO_2, priors with growth and a 14-site distribution of CO_2, priors plus rates of growth of concentration of CO_2 and $\delta^{13}C$, priors with growth and a 14-site distribution of CO_2 and rate of growth of $\delta^{13}C$. The inversions use the basis described in Box 10.1, with data for 1985–95. Compared with earlier inversions with this model and basis set [143, 145], the prior estimates of the isofluxes have been revised to reflect new information cited by Trudinger [487].

13.3 Network design

The 'experimental-design' calculations described in the previous section considered the improvements that would be achieved by making specific additional measurements at a specified precision. In practice, a more common question is 'where should we measure (or sample)?'.

Rayner *et al.* [400] extended the 'experimental-design' approach to analyse the cases of optimally re-locating the whole CO_2-observing network. This published analysis should be regarded as a 'proof of concept'. Operational use of the technique will require significant refinement of the underlying statistical modelling; the discussion below indicates some of the needs.

The evaluation of alternative networks and selection of the best requires an evaluation criterion. Our experimental-design approach uses criteria based on reducing the variance of estimates. For a specific network-design calculation, a specific choice of objective function is required. Usually, a Bayesian approach is essential. The network-design calculations need to be in the form of how much new information is added by new or changed measurements, relative to what is already known. Designing a network to determine, for example, the North-Atlantic flux of CO_2, while assuming the possibility of an arbitrarily large and varied flux everywhere else, will not lead to a useful or even meaningful result.

The network-design calculations involve two components:

- an inversion technique that can calculate posterior uncertainties of a chosen estimate, i.e. output, given specified uncertainties in the input; and
- a procedure for iteratively applying this inversion with a sequence of different networks, seeking to minimise the calculated uncertainty in the chosen estimate.

A schematic representation of this, with the optimisation as an outer shell around a self-contained inversion procedure, is given in Figure 7.8.

In the initial study [400], the inversion technique was the steady-state (i.e. cyclo-stationary) Bayesian synthesis inversion of CO_2, based on those undertaken by our group [143, 145], using the formalism described in Section 10.3 above. In order to simplify the interpretation, the initial 'network-design' calculations used only CO_2-concentration data, excluding ^{13}C data. The iterative adjustment used a procedure known as simulated annealing (one of the Monte Carlo algorithms described in Box 7.1). Simulated annealing is particularly suitable for this type of problem when we can expect that many different networks will lead to similar minima. The technique cannot be guaranteed to find the absolute minimum, but can be expected to find a network with performance very close to the best possible. Although our use of simulated annealing as the optimisation procedure in network design derives from work by Hardt and Scherbaum [194] on the design of seismic networks, a simulated annealing approach to network optimisation had been described earlier by Barth and Wunsch [24]. There have been analogous network-design studies in other branches of earth science. The main case considered by Rayner *et al.* was that of optimising the global network, restricted to 20 stations, in order to estimate the net global exchange of CO_2 between the atmosphere and the oceans. Relative to a network of 20 existing stations, the uncertainty (1 SD) for the net air–sea exchange of CO_2 could be reduced from 1.2 to 0.7 GtC yr^{-1} with the optimised network.

This initial network-design work [400] was intended as a set of exploratory 'proof-of-concept' calculations. There are several limitations to be overcome before the calculations would form a credible basis for designing operational systems. These are discussed in some detail in the following section. Nevertheless, it seems that some general conclusions can be drawn from these early calculations.

First, as described in the previous section, the measured rate of growth in concentration of CO_2 constrains the total net CO_2 source more tightly than do other data. This results in the various source-component estimates being generally negatively correlated. Consequently, for a large number of source components, the measurements that give most improvement to the estimates are those that most directly relate to the most poorly determined source component. This was particularly true for networks chosen to optimise the determination of the net air–sea flux of CO_2.

Secondly, the type of optimised network that is produced depends very greatly on the choice of objective function.

Other cases that have been analysed are the following.

The optimal location for a single additional site. This calculation does not need an iterative-adjustment approach such as simulated annealing. If only one additional site is to be considered, then all grid points can be tried. The results of this type of calculation can be communicated by plotting maps, showing the objective function that results from adding new measurements at each particular point [400: Figure 7].

Sums of variances. For the global net uptake of CO_2, the variance for estimates of this sum is obtained as

$$C_{\text{inferred:ocean}} = \sum_{\mu, \mu' \in \text{ocean}} C_{\text{inferred:}\mu\mu'}$$

In contrast, the use of the objective

$$J_{\text{inferred:ocean}} = \sum_{\mu \in \text{ocean}} C_{\text{inferred:}\mu\mu}$$

aims to produce networks that resolve the exchanges of CO_2 for each of the individual ocean regions. With this form of objective function, the global constraint on the CO_2 budget does not constrain the variability and so the greatest reductions are found when the observational sites are concentrated over the oceans.

This criterion was used in a more detailed analysis of network design by Gloor et al. [181]. They concluded that the GlobalView network gave reasonable resolution of latitudinal differences; for resolving longitudinal differences, on locating observing sites nearer to source regions the advantage of a stronger 'signal' outweighed the disadvantages of higher 'noise'; and use of aircraft profiles could lead to large reductions in uncertainties in fluxes.

The initial network-design calculations presented by Rayner et al. [400] were intended as a proof-of-concept calculation rather than as a proposal for changes to operational networks. It is possible to identify a number of areas in which refinement is needed before the technique can be considered for operational use. In general terms, what is required is an improvement in modelling of errors. The specific areas of concern are

- the need for a quantitative analysis of the effects of limited resolution in the model;
- quantification of other model errors;
- improvement of analysis of the problems involved in matching observations to models.

Other chapters of this book have addressed many of these issues. In many cases partial solutions exist, but these are often not in a form that is sufficiently simple to be included in a calculation as complex as network design. One aspect of phase 3 of the TRANSCOM intercomparison is the attempt to choose a network in such a way as to minimise model error. The difficulty that is common to all these aspects of error modelling is the extent to which model errors are correlated.

Further reading

- Our work on network design [400] was based on analogies with seismology [194], but there is also significant application of such techniques in oceanography [24].

Exercises for Chapter 13

1. Problem 2 of Chapter 1 was to calculate the variance of the estimated slope in linear regression from Box 1.4, with unit data spacing. If the amount of data is changed from n to kN, how does the variance of the estimated trend change (a) for extending the data at the data density and (b) on increasing the density of data for the same range on the independent variable?

2. The usual role of the Bayesian approach is to introduce additional information through the priors as a 'zeroth' data point. Use the recursive regression approach of Box 4.4 to show how much a measurement (with variance R) of a quantity $\alpha = \mathbf{a} \cdot \mathbf{x}$, will reduce the uncertainty in α relative to the 'final' uncertainty, determined by $\hat{\mathbf{P}}(n)$, when the measurement of α is treated as an $n + 1$th observation.

Part B

Recent applications

Chapter 14

Global carbon dioxide

Thus human beings are now carrying out a large scale geophysical experiment of a kind that could not have happened in the past nor be reproduced in the future.

Roger Revelle and Hans Suess [406].

14.1 Background

CO_2 is the most important of the anthropogenic greenhouse gases. Its concentration in the atmosphere has increased by about 30% during the industrial period and is almost certain to reach double pre-industrial concentrations in the second half of the twenty-first century.

On time-scales of millions of years, geological processes dominate the carbon cycle. Atmospheric concentrations of CO_2 are determined by the balance among erosion, sedimentation and volcanic emissions. Sundquist [456] has described a hierarchical classification of geological exchanges of carbon on the basis of time-scale. On any particular time-scale, any faster processes make the reservoirs involved appear to be in a well-mixed equilibrium state whereas any slower processes appear as almost-fixed boundary conditions. Geological processes act too slowly to cause significant changes to atmospheric concentrations of CO_2 on time-scales less than millennia. On time-scales of centuries or less, the dominant fluxes of CO_2 to and from the atmosphere come from the oceans and the terrestrial biosphere, i.e. living biota, dead and decaying biota and active carbon reservoirs in soil. It is these 'active' reservoirs that we need to consider when we are analysing anthropogenic changes in the carbon cycle. Figure 14.1 gives a schematic representation of the exchanges between these reservoirs, showing the large natural cycle in approximate balance and a comparatively small anthropogenic

233

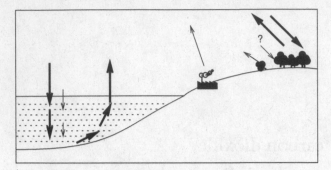

Figure 14.1 The carbon cycle, showing processes affecting the atmosphere on time-scales of decades to centuries. Heavy arrows represent the natural cycle and light arrows represent the anthropogenic perturbation. The net uptake by terrestrial ecosystems is identified mainly by multi-species budgeting but the common attribution to CO_2-enhanced growth is less well established.

perturbation. The net changes, both anthropogenic and natural, can be quantified in terms of the atmospheric carbon budget, which is expressed as

$$\frac{d}{dt}M_{A:C} = \Phi_F + \Phi_O + \Phi_B \tag{14.1.1}$$

In this equation, $M_{A:C}$ is the amount of carbon in atmospheric CO_2, i.e. essentially all carbon in the atmosphere. It has become conventional to analyse the CO_2 and the carbon cycle in terms of mass of carbon rather than mass of CO_2, in order to emphasise the role played by conservation of mass in studying the exchanges of carbon. (The main exception to this convention is in global-warming potentials, see Box 6.2.) The net fluxes are defined with fluxes to the atmosphere being taken as positive: Φ_F is the flux of fossil carbon, Φ_O is the net sea–air exchange of carbon and Φ_B is the net transfer of carbon from the terrestrial biota. This representation (14.1.1) is known as a 'flux budget'.

   An alternative characterisation of atmospheric CO_2 is in terms of 'storage' budgets. These come from the perspective of regarding the fossil flux as an input into an 'atmosphere–biosphere–ocean' system that is otherwise essentially closed. Thus

$$\frac{d}{dt}(M_{A:C} + M_{B:C} + M_{O:C}) = \Phi_F \tag{14.1.2}$$

where $M_{O:C}$ and $M_{B:C}$ are the masses of carbon in the oceans and the terrestrial biota. If these two reservoirs exchanged carbon only via the atmosphere, then (14.1.2) would be simply a re-ordering of (14.1.1) with $(d/dt)\, M_{B:C} = -\Phi_B$ and $(d/dt)\, M_{O:C} = -\Phi_O$. However, Sarmiento and Sundquist [420] pointed out that there is a flux, Φ_{BO}, of carbon from terrestrial ecosystems into the oceans via rivers. This leads to a distinction between the 'flux budget' (14.1.1) and (14.1.2) (which we term a 'storage budget'). The two budgets are related by

$$\frac{d}{dt}M_{B:C} = -\Phi_B - \Phi_{BO} \tag{14.1.3a}$$

and

$$\frac{d}{dt}M_{O:C} = -\Phi_O + \Phi_{BO} \tag{14.1.3b}$$

In characterising the carbon cycle on a regional scale, it is also necessary to distinguish fluxes of CO_2 from fluxes of carbon (see Box 14.1), although on the global scale virtually all carbon entering the atmosphere in other gases ends up as CO_2.

The use of atmospheric-transport modelling in interpreting the carbon cycle needs to be appreciated in the context of how our understanding of the carbon cycle has evolved over the last four decades. Some of the key steps have been the following.

Pre-1958. Over a century ago, Arrhenius [14] had produced quantitative estimates of the global warming to be expected from doubling the atmospheric concentration of CO_2. However, prior to the commencement of high-precision CO_2 measurement by Keeling *et al.* [248], it was not known whether significant amounts of fossil carbon remained in the atmosphere.

Post-1958. The atmospheric concentration of CO_2 was found to be increasing at a rate implying that about half the fossil CO_2 emitted was remaining in the atmosphere.

From 1975. Rates of increase of atmospheric concentration of CO_2 calculated by using ocean models calibrated by ^{14}C (particularly ^{14}C from nuclear testing) were higher than those observed, implying the existence of what came to be known as the 'missing sink'.

1980s. Estimates of large releases of CO_2 from deforestation exacerbated these apparent discrepancies. Even though the extremely high initial estimates (as much as 20 GtC yr^{-1}) have been discounted, even current estimates of 1–2 GtC yr^{-1} would have implied significant problems with the understanding of the atmospheric carbon budget.

1983. Oeschger and Heimann [343] used a response-function analysis (c.f. Section 6.6) to demonstrate that ocean models implied a lack of independence between carbon budgets at different times – the 1980s budget could not be considered in isolation from previous changes.

1985 onwards. The availability of a time-history of atmospheric concentrations of CO_2 from measurements of gas trapped in polar ice provided the time-history that Oeschger and Heimann had identified as necessary. The inverse calculation of relating the measured concentrations to net fluxes of CO_2 was developed by Siegenthaler and Oeschger [434].

1990. By combining ocean p_{CO_2} data and atmospheric-transport modelling, Tans *et al.* [467] estimated a net uptake of CO_2 by the ocean that was much smaller than that estimated from ocean models.

1992. Sarmiento and Sundquist [420] identified the distinction between flux and storage budgets (see equations (14.1.3a) and (14.1.3b)). This was part of their re-analysis that reduced the discrepancy between ocean models and the result of Tans *et al.* [467]. They found three 'corrections': the flux-versus-storage distinction, a correction to the p_{CO_2} data due to ocean-surface skin-temperature effects [409] and a correction to the atmospheric-transport modelling to take account of carbon transported as CO rather than CO_2 (see Box 14.1).

1990s. New approaches using ^{13}C data produced disparate answers. Quay *et al.* [387] used changes in the ocean ^{13}C inventory to estimate an average accumulation of carbon by the ocean of 2.1 ± 0.7 GtC yr^{-1} over 1970–90 (a storage budget – see Figure 14.4). Tans *et al.* [469] used ocean-surface ^{13}C measurements to estimate a net air-to-sea flux in the range 0.2–1.1 GtC yr^{-1} for 1970–90 (a flux budget – see Figure 14.3). It was during the 1990s that inversions of the spatial distributions of CO_2 started to provide information about the details of the carbon budget.

1995. Modelling studies reported in the IPCC Radiative Forcing report identified uncertainty in the current atmospheric budget as the largest cause of uncertainty in projecting future concentrations of CO_2 [144, 422].

1995. Analysis of long time-series of atmospheric concentrations of CO_2 and δ^{13}C implied that there is a large interannual variability in net fluxes of CO_2 to the oceans and terrestrial biota, but with significant differences between the CSIRO record [159] and the Scripps record [250] until the late 1980s.

Mid-1990s. The development of techniques to measure changes in atmospheric concentration of oxygen ($O_2 : N_2$ ratio) by interferometry [251] and mass spectrometry [27] provided additional constraints.

Inversion studies using transport modelling to deduce fluxes of CO_2 began in the late 1980s. CO_2 inversions have motivated many of the technical developments in inversion methodology described in Section 10.4. Section 14.4 presents results of some CO_2 inversions in relation to other approaches to atmospheric carbon budgeting.

In parallel with these steps, there has been significant development in 'process models' that calculate exchanges of CO_2 with terrestrial and oceanic systems. This development has been an on-going process of refinement, rather than being marked by major changes in understanding. The capabilities are developing towards the interactive form required for earth-system modelling.

A wider perspective on our increasing knowledge of the carbon cycle is given by Heimann [201], who reviewed the current (1998) understanding relative to the account given in 1896 by Arrhenius [14] (which drew extensively on the work of Högbom). Arrhenius' interest was in the possible CO_2–climate connection over glacial–interglacial cycles and mainly concerned the role of the lithosphere in determining atmospheric levels of CO_2. Failing to anticipate the near-exponential rate of growth of emissions during the twentieth century, he estimated that only a sixth of emissions would remain in the atmosphere. The factor $\frac{1}{6}$ is appropriate when emissions grow sufficiently slowly for complete equilibration between the atmosphere and the whole of the oceans and the total amount released does not greatly exceed the pre-industrial carbon content.

Classification of fluxes

Figure 14.2 shows some of the complexities that arise for classifying fluxes of CO_2 to and from the atmosphere. The grouping on the left-hand side is initially by reservoir:

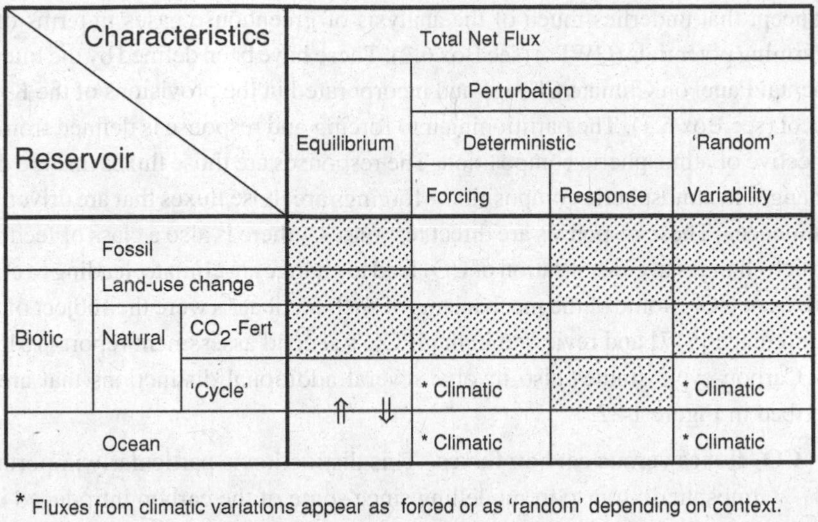

Characteristics / Reservoir	Total Net Flux			
		Perturbation		
	Equilibrium	Deterministic		'Random'
		Forcing	Response	Variability
Fossil	▒		▒	▒
Land-use change	▒		▒	▒
Biotic — Natural — CO_2-Fert	▒	▒		▒
Biotic — Natural — 'Cycle'	⇑ ⇓	* Climatic	▒	* Climatic
Ocean		* Climatic		* Climatic

* Fluxes from climatic variations appear as forced or as 'random' depending on context.

Figure 14.2 Classification of net fluxes of CO_2 to and from the atmosphere according to reservoir and 'functional role'. The shading shows combinations that make little if any contribution to the atmospheric carbon budget. CO_2-Fert denotes the natural response of terrestrial systems to elevated concentrations of CO_2, the net effect of so-called CO_2-fertilisation. The arrows represent a natural equilibrium cycle involving transfer of carbon from terrestrial systems to the ocean via rivers, with carbon returning from the oceans via the atmosphere.

fossil, ocean or the terrestrial biota. These are associated with specific regions and so this division is (potentially) addressable by transport modelling. For the terrestrial biota, there is a further distinction between managed and never-managed systems. In principle this is a spatial distinction but in practice differences can occur on scales much too small for tracer inversions to provide direct information about the partitioning between managed and unmanaged ecosystems. Changes in managed systems can be regarded as anthropogenic, but unmanaged systems can still change by responding to influences such as increases in atmospheric concentration of CO_2, variation of climate (both natural and anthropogenic) and other natural disturbances.

The 'functional' classification across the top of Figure 14.2 builds on the description outlined in Chapter 6. The primary division is into equilibrium fluxes and variations from equilibrium. The main case of a non-zero equilibrium flux (identified by the arrows) is the cycling of carbon from terrestrial ecosystems into rivers and thence to the oceans from which it is outgassed and returned from the atmosphere to the biota via photosynthesis.

The variation can be subdivided into 'deterministic' and 'random'. This classification is context-dependent – as emphasised previously, the random component has to include anything that cannot be modelled deterministically with the available data. Natural variability in fluxes of CO_2 can be (and often is) treated as random. However, if the requisite forcing functions are known, such fluxes can also be treated as a deterministic response to climatic variation, as in the work of Dai and Fung [93]. The deterministic component can be further divided into 'forcing' and 'response'. This division is the concept that underlies much of the analysis of greenhouse gases in terms of global-warming potentials (GWPs) (see Box 6.2). These have been defined by the Intergovernmental Panel on Climate Change and incorporated in the provisions of the Kyoto Protocol (see Box 6.1). The partitioning into forcing and response is defined from the perspective of atmospheric composition. The responses are those fluxes that are driven by changes in atmospheric composition – forcings are those fluxes that are driven by other influences. These responses are direct responses – there is also a class of feedbacks for which changes in concentration of CO_2 lead to changes in climate, leading to changes in fluxes of CO_2. Some of the mechanisms of such feedbacks were the subject of an IPCC conference [517] and reviewed in the IPCC's second assessment report [101, 315].

Carbon-cycle studies also involve several additional distinctions that are not described in Figure 14.2.

> **CO_2 fluxes versus carbon fluxes.** This distinction is particularly important in atmospheric-transport modelling since some of the carbon introduced into the atmosphere is transported in forms other than CO_2. These are mainly carbon monoxide (CO) and compounds that oxidise to CO_2 via CO (see Box 14.1).

> **Flux budgets versus storage budgets.** Most changes in carbon content of the oceans and biota occur via fluxes to and from the atmosphere. However, as described by equations (14.1.3a) and (14.1.3b), there is a small additional component that by-passes the atmosphere and leads to a closed cycle of non-zero fluxes in equilibrium: land biota to rivers to ocean to atmosphere to land biota. Sarmiento and Sundquist suggest that this flux Φ_{BO} should be about $0.5\,\mathrm{GtC\,yr^{-1}}$.

> **Net fluxes versus gross fluxes.** The anthropogenic perturbation and other net fluxes categorised in Figure 14.2 represent small changes to a large natural cycle. This is described in terms of gross fluxes, $\Phi_O^- \approx \Phi_O^+$ and $\Phi_B^- \approx \Phi_B^+$, to and from oceans and biota. The net fluxes of carbon $\Phi_O = \Phi_O^+ - \Phi_O^-$ and $\Phi_B = \Phi_B^+ - \Phi_B^-$ are small residuals of the two opposing gross fluxes. This notation is an exception to the general sign convention used in this book, of fluxes to the atmosphere being positive – for fluxes Φ_X^- a positive value is a flux *from* the atmosphere to reservoir X. As described in Chapter 10, the consideration of gross fluxes is of particular importance in isotopic studies.

In the budget equations (14.1.1) and (14.1.2) the oceanic and biospheric fluxes (or changes) are poorly known relative to the other terms. In order to determine these two 'unknowns' what is required is a second budget equation that captures an independent relation between these variables. The use of multi-tracer budgeting, which is outlined in Section 10.5, using ^{13}C and/or oxygen provides such an approach. Applications are described in the following section.

The use of atmospheric-transport models becomes important when one is analysing variations on time-scales of less than 2 or 3 years because, on these relatively short time scales, observations at a small number of sites need not be representative of the atmosphere as a whole. Atmospheric-transport modelling can also be used to study changes on longer time-scales, so long as sufficient data are available.

Box 14.1. **Carbon monoxide**

Most but not all of the carbon in the atmosphere enters as CO_2. There are significant inputs of carbon as methane (CH_4) and carbon monoxide (CO), which are oxidised to CO_2 in the free atmosphere, away from the surface. This has two consequences for inverse calculations of surface fluxes. The first is a small error in inversions of CO_2 data that assume that only surface sources occur. Secondly, estimates of fluxes of CO_2 will differ from fluxes of carbon both for natural and for anthropogenic processes.

The conversion from zonal CO_2 budgets to estimates of zonal carbon budgets was described by Enting and Mansbridge [137]. A simplifying assumption was that all non-CO_2 carbon entered the atmosphere either as CO or in a form that was oxidised to CO.

The correction for error in inversion can be understood by partitioning the sources into surface sources (with magnitudes s_μ) and free-atmosphere sources (with magnitudes s'_ν) and writing the Green-function relation for CO_2 as

$$c_j = \sum_\mu G_{j\mu} s_\mu + \sum_\nu G'_{j\nu} s'_\nu \qquad (1)$$

There are several possible approaches to considering CO in CO_2 inversions.

- ■ Calculate one or more responses, $G'_{j\nu}$, and estimate the coefficient(s), s'_ν, as part of the synthesis inversion, e.g. [145].
- ■ Run a mass-balance inversion with oxidation of CO modelled explicitly.
- ■ For a known source of CO, s', calculate $\chi = G's'$ and invert $c - \chi$, estimating surface sources by any of the inversion techniques.
- ■ Ignore χ as small, relative to other errors.

Once the CO_2-flux estimates have been obtained, they can be converted to carbon-flux estimates by the relation

$$s^{(C)} = s^{(CO_2)} + s^{(CO)} + s^{(CH_4)*} - d^{(CO)} - d^{(CH_4)} \qquad (2)$$

where $d^{(CO)}$ and $d^{(CH_4)}$ are the rates of destruction of CO and CH_4 at the earth's surface and $s^{(CO)}$ and $s^{(CH_4)*}$ are the CO source and the 'effective' methane source which excludes the proportion of methane which is oxidised to compounds other than CO, which are assumed to be re-deposited near the source.

14.2 Multi-tracer analysis of global budgets

Section 10.5 described the way in which multi-species budgeting can provide constraints that can help distinguish among processes. For atmospheric carbon, this approach has been used on long time-scales using ^{14}C and on shorter time-scales using ^{13}C. More recently, changes in concentration of oxygen have also been used.

The oxygen budget

The use of oxygen in atmospheric carbon budgeting is illustrated by Figure 10.1. The principles have been discussed by Keeling *et al.* [252], following the analysis by Keeling [251],

$$\dot{M}_{A:C}[1, 0] + \dot{M}_{A:O2}[0, 1] = [1, -1.3]\Phi_F + [1, -1.1]\Phi_B + [1, 0]\Phi_O \quad (14.2.1)$$

where $M_{A:O2}$ is in molar equivalents to mass of CO_2.

The terms in the budget are given below.

Quantity	Symbol	GtC yr^{-1} (equivalent)	[C, O] factors
Change in concentration of CO_2	$\dot{M}_{A:C}$	3.1	[1, 0]
Change in concentration of oxygen	$\dot{M}_{A:O2}$	7.5	[0, 1]
Fossil emission	Φ_F	5.6	[1, −1.38]
Estimates			
Ocean	$\hat{\Phi}_O$	−2.3 ± 0.7	[1, 0]
Terrestrial	$\hat{\Phi}_B$	−0.2 ± 0.9	[1, −1.1]

The values [269] represent annual averages over a 19-year period, 1978–97, with the change in concentration of oxygen measured in selected tanks of air archived at CSIRO Atmospheric Research [268]. As noted in Section 10.5, the C : O ratios used in this budget reflect the difference between oxidised and reduced carbon, i.e. CO_2 dissolved in the oceans versus CO_2 transformed into carbohydrate and other organic matter through photosynthesis. The use of [1, −1.1] rather than [1, −1] is due to the departures from the CH_2O stoichiometry in plant organic matter. Since some organic carbon is transported to the oceans via rivers, this carbon will eventually re-enter the atmosphere as a flux from the ocean. Thus the 'terrestrial' component in the C–O budget does not refer to fluxes but rather to net changes in the amount of reduced carbon, i.e. effectively to changes in the amount of carbon *stored* in terrestrial ecosystems.

An additional point to note is that the budget described above is a carbon budget, not a CO_2 budget. The assumption in using the 1 : 1.1 C : O_2 ratio from plant tissue is that virtually all carbon returned to the atmosphere is oxidised to CO_2 on time-scales comparable to those used in global budgeting – an assumption that is reasonable for a 19-year average. For inversion calculations, the shorter time-scales may require the use of the CO_2 : O_2 ratio and explicit treatment of the non-CO_2 component of the carbon cycle.

Isotopic flux budgets

Carbon occurs naturally as the three isotopes ^{12}C (about 99%), ^{13}C (about 1%) and ^{14}C (about one part in 10^{12}). The importance of ^{14}C in carbon-cycle studies is that it is radioactive with a half-life of 5730 years. Therefore its decay after production in the stratosphere provided a 'clock' that measured the amount of time that the carbon had

---

Box 14.2. The isotopic composition of CO_2

The carbon isotopes are ^{12}C (about 99% of carbon), ^{13}C (about 1%) and ^{14}C (known as radiocarbon), a radioactive isotope with a half-life of 5730 years. Naturally, ^{14}C is about one part in 10^{12} of atmospheric carbon. However, this proportion was approximately doubled by nuclear testing.

As noted in Box 10.2, the measured isotopic ratios are generally expressed in δ notation with

$$\delta^{13}C = \left(\frac{[^{13}C]/[^{12}C]_{\text{obs}}}{[^{13}C]/[^{12}C]_{\text{standard}}} - 1 \right) \times 1000 \quad \text{in } ‰$$

The standard $[^{13}C]/[^{12}C]_{\text{standard}}$, denoted r_s, has been measured as 0.011 237 2. However, measurements of $\delta^{13}C$ are made by comparison of a sample with a working standard and so in practice the accuracy of the measurement lies in the extent to which the working standard can be linked back to the international standard. For ^{14}C an isotopic anomaly $\delta^{14}C$ is defined analogously to $\delta^{13}C$. In addition, a quantity $\Delta^{14}C$ is defined, to convert measured $\delta^{14}C$ into an effective atmospheric level, using measured $\delta^{13}C$ to define the correction, which notionally includes both fractionation in the formation of biological material and fractionation in the process of measurement.

The oxygen isotopes ^{16}O (99.63%), ^{17}O (0.037%) and ^{18}O (0.204%) have also contributed to carbon-cycle studies. The oxygen-isotope ratios in CO_2 reflect gross photosynthetic exchange since this process couples the distribution of oxygen isotopes in CO_2 to the distribution of oxygen isotopes in leaf water [158, 68, 69].

---

been isolated from the atmosphere. More recently, the amount of ^{14}C in the atmosphere was nearly doubled by production of ^{14}C from the testing of nuclear weapons. Both the natural and bomb-produced ^{14}C distributions play important roles in ocean tracer modelling and inversion, but the use of ^{14}C in purely atmospheric studies is much more restricted.

Biogeochemical processes induce small variations in the $[^{13}C]/[^{12}C]$ ratio due to the fact that many exchange processes discriminate against the heavier isotope. This has established differences such that the biotic and fossil-fuel reservoirs are about 18‰ lower than the atmosphere and the oceans are about 7‰ higher.

Isotopic budgeting was introduced in Chapter 10, using CH_4 as an example. For CO_2 it was noted that isotopic budgeting involved an additional complication due to the atmosphere having two-way exchanges with the oceans and terrestrial biota.

Following the analysis in Section 10.5 (see Figure 10.3), we write the net fluxes as a difference between gross fluxes:

$$\Phi_O = \Phi_O^+ - \Phi_O^- \tag{14.2.2a}$$
$$\Phi_B = \Phi_B^+ - \Phi_B^- \tag{14.2.2b}$$

The anomaly fluxes are

$$\Phi_O^+ \delta_O^+ - \Phi_O^- \delta_O^- = (\Phi_O^+ - \Phi_O^-)\delta_O^+ + \Phi_O^-(\delta_O^+ - \delta_O^-) \tag{14.2.3a}$$

and

$$\Phi_B^- \delta_B^+ - \Phi_B^- \delta_B^- = (\Phi_B^+ - \Phi_B^-)\delta_B^+ + \Phi_B^-(\delta_B^+ - \delta_B^-) \qquad (14.2.3b)$$

The isotopic anomaly δ refers to the fluxes, not the reservoirs. In equations (14.2.3a) and (14.2.3b) the first term on the right-hand side vanishes if the net flux from reservoir X vanishes. The second term on the right-hand side (often termed the isoflux) would vanish if reservoir X were in isotopic equilibrium with the atmosphere. However, neither the oceans nor the terrestrial ecosystems are in isotopic equilibrium with the atmosphere. The $[^{13}C]/[^{12}C]$ ratio of the atmosphere has changed due to the input of fossil carbon, which is depleted in ^{13}C. The isotopic compositions of the oceans and biota have followed this on-going depletion, but with a time lag, so that the atmosphere remains out of isotopic equilibrium with these reservoirs. For the purposes of carbon-budgeting studies, these disequilibrium effects are 'nuisance' variables since the effects are not related to the net fluxes.

For CO_2, the joint CO_2–$^{13}CO_2$ budget combines the carbon budget

$$\frac{d}{dt}M_A = \Phi_F + \Phi_O + \Phi_B \qquad (14.2.4a)$$

and the ^{13}C anomaly budget

$$\frac{d}{dt}(M_A\delta_A) = \delta_F\Phi_F + \delta_O^+\Phi_O + \delta_B^+\Phi_B + \Phi_O^-(\delta_O^+ - \delta_O^-) + \Phi_B^-(\delta_B^+ - \delta_B^-)$$

$$(14.2.4b)$$

Using the vector representation described in Section 10.5, the vector sum equivalent to (14.2.4a) and (14.2.4b) is

$$+\Phi_F[1, \delta_F] - \dot{M}_A[1, \delta_A] + \Phi_B[1, \delta_B^+] + \Phi_O[1, \delta_O^+]$$
$$+[0, 1]\Phi_O^-(\delta_O^+ - \delta_O^-) + [0, 1]\Phi_B^-(\delta_B^+ - \delta_B^-) - [0, M_A\dot{\delta}_A] = [0, 0] \qquad (14.2.4c)$$

A graphical representation of this relation is shown in Figure 14.3. The order of terms in (14.2.4c) follows the lines anti-clockwise from the origin.

Tans et al. [469] introduced this approach to carbon budgeting. The most important uncertainty in their calculation was in the value of the ocean isoflux which they determined from observations. An accurate summation is extremely difficult because the net global isoflux is a sum over both positive and negative contributions. Their original estimate of oceanic uptake for 1970–90 was 0.2 GtC yr^{-1}, which they rejected as being unphysical. The argument was that, on time-scales of more than a few years, the mixed-layer partial pressure should track the atmospheric changes as

$$-\Phi_B \approx \dot{M}_A[M_{mixed}/(\xi_{buffer}M_A)] - \Phi_{deep} \qquad (14.2.5a)$$

where $\Phi_{deep} < 0$ represents loss to the deeper ocean layers. Tans et al. interpreted this as implying that $\Phi_B < -0.2$ GtC yr^{-1}. However, the correct form of the constraint is

$$-\Phi_B + \Phi_{BO} \approx \dot{M}_A[M_{mixed}/(\xi_{buffer}M_A)] + \Phi_{deep} \qquad (14.2.5b)$$

or approximately $\Phi_B < -0.2 + \Phi_{BO}$, which is a much weaker constraint than that used by Tans et al.

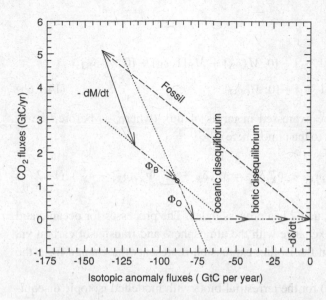

Figure 14.3 A vector representation of the balance of CO_2 and ^{13}C fluxes. The components follow the order of equation (14.2.4c) anti-clockwise from the origin: – –, fossil; ——, $\dot{M}$; — — ·, Φ_B; — — ··, Φ_O; — · — ·, ocean isoflux; — ··· — ···, biotic isoflux; and ······, $-\delta$. The solution for Φ_B and Φ_O is given by the intersection of the dotted lines to give the vectors shown.

However, although the result of Tans et al. does not violate the constraint, it should be regarded as an under-estimate because they used the value $12\%_0$ GtC yr^{-1} for the terrestrial isoflux. This was done to provide a consistent comparison with the ^{13}C-storage-budget technique of Quay et al. described below. However, as can be seen by comparing the vector plots in Figures 14.3 and 14.4, the technique of Tans et al. is more sensitive to errors in the isoflux than is the approach of Quay et al.

A biotic isoflux of $26.5\%_0$ GtC yr^{-1} is obtained from a calculation with a five box terrestrial model [143] and a similar result has been obtained with high-resolution terrestrial modelling [166]. This revised isoflux changes the budget estimates to those given below.

Quantity	Symbol	GtC yr^{-1}	$\%_0$ GtC yr^{-1}	$[C, \delta \times C]$
Fossil	Φ_F	5.1		$[1, -27.2\%_0]$
Atmospheric concentration of CO_2	$\dot{M}_A$	2.9		$[1, -7.56]$
Ocean isoflux	$\Phi_O^-(\delta_O^+ - \delta_O^-)$		36.5	$[0, 1]$
Biotic isoflux	$\Phi_B^-(\delta_B^+ - \delta_B^-)$		26.5	$[0, 1]$
	$M_A\dot{\delta}_A$		14.8	$[0, 1]$
Estimates				
Ocean flux	$\hat{\Phi}_O$	−1.2		$[1, -9.6\%_0]$
Terrestrial flux	$\hat{\Phi}_B$	−1.0		$[1, -26.6\%_0]$

Isotopic inventory budgets

An alternative approach to CO_2–$^{13}CO_2$ budgeting was introduced by Quay et al. [387]. It is based on pairing (14.1.2) with the ^{13}C anomaly budget for storage of carbon:

$$\frac{d}{dt}(M_A\delta_A + M_B\delta_B + M_O\delta_O) = \Phi_F\delta_F \qquad (14.2.6a)$$

to give the vector equation

$$\Phi_F[1, \delta_F] = \frac{d}{dt} M_A[1, \delta_A] + [0, M_A\dot{\delta}_A] + \dot{M}_B[1, \delta_B] + [0, M_B\dot{\delta}_B]$$
$$+ \dot{M}_O[1, \delta_O] + [0, M_O\dot{\delta}_O] \qquad (14.2.6b)$$

The isotopic changes can be expressed in terms of flux (where, as before, the + superscript represents fluxes to the atmosphere)

$$\frac{d}{dt} M_X\delta_X = \sum_j \Phi_{Xj}^- \delta_{Xj}^- - \Phi_{Xj}^+ \delta_{Xj}^+ = \dot{M}_X\delta_X^- + \sum_j \Phi_{Xj}^+ (\delta_{Xj}^- - \delta_{Xj}^+) \quad (14.2.6c)$$

where X is a reservoir index and j is a process index. The processes for oceanic and biotic reservoirs are direct exchange with the atmosphere and transfer of carbon via rivers. The change in sign in the isoflux expression represents the change from the atmospheric perspective.

Quay *et al.* used (14.2.6c) for the terrestrial biota with modelled isotopic disequilibrium. For the ocean, they used the form

$$\frac{d}{dt} M_X\delta_X = \dot{M}_X\delta_X + M_X\dot{\delta}_X \qquad (14.2.6d)$$

and used measurements of ocean $\delta^{13}C$ to determine the $M_O\dot{\delta}_O$ term as

$$M_O\dot{\delta}_O = C \frac{d}{dt} \int \delta_O \, dV \qquad (14.2.7)$$

where C is the concentration of carbon in the ocean (in the upper layers where the $\delta^{13}C$ was changing) and the volume integral is taken over these upper layers.

The terms in the budget equation, with magnitudes for 1970–90 taken from Quay *et al.*, apart from the larger biotic isoflux based on a five-box terrestrial model [143], are given below.

Quantity	Symbol	GtC yr^{-1}	‰GtC yr^{-1}	[C, $\delta \times$C]
Fossil	Φ_F	5.1		[1, -28‰]
Atmospheric concentration of CO$_2$	$\dot{M}_A$	2.9		[1, -7.5]
Ocean ^{13}C anomaly	$M_O\dot{\delta}_O$		-89.9	[0, 1]
Biotic isoflux	$\Phi_B^-(\delta_B^+ - \delta_B^-)$		-26.5	[0, 1]
	$M_A\dot{\delta}_A$		-14.8	[0, 1]
Estimates				
Ocean	$\hat{\Phi}_O = -\dot{M}_O$	-2.6		[1, 1.7‰]
Terrestrial carbon	$\hat{\Phi}_B = -\dot{M}_B$	0.4		[1, -26.6‰]

Figure 14.4 gives a graphical representation of how this relation can be used to determine this 'storage budget'. Note that part of the difference between this budget and the result of Tans *et al.* (as modified above) is that one is a storage budget whereas

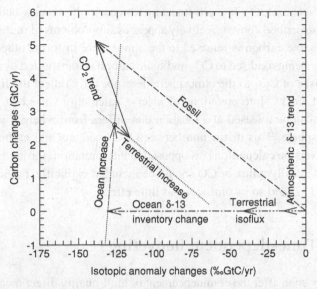

Figure 14.4 A vector representation of the balance of storage for CO_2 and ^{13}C. Following the components anti-clockwise from the origin, the fossil-carbon flux, $--$, balances the other changes in the atmosphere–ocean–biosphere system. These are —, $\dot{M}$; — —, M_B; – – · · – –, M_O; — · — ·, change in oceanic $\delta^{13}C$ inventory; · · · — · · ·, biotic disequilibrium; and · · · · · ·, δ. The solution for the net change in biotic and oceanic carbon content is given by the intersection of the dotted lines to give the vectors as shown.

the other is a flux budget. Nevertheless, the difference is greater than would be expected for a river carbon flux of order 0.5 GtC yr^{-1}.

Heimann and Maier-Reimer [202] produced a combined budget estimate based on combining the information from the two forms of joint CO_2–$^{13}CO_2$ budget, (14.2.4c) and (14.2.6b), and a third approach based on the long-term trends in concentration of CO_2 and $^{13}CO_2$. This involved simultaneously fitting the carbon-budget equation, the trend constraint, the ^{13}C-flux budget and the oceanic ^{13}C-change budget, with Bayesian constraints on the adjustable parameters, including the river flux term that converts flux budgets into storage budgets. The reason for simultaneous fitting (rather than combining the three estimates) is that many of the data trends and parameters are common to more than one of the techniques. Therefore estimates from the three techniques will not be independent. Heimann and Maier-Reimer obtained a final budget with acceptable agreement with all the data. Their estimate of the terrestrial isoflux was 23.4‰ GtC yr^{-1}. Fung *et al.* [166] undertook a spatially explicit analysis of disequilibria in the terrestrial biota and estimated a flux-weighted mean disequilibrium of 0.33‰ for 1988.

Implications for inversions

The discussion of multi-species budgeting in Chapter 10 noted that, in inversions, most of the constraint on the global budget came from the global mean concentration data (see also Section 13.1). In multi-tracer inversions, multi-species-budgeting analyses of the types described above effectively appear as subproblems of multi-tracer inversions.

Since some carbon is released to the atmosphere in forms other than CO_2 before ultimately being oxidised to CO_2 and some carbon is transported by rivers, the location of net fluxes of CO_2 to the atmosphere does not fully reflect the location of changes in carbon stocks. Interpretations of inverse calculations need to allow for the fact that CO_2 can be released at a location that differs from where it was taken up from the atmosphere. This distinction between CO_2 budgets and carbon budgets mainly affects inversion calculations (as opposed to interpretation) through the choice of prior estimates. Ideally a flux of CO should be included explicitly, but its spatial signature is small [137] and so its omission has little effect.

14.3 **Time-dependent budgeting**

Relatively soon after the commencement of high-quality direct measurements of concentrations of CO_2, it was appreciated that the long-term rate of growth had significant year-to-year variation. Bacastow [18] found a negative correlation between the rate of increase in concentration of CO_2 (after removing seasonality) and the southern-oscillation index (SOI). This connection has been explored in many later time-series studies, e.g. [119, 120, 474].

The processes causing this interannual variability have been investigated by using the joint CO_2–^{13}C budgeting – the formalism of Tans *et al.* [469] – described by equation (14.2.4c), applied to the time-evolution of the carbon cycle. A multi-isotope time-domain analysis was developed by Francey *et al.* [159] and Keeling *et al.* [250]. They produced estimates that disagreed, mainly due to the differences between the CSIRO and Scripps $\delta^{13}C$ records prior to 1990. This led to a number of claims.

- The estimates of Keeling *et al.* were regarded as implausible because they involved a large anti-correlation between terrestrial and oceanic fluxes. The mechanisms for forcing such coherent variation seemed weak, but such anti-correlated estimates could readily arise from errors in the ^{13}C data. It was suggested that a key measurement requirement was consistency in the isotopic composition of the reference sample used in the mass spectrometer.
- It was suggested that the CSIRO Cape Grim record was too far from being globally representative to be used for this type of budget analysis.

Enting [131] revisited this joint-budget analysis, pointing out the need for a comprehensive statistical analysis, including the requirement that the various terms in equation (14.2.4c) need to have the same time-averaging applied if the equation is to remain consistent. Apart from a need to review the statistics of the earlier studies, an additional motivation was to develop a 'prototype' for improved statistical modelling in time-dependent synthesis inversions.

Formally, the solution of the vector equation (14.2.4c) is

$$\Phi_O(\delta_O^+ - \delta_B^+) = \dot{M}(\delta_A - \delta_B^+) + M_A\dot{\delta}_A - \Phi_F(\delta_F - \delta_B^+) - \Phi_O^-(\delta_O^+ - \delta_O^-)$$
$$-\Phi_B^-(\delta_B^+ - \delta_B^-) \qquad\qquad\qquad (14.3.1a)$$

However, a form more suitable for uncertainty analysis is

$$\Phi_O(\delta_O^+ - \delta_{\text{ref}}) = \dot{M}_A(\delta_A - \delta_{\text{ref}}) + M_A\dot{\delta}_A - \Phi_F(\delta_F - \delta_{\text{ref}})$$
$$-\Phi_O^-(\delta_O^+ - \delta_O^-) - \Phi_B^-(\delta_B^+ - \delta_B^-) - \Phi_B(\delta_B^+ - \delta_{\text{ref}}) \quad (14.3.1b)$$

This corresponds to re-expressing the δ values relative to a reference ratio whose $\delta^{13}C$ value is fixed, i.e. not subject to uncertainty, but numerically close to δ_B^+. The form (14.3.1b) deals with non-linearity associated with uncertainties in δ_B^+ by confining it to one term involving Φ_B.

The autocovariance in the estimate, $\hat{\Phi}_O(\delta_O^+ - \delta_{\text{ref}})$, was then obtained by summing the autocovariance functions of each term on the right-hand side of (14.3.1b). The derivative terms have negative autocovariances on time-scales comparable to the smoothing and so the autocovariance of the flux estimates has a local minimum near that time-scale [131].

Battle *et al.* [25] have used both oxygen and ^{13}C data in carbon budgeting. They expect oxygen to constrain the long-term partitioning but to have short-term uncertainty in its concentration due to variations in oceanic uptake. For ^{13}C, the short-term variations were regarded as reliable but the long-term partitioning was regarded as less certain due to uncertainties in the isofluxes. Their main result was that terrestrial systems were close to neutral over 1970–90, but became a net sink over the period 1990–97. The oceans appeared as about a 2-GtC yr^{-1} sink early in the 1990s, but this dropped to about 1 GtC yr^{-1} around 1996–97, with a possible subsequent increase starting to become apparent.

In steady-state synthesis inversion [143, 145 and Figure 13.1], it was apparent that most of the information about the global total fluxes came from the global mean concentration data. In time-dependent analyses this will be true on the longer time-scales, but less applicable on shorter time-scales. The time-scales for transport add an extra contribution to the multiple time-scales associated with the uncertainties in the multi-species budgeting illustrated in this section.

A new approach to time-dependent global-scale budgeting is described by Trudinger [487]. This used the Kalman-filtering formalism to implement a statistical version of double deconvolution. The idea is to use model responses to calculate the isoflux terms. In the sense that it is making use of the dynamics of the time-evolution, double deconvolution and its statistical generalisation can be regarded as a time-dependent analogue of the 'dynamic constraint' approach of Heimann and Maier-Reimer [202]. This should be an improvement over previous approaches that incorporated only slowly varying components in the isoflux terms.

14.4   Results of inversions

Requirements

The complexity of the carbon cycle makes the communication of atmospheric carbon budgets particularly complicated. The fluxes vary greatly in space and time while the resolution of the inversions has not been characterised well. Among the characteristics that need to be captured in any description are the following.

The space–time dependence. The space–time dependence is the primary output of inversion calculations.

The partitioning into processes. This is the primary aim of CO_2 inversions: to use the space–time distribution of concentrations to determine the space–time distribution of fluxes and thus provide information about processes. Multiple tracers (adding O_2 and/or $^{13}CO_2$ to the CO_2 budget) can provide valuable extra information about the partitioning of the budget. As noted above, uncertainties in the current carbon budget have been assessed as being the major contributions to uncertainties in projections of future concentrations of CO_2 from proposed emission scenarios.

The time-averaging. Specifying the degree of time-averaging is important because both the anthropogenic inputs and the responses to these inputs are changing over time. In addition, there is a degree of natural year-to-year variability that is comparable to (and often larger than) the annual changes from anthropogenic causes.

The spatial resolution. Section 8.3 has discussed the various aspects of resolution. In describing the resolution of estimates and the uncertainties that occur at different resolutions, it is important to distinguish the 'target' resolution (the resolution at which the results are presented) from the inherent resolution imposed by the signal-to-noise characteristics of the data.

The uncertainties. As is emphasised throughout this book, assessment of uncertainty is an important part of presenting the science, particularly for ill-conditioned problems that are subject to very great amplification of errors. In many cases, consideration of the covariance structure of the estimated budgets will be essential.

In considering the partitioning into processes, the relevant question is 'what information does the inversion contribute?'. More specifically, what does the inversion determine, beyond what is known from the global budgeting *and the prior estimates*? Inversion calculations using CO_2 and ^{13}C have generally taken the form of extending the formalism of Tans *et al.* [469] to be spatially explicit. Time-dependent inversions with ^{13}C represent a spatially explicit form of the multi-tracer budgeting described in Section 14.3.

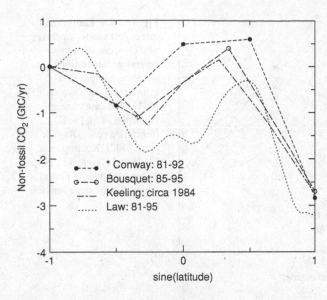

Figure 14.5 Latitudinally integrated fluxes of non-fossil CO_2 from various inversion calculations. The * denotes two-dimensional calculations. The cases are Conway *et al.* [78], Bousquet *et al.* [41] (reference case), Keeling *et al.* [249] and Law [274: case 2]. Years indicate the period to which the estimate applies.

Spatial distributions

In order to focus on the spatial aspects of the inversions, the first point of comparison is the distribution of long-term average fluxes. Following the discussion in Chapter 8, these are presented as fluxes integrated over latitude. The reasons for this are (a) to aid comparisons of model results presented on different grids and (b) to reduce the degree of spatial autocorrelation (c.f. equations (8.2.2b) and (8.2.2c)). Figure 14.5 shows the estimates of the non-fossil flux of CO_2 and Figure 14.6 shows estimates of the oceanic component.

These diagrams suggest that the inversion calculations are giving a consistent view of the large-scale fluxes and that a significant amount of the uncertainty should be regarded as ambiguity in the precise location of the flux, rather than in the size of the flux.

The main features in the combined non-fossil estimates are the following.

- A southern-latitude sink of order 1–1.5 $GtC\,yr^{-1}$. Some of the analyses suggest that this is the net result of a high-latitude source (of order 0.5 $GtC\,yr^{-1}$) and a mid-latitude sink of order 2 $GtC\,yr^{-1}$. Other analyses use insufficient resolution to detect such a source *or* fail to report in sufficient detail *or* explicitly fail to find such a source.

- A strong northern-latitude sink of as much as 2–3 $GtC\,yr^{-1}$. The inversions differ in the range of latitudes to which this source is ascribed. In part this represents differences in the reporting of the results. This estimate needs to be qualified with the comment that most of these calculations obtain this sink by subtracting an estimated fossil source from the net flux, without correcting for the component of fossil emissions released as CO.

- Comparing inversion estimates *of fluxes of CO_2* calculated without CO with those that included CO (through lower amounts of fossil CO_2 and an explicit source of

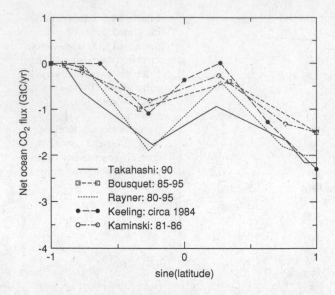

Figure 14.6 Latitudinally integrated sea-to-air fluxes of CO_2 from various inversion calculations. Also shown are direct flux estimates from Takahashi [461]. The cases are Bousquet *et al.* [41] (reference case), Rayner *et al.* [402], Keeling *et al.* [249] and Kaminski [239]. Years indicate the period to which the estimate applies.

CO_2 from CO) indicate that ignoring CO leads to mainly higher (≈ 0.5 GtC yr^{-1}) estimated tropical fluxes of CO_2 (i.e. some of the oxidation of CO is misinterpreted as a surface source). Of course, interpretations in terms of *net carbon fluxes* or *net carbon storage* will differ from those in terms of fluxes of CO_2.

■ A balancing source somewhere between 30° S and 30° N. This is poorly resolved, as might be expected from the fact that the Hadley circulation transports much of the tropical CO_2 signal away from the surface. This problem is exacerbated by the small number of stations in the tropics. In particular, the toy-model analysis using robust estimation [130] implies a strong sensitivity to data from Samoa, a station that is strongly influenced by ENSO-related interannual variability.

For the ocean fluxes, the main features of the estimates are the following.

■ An uptake of 1–1.5 GtC yr^{-1} in the extra-tropical northern hemisphere. This is in general agreement with recent estimates from Takahashi [461], based on p_{CO_2}, but larger than the estimates from Tans *et al.* [467] which were based on older p_{CO_2} data. In many of the inversions, this agreement is, of course, due in part to the use of p_{CO_2} data for the prior estimates.

■ A southern hemisphere sink of about 1 GtC yr^{-1}, which is rather less than the estimates by Takahashi.

■ A tropical source that is poorly determined for the same reasons that the total flux is poorly determined. There is a strong negative correlation between estimates for the tropics and the southern hemisphere.

Figure 14.7 summarises some of the information from a single inversion. It shows the covariances between the estimates of various components of the global budget for 1980–90. The latitudinal groupings were obtained by linear combination of the basis

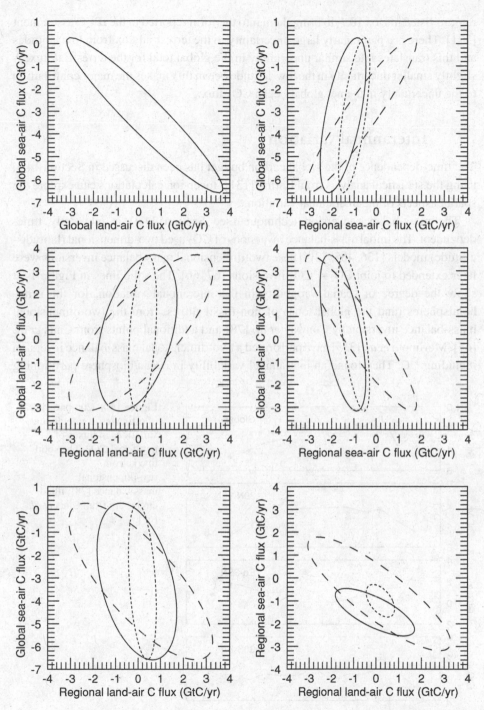

Figure 14.7 Covariances between estimates of components of the carbon budget. For cases involving multiple zones, the solid, dashed and dotted curves are for north of 30° N, 30° S to 30° N and south of 30° S, respectively.

regions listed in Box 10.1, to bring them into the form reported in the IPCC assessment [371]. There is a particularly large uncertainty in the terrestrial flux from low latitudes and this translates into similar uncertainty in the global total for the terrestrial flux. A slightly smaller uncertainty in the low-latitude ocean flux makes the major contribution to the uncertainty in the net global ocean CO_2 flux.

Interannual variation

The time-dependence of the global carbon budget has been discussed in Section 14.3, using the statistical analysis from Enting [131]. Inversion calculations can explore the spatial aspects of the interannual variation.

The mass-balance-inversion technique (see Section 11.1) is inherently time-dependent. The initial mass-balance inversions of CO_2 used two-dimensional (latitude–altitude) models [136, 465, 78]. These two-dimensional mass-balance inversions were later extended to joint CO_2–$^{13}CO_2$ inversions [67, 66]. The solid lines in Figure 14.8 show the degree of spatial variability in the interannual variation, for the semi-hemispheres (and the global total) of non-fossil fluxes, from the two-dimensional mass-balance inversion by Conway *et al.* [78] and additional points from Ciais *et al.* [67]. Morimoto *et al.* [325] have performed a two-dimensional mass-balance inversion including ^{13}C. They found an interannual variability in ocean/biosphere partitioning

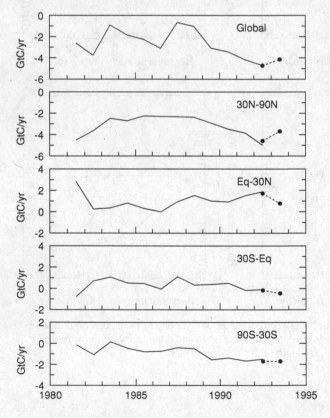

Figure 14.8 Comparisons of estimates of interannual variations in fluxes of non-fossil CO_2. The solid line is from two-dimensional mass-balance [78]; filled circles are from Ciais *et al.* [67].

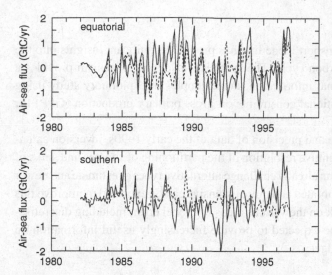

Figure 14.9 Interannual variations in ocean fluxes of CO_2 from time-dependent synthesis. For the equatorial region, the solid line is for all oceans, whereas the dotted line is for the east Pacific. For the southern regions, the dotted line is for the southern ocean, whereas the solid line is for the southern ocean plus southern gyres of all oceans. (Data from P. J. Rayner, personal communication, 2001).

that combined features of the global estimates by Francey [159] and Keeling *et al.* [250]. They identified tropical biota as the main region of ENSO–non-ENSO differences.

To investigate interannual variability in fluxes of CO_2, Rayner *et al.* [401] compared three inversion techniques: Bayesian synthesis and mass-balance using two different ways of producing the concentration fields. They found that the sources in the zone 15° S to 30° N exhibited a strong similarity to the SOI, but that a correlation analysis showed the flux of CO_2 leading by several months. Comparing the time-evolution of fluxes of CO_2 with the SOI over several events showed that the fluxes followed the SOI through the onset of an El Niño event but returned to near normal after about 6 months, so that the flux 'pulses' were completed sooner than were the SOI pulses.

Figure 14.9 shows ocean fluxes from a time-dependent synthesis [P. J. Rayner, personal communication]. This particular calculation explored the consequences of modifying the statistical models to include 'bias' terms. These were given a correlation of unity over time and were intended to represent model error that would not behave as $1/\sqrt{N}$ for a time series of length N. Both the observations and the priors were modelled in this way. The main effect of this refinement of the model is on the long-term means. As with the earlier study [401], the pattern of time-dependence seems relatively insensitive to the choice of inversion technique on time-scales of seasons to years.

Results of two recent studies have suggested that there is relatively low interannual variability of the air–sea flux: Lee *et al.* [278] analysed the interannual variability of p_{CO_2} and wind-speed to find a year-to-year variation of 0.4 GtC yr^{-1}(1 SD), with most of the variability coming from the equatorial east Pacific. Le Quéré *et al.* [279] found a similar result from their ocean model. Given the difficulties in resolving tropical fluxes and the uncertainties associated with ^{13}C, land–sea partitioning of tropical fluxes in inverse calculations must still be regarded as tentative. This is, however, an extremely active area of research.

Summary

The use of atmospheric-transport modelling has provided significant insights into the behaviour of the global carbon cycle. The synthesis-inversion formalism provides a basis for including additional information. For example, an exploratory study using ^{18}O [68, 69] provided additional constraints on gross primary production (GPP) for the tropics.

At the levels of coverage and precision of data of the early 1990s, inversion calculations do not provide a definitive resolution of the partitioning of sources and sinks of CO_2. However, the results can already challenge alternative types of estimate and newer data that are better inter-calibrated and more extensive can be expected to improve this situation. With further work on the various areas of model error, including discretisation error, these data can be expected to provide increasingly useful information on the atmospheric carbon budget.

Further reading

- The assessments by the Intergovernmental Panel on Climate Change (IPCC) include chapters on CO_2 [499, 422, 371].

- Several volumes from 'summer schools' [38, 200, 507] provide valuable introductions, as do special assessment reports such as those by the US Department of Energy [483]

- The JGOFS mid-term synthesis includes a useful summary of the ocean carbon cycle [294].

- The textbook *Greenhouse Puzzles* [54] places the carbon cycle in the context of ocean biogeochemistry and includes a number of exercises.

- Sundquist [456] provides a view of the carbon cycle on geological time-scales.

- Aspects of carbon-cycle science are also topics of periodic reviews in publications such as *EOS*, *Nature* and *Science*.

Chapter 15

Global methane

The biosphere retains in its midst gases such as hydrogen and methane that otherwise would long ago have been lost from the Earth by cosmic processes. It is a souvenir of itself.

Lynn Margulis and Dorion Sagan: *Microcosmos: Four Billion Years of Evolution from Our Microbial Ancestors* (1986) (Summit Books: New York).

15.1 Issues

Methane (CH_4) is the second most important of the anthropogenic greenhouse gases in terms of its current contribution to radiative forcing. Mole for mole, the radiative forcing is 21 times that of CO_2, but the climatic impact from methane is smaller than that from CO_2 due to the lower concentrations which result from smaller sources and faster removal from the atmosphere.

Over the period of direct observations, the atmospheric concentration of CH_4 increased for several decades, but during the late 1980s and the 1990s the rate of growth dropped considerably [448, 112, 111]. On longer time-scales, ice-core records [147] show that the atmospheric concentration has more than doubled over the last 400 years. On still longer time-scales, concentrations of CH_4 are closely coupled to glacial/interglacial cycles [355]. However, the current budget of atmospheric CH_4 remains rather uncertain (see Section 15.2 and Table 15.1 below).

Analysis of the behaviour of CH_4 is in many ways more complicated than that for CO_2 because the main process by which CH_4 is lost from the atmosphere is by oxidation through the OH radical in the free atmosphere, rather than by gas exchange

Table 15.1. *The global methane budget for the 1980s from the IPCC [369], with isotopic values from Houweling et al. [217] (negative changes are decreases).*

Process	Change Tg CH_4 yr^{-1}	‰
Atmospheric increase	35–40	
Stratospheric loss	−32 to −48	δ_A + 12
Tropospheric OH	−360 to −530	δ_A + 5.4
Soil loss	−15 to −45	δ_A + 22
Wetlands	55–150	−60
Termites	10–50	−57
Oceans	5–50	−40
Other natural sources	10–40	
Fossil-fuel-related	70–120	−40
Enteric fermentation	65–100	−62
Rice paddies	20–100	−64
Burning of biomass	20–70	−25
Animal waste	20–30	−55
Domestic sewage	15–80	−55

at the earth's surface. Additional sinks occur through photolytic destruction in the stratosphere and through interactions in soils. These sinks give CH_4 an atmospheric lifetime of about 10 years. The relatively short lifetime, compared with those of the other main greenhouse gases, means that concentrations of CH_4 will respond quite rapidly to changes in sources.

An important aspect of the behaviour of methane is its effect on atmospheric chemistry. As one of the main compounds that reacts with the OH radical, methane can greatly affect the concentration of OH. This raises a number of issues.

■ Changes in concentration of methane can change the overall oxidising capability of the atmosphere, i.e. the capacity of the atmosphere to break down a large range of pollutants.

■ Other compounds that react with OH have the potential to affect the atmospheric budget of methane. Therefore emission scenarios for greenhouse gases need to include emissions of gases such as CO and volatile organic compounds in order to determine the anthropogenic influence on radiative forcing.

■ Combining the two preceding points, changes in the concentration of methane will affect its own atmospheric budget. For small changes this can be characterised by an 'adjustment time' that differs from the atmospheric 'lifetime' or turnover time, as described below. For large changes, the whole character of the atmosphere can be changed. Guthrie [186] has used a 'toy model' to suggest that the chemical balance is unstable outside a limited range of magnitude of methane sources. Outside this range lie the extremes of having virtually no methane (if the OH source were more than was needed to remove all CO and CH$_4$ as fast as it was produced) and having unbounded growth in CH$_4$ concentrations (if the CO source were more than strong enough to remove all OH as fast as it was produced). As illustrated by the toy model described in Box 1.5, the inclusion of a number of other minor processes prevents such extreme instability, but it is still true that the present balance is in a region of relative local stability outside which quite different conditions can occur.

■ Any interpretation of past concentrations recovered from ice-cores needs to take into account the extent to which the chemistry of the atmosphere may have changed.

The concept of the 'adjustment time' is illustrated by Figure 15.1. This shows the way in which atmospheric levels of CH$_4$ and OH respond to an input pulse of CH$_4$, relative to a steady background rate of emission. The contribution from 'new' methane is shown by the dark shading and the contribution of methane from a fixed set of background sources is shown by the stippling. The solid line shows levels of OH. The response to an increase in the total amount of methane is a greater loss through oxidation by OH. This leads to a rapid drop in levels of OH by an amount proportional to the input pulse. The concentration of 'new' methane drops at a rate that (for a small

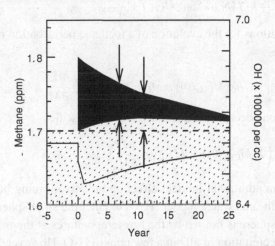

Figure 15.1 The difference between lifetime and adjustment time for methane perturbations, illustrated by an addition of a pulse to a background due to fixed sources. The pulse (dark shading) decays at a rate characterised by an atmospheric lifetime so that 50% is lost after 6 years (left-hand arrows). The increase in concentration of methane causes a decrease in concentration of OH, so that the concentration of methane from the fixed sources (stippling) increases because of the smaller sink and hence the net methane perturbation decays more slowly (on a time-scale characterised by an 'adjustment time') than does the methane from the pulse alone. It is only after 11 years (right-hand arrows) that 50% of the perturbation has decayed.

pulse) is defined by a lifetime that is very close to the bulk lifetime for the initial state. In the example, as shown by the left-hand pair of arrows, 50% of the pulse is lost after about 6 years. However, because of the initial increase in concentration of methane and decrease in concentration of OH, the 'original' sources are no longer balanced by the OH sink. The amount of imbalance is proportional to the pulse input and the concentration due to the 'original' sources increases. Therefore, the net perturbation decays more slowly than does the concentration of methane that comes directly from the pulse. In Figure 15.1, as shown by the right-hand pair of arrows, it is 11 years until 50% of the net perturbation has decayed.

A simplified derivation follows the form of the toy chemical model (Box 1.5) with

$$\frac{d}{dt}M_{CH_4} = S_{CH_4} - k_1 M_{OH} M_{CH_4} - \lambda_{other} M_{CH_4} \qquad (15.1.1a)$$

If we regard OH and CO as being in an effective equilibrium, then the concentration of OH depends on the source strengths of OH and CO and the concentration of methane. We write this as

$$M_{OH} = f(S_{CO}, S_{OH}, M_{CH_4}) \qquad (15.1.1b)$$

Taking the other quantities as fixed, we write the methane balance (15.1.1a) as

$$\frac{d}{dt}M_{CH_4} = S_{CH_4} - \lambda(M_{CH_4})M_{CH_4} \qquad (15.1.1c)$$

with

$$\lambda(M_{CH_4}) = k_1 f(S_{CO}, S_{OH}, M_{CH_4}) + \lambda_{other} \qquad (15.1.1d)$$

We write the equations for the evolution of a methane perturbation relative to concentration $M_{CH_4}(0)$ as

$$\frac{d}{dt}\Delta M_{CH_4} = -\lambda(M_{CH_4}(0))\,\Delta M_{CH_4} - M_{CH_4}(0)\frac{\partial\lambda}{\partial M}\,\Delta M_{CH_4} \qquad (15.1.1e)$$

so that perturbations decay with a characteristic time-scale

$$\tau_{adjust} = \left(\lambda(M_{CH_4}(0)) + M_{CH_4}(0)\frac{\partial\lambda}{\partial M}\right)^{-1} \qquad (15.1.1f)$$

Methane has an additional indirect effect on radiative forcing because it provides a pathway for additional water vapour (H_2O) to enter the stratosphere. Direct input of H_2O to the stratosphere is limited by the low temperatures at the tropical tropopause, which leads to precipitation of all but a few ppm of H_2O. However, *in situ* production of H_2O through oxidation of CH_4 in the stratosphere is not temperature-limited in this way.

An improved understanding of atmospheric methane is important for several reasons. First, current predictions of future concentrations of CH_4 are quite sensitive to uncertainties in inputs of other trace species, e.g. [265, 260]. Secondly, there is significant scope for decreasing anthropogenic emissions. In particular, emissions of methane from coal-mining operations and methane from processing of domestic sewage are increasingly being trapped to produce energy.

Some of the most important scientific questions regarding atmospheric methane are the following.

- What is the atmospheric budget?
- In particular, what are the anthropogenic contributions?
- Is the sink strength changing?
- What is the reason for the slowdown in the rate of growth during the 1990s?
- To what degree are projections of future concentrations sensitive to inputs of other gases?

Table 15.1 lists the atmospheric budget of CH_4 during the 1980s as estimated in the IPCC Radiative Forcing Report [369]. It will be seen that there are very large uncertainties. The budget is, like most other greenhouse-gas budgets, expressed in masses of gas (Tg CH_4 in this case), but, unlike CO_2 budgets, which are conventionally expressed in terms of mass of carbon. Compared with that for CO_2, a notable feature of the CH_4 budget is the extent to which the anthropogenic emissions (the items in the lowest group in Table 15.1) are poorly known.

Examples of the complexities in the methane budget are

- emissions of CH_4 from rice paddies are strongly dependent on cultivation techniques;
- enteric fermentation depends strongly on the type of feed;
- there is great variation in the $^{13}C : {}^{12}C$ ratio in natural gas and to a lesser extent other biogenic sources; and
- there is a possibility of other poorly known sources such as the action of termites and oceanic clathrates.

15.2   Inversion techniques

Compared with the case for CO_2, inversions of CH_4 are complicated by the sink in the free atmosphere. The computational difficulty is that the sink depends on the concentration of methane. The transport equation takes the form

$$\frac{\partial}{\partial t}\mathbf{m} = \mathcal{T}[\mathbf{m}] - \lambda(\mathbf{r}, t)m(\mathbf{r}, t) + s(\mathbf{r}, t) \qquad (15.2.1)$$

but, as indicated by (15.1.1c), $\lambda(\mathbf{r}, t)$ will depend on the concentrations of methane and reactive species.

These difficulties can be addressed in a number of different ways.

Mass-balance with a chemical transport model. In this case, equation (15.2.1) is integrated, using an atmospheric model consisting of a transport module (to evaluate $\mathcal{T}[.]$) and a chemistry module (to evaluate λ). At the surface, the mass-balance equation is used to estimate the surface sources, given the specified surface concentrations. This is just a variation of the general principle that mass-balance inversions estimate sources at locations where the concentrations are known but require known sources at locations where the concentrations have to be calculated. The specified sources do not have to be zero (although this is common) and may depend on the concentrations, as in equation (15.2.1). Furthermore, when the source/sink term depends on the concentration, there is no requirement for a linear dependence. Saeki *et al.* [418] used a mass-balance inversion in a two-dimensional model with the distribution of OH derived from a three-dimensional chemical model. Law and Vohralik [276] have performed a mass-balance inversion of distributions of CH_4 using a three-dimensional transport model and with the sink specified using production/loss chemistry derived from a two-dimensional chemical model.

A modification of this approach involves using a synthesis of perturbations in surface fluxes about a 'current best guess', which is used in a chemical model to calculate the chemical sink. This approach was analysed, using a two-dimensional model, by Brown [56], using a decomposition of fluxes into latitude bands. A later paper [57] refined the data-fitting procedure by using SVD truncation as a regularisation technique and also considered isotopic inversions.

A specified sink. The synthesis-inversion study of CH_4 by Fung *et al.* [165] used a specified sink. In the term $\lambda(\mathbf{r}, t)m(\mathbf{r}, t)$, the variation in $\lambda(\mathbf{r}, t)$ was based on model calculations of distributions of OH. The distribution of $m(\mathbf{r}, t)$ was fixed and included only the interhemispheric difference based on observations. This distribution was not further adjusted to be consistent with the results of the inversion. The justification for this is the great attenuation of the more variable modes of sink variation as reflected in surface concentrations. This study preceded the development of formal estimation procedures in synthesis inversion; the results were presented as a set of possible scenarios.

Scaling a specified sink. In this case, the sink is represented as $-\alpha\lambda(\mathbf{r}, t)m_0(\mathbf{r}, t)$, where $m_0(\mathbf{r}, t)$ is taken from a chemical model calculation. The sink is adjusted to fit the observation by varying α but not otherwise changed. This approach was used by Hein and co-workers [206, 207] in a Bayesian synthesis inversion involving simultaneous estimation of α and the strengths of the surface source components.

Scaling a specified distribution of OH. A study by Houweling and co-workers [214: Chapter 3; 215] used an iterative Green-function approach to inverting (15.2.1). The source was written as

$$s(\mathbf{r}, t) = -\alpha\lambda(\mathbf{r}, t)\, m(\mathbf{r}, t) + s(\mathbf{r}, t) \tag{15.2.2a}$$

where the space–time dependence of λ was taken from a chemical model and a single scale factor α was fitted to the CH_4 data. The prior value of α was based on fitting CH_3CCl_3. The iterative Green-function technique was based on the approximation

$$s'(\mathbf{r}, t) = -\alpha\lambda(\mathbf{r}, t)\tilde{m}(\mathbf{r}, t) + s(\mathbf{r}, t) \qquad (15.2.2b)$$

where $\tilde{m}(\mathbf{r}, t)$ was a fixed 'first-guess' concentration distribution. Using this approximation, the problem of estimating α and $s(\mathbf{r}, t)$ was linear. The estimated sources were used to calculate an improved guess of $m(\mathbf{r}, t)$, which was used to replace $\tilde{m}(\mathbf{r}, t)$ in (15.2.2b) and the linear-inversion calculation was repeated.

Other noteworthy features of the study by Houweling *et al.* were the use of a high-resolution representation of the sources, which was made possible by determining the Green-function matrix from the adjoint of the transport model (see Section 10.6). Additional calculations explored the degree of sensitivity to assumptions about correlations in the prior estimates.

Extended Green's functions. For the case of production/loss chemistry, for which the strength of the free-atmosphere sink is specified independently of the concentration, the relation (15.2.1) can be written

$$\frac{\partial}{\partial t}m(\mathbf{r}, t) - \mathcal{T}[m(\mathbf{r}, t)] + \lambda(\mathbf{r}, t)\, m(\mathbf{r}, t) = \mathcal{L}m(\mathbf{r}, t) = s(\mathbf{r}, t) \qquad (15.2.3)$$

and a Green function can be defined as the inverse of the operator $\mathcal{L}$. This implies that Green-function inversions can be developed. These could be the steady-state or the time-dependent forms described in Chapter 10 or the time-stepping forms described in Chapter 11. The effect that is neglected in this approach is the influence of CH_4 on its own rate of destruction. For time-dependent calculations, the use of production/loss chemistry will result in the decay time of perturbations being given (incorrectly) by the mean turnover time rather than the adjustment time.

In addition to the global-scale inversions described above, there has also been an inversion estimate of methane sources in north-eastern Europe, which was based on a 'synthesis' technique [450]. This is discussed in Section 17.3. On these regional scales, the local rate of destruction is negligible and the effect of global-scale destruction of CH_4 can be characterised using boundary conditions based on observed concentrations.

15.3   Chemical schemes

The various inversion schemes described in the previous section all rely in some way on results from a chemical transport model. Modelling atmospheric chemistry involves some major computational difficulties. These stem from three main causes:

■ the large number of compounds involved;

■ the importance of photochemical reactions, requiring a careful treatment of the radiative budget; and

■ reaction rates that span many orders of magnitude. Furthermore, the rates of particular reactions can vary greatly in space and time due to differences in temperature and radiation.

Specialised integration techniques are needed in order to handle the disparate reaction rates [228]. This is an on-going area of research and new integration techniques are being developed from time to time [531].

To deal with the range of compounds, several approximations need to be made. Important choices are

■ deciding which compounds are included;

■ deciding which of these compounds have their time-evolutions calculated (and are thus subject to transport) and which are treated as being in instantaneous equilibrium;

■ deciding which, if any, groups of compounds are treated using a proxy for a 'family' of compounds with similar behaviours;

■ the reaction rates; and

■ deciding which processes are treated by parameterisation rather than explicit chemical calculation.

Some of the widely used cases are the following.

Production/loss. This approach (P/L) specifies the production and loss of individual compounds of interest, generally with an absolute specification for production and loss specified as proportional to concentration, i.e. through a local lifetime. This approach considers one or more compounds in isolation and so it is largely restricted to diagnostic modelling rather than prognostic modelling. An application to methane inversions [276] is described above.

TC1. This is a basic tropospheric CH_4–CO–O_3–NO_x–OH scheme, involving 13 species, seven of which are transported. It has 25 chemical reactions including seven photolysis reactions. A description of reactions identifying key chains of reactions is given by Crutzen and Zimmermann [87]. The reaction rates were taken from the JPL compilation of DeMore *et al.* [100]. The TC1 scheme has been used in studies of CH_4 [207] and CO [30]. Wang *et al.* [497] used the TC1 mechanism, augmented by three tropospheric sulfur reactions, nine stratospheric reactions for CFCs and N_2O and 12 aqueous reactions.

The carbon-bond mechanism (CBM). This is a type of lumped scheme that works (in part) with 'species' that represent groups of compounds. A commonly used recent version is CBM-4 [175]. It includes 87 reactions for 33 species. It was evaluated against an expanded scheme, CBM-EX with 204 reactions describing 87 species.

Houweling [214] and co-workers used a form of CBM-4 modified to include additional background reactions. Zaveri and Peters [528] recently described an extension, CBM-Z, which is designed to give a better performance for large scales.

RADM. This scheme is designed for regional pollution studies. It is often used as a reference case for testing less comprehensive schemes. The original version [451] included 80 reactions. Updated versions are RADM2 [452] and RACM, the Regional Air Chemistry Model, which includes reactions relevant to the troposphere as a whole [453].

These schemes are modified from time to time as improved data on reaction rates become available. Other photo-chemical schemes have been described by Thompson and Cicerone [473] and Lurmann *et al.* [293].

Comparisons of photo-chemical schemes have been noted in the IPCC Radiative Forcing Report [369] and later reported in detail [346]. They focused on ozone chemistry and found a $\pm 10\%$ spread for moist unpolluted marine conditions and rather larger differences for polluted mid-troposphere conditions. Details of the treatment of radiation were found to be important.

15.4 Results of inversions

In Section 15.1, some of the current scientific questions were identified as those involving the budget and the present and future strength of the tropospheric sink. In this section we consider the extent to which global and regional budgeting contribute to answering these questions.

Estimation of the CH_4 sink due to oxidation by OH poses considerable difficulties. The concentration of OH radical is highly variable and is extremely difficult to measure. Estimates from chemical models suffer from the sensitivity to the large number of reactions that affect OH. The mean sink due to OH can be calibrated from budgets of compounds for which there are good data on the concentrations and knowledge of all sources and sinks other than OH. This estimation problem is discussed in Section 16.3. Such estimated concentrations of OH are averages whereby the concentration of OH is averaged over space and time with a weighting involving the concentration and rate of reaction of the calibration tracer. For estimating the CH_4 sink, budgets of CH_3CCl_3 are particularly valuable because the rates of reaction of CH_4 and CH_3CCl_3 have quite similar dependences on temperature.

In Chapter 10 a methane-budget estimate [386] based on isotopic data was used as an example of multi-species budgeting. This was presented as a CH_4–$^{13}CH_4$ budget. The use of $^{14}CH_4$ was implicit through its role as a constraint on the fossil component.

A more comprehensive global budgeting analysis was presented by Kandlikar [242]. Data for ^{12}C, ^{13}C and ^{14}C, as well as pre-industrial values from ice-cores (to help estimate the natural sources), were used in a Bayesian Monte Carlo inversion. The estimates of 12 source components were presented as cumulative distributions for the source strength. The results suggested that the 1992 IPCC budget over-estimated the contributions from rice growing and wetlands. However, this study used priors based on the IPCC budget and so it is not clear how much 'double counting' is involved, i.e. to what extent the same atmospheric data were used in the inversion and in the IPCC budget that was used as a prior. In a more recent analysis, considering the ^{13}C, ^{14}C and D in methane, Quay et al. [388] concluded that the isotopic budgets could not uniquely identify the origin of the slowdown in rate of growth. They suggested that isotopic budgeting could be improved with a better knowledge of emissions of ^{14}CH$_4$ from nuclear reactors and improvements in measurements of the fractionation in methane-oxidation reactions. As with CO$_2$, global-scale budgets derived from global mean data serve as a reference point for identifying the extent to which that analysis of spatial distributions provides additional information.

Hein and co-workers [206, 207] applied the steady-state Bayesian synthesis-inversion formalism to CH$_4$. They used CH$_4$ data from 22 sites and ^{13}CH$_4$ data from three sites, using monthly means in each case, as well as global trends for CH$_4$ and ^{13}CH$_4$. They fitted source strengths for 11 surface sources and one surface sink from soil as well as sinks in the stratosphere and the troposphere. These components had specified spatial and temporal variations. Measurements of concentrations of ^{14}C were incorporated through the prior uncertainty assigned to the fossil component. In results similar to those of Kandlikar, they found that the data provided a significant reduction in the uncertainties of the rice-cultivation and wetland sources. They also found a reduction from 25% to 8% in the uncertainty in the strength of the tropospheric sink. This study prompted the cautionary note by Tans [463], discussed in Chapter 10, concerning the slow equilibration of isotopic ratios.

Houweling and colleagues [214, 215] used the adjoint of the TM3 transport model so that they could estimate sources at every grid point. This means that the estimated source strengths are expressed as spatial distributions rather than being attributed to processes. However, they did find that their inversion could roughly halve the uncertainty in the flux from a region bounded by 10° N, 40° N, 75° E and 135° E, containing the main rice-growing areas.

They also investigated the effect of assuming temporal correlations in the concentration data and spatial correlations in the uncertainties used for prior flux estimates. Applying spatial correlation over distances of 2000 km has a significant effect on the flux estimates for particular months, but rather less effect on the annual mean. Saeki et al. [418] used a two-dimensional mass-balance inversion for the period 1984–94. In their analysis of trends for emissions in the four semi-hemispheres, they found a decrease in the zone from 30° N to 90° N and a matching increase in the zone from the equator to 30° N.

Law and Vohralik [276] performed a series of three-dimensional mass-balance inversions using zonally uniform sinks, with no interannual variability, which they derived from a two-dimensional CTM. Two different transport models were used

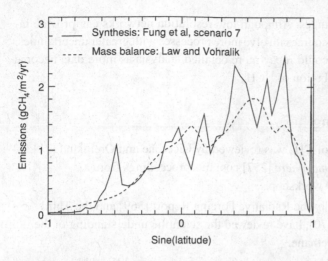

Figure 15.2 The latitudinal distribution of emissions of methane estimated by two inversion calculations: synthesis inversion by Fung *et al.* [165: scenario 7] with total emissions of 489.9 Tg CH_4 yr^{-1}; and mass-balance inversion by Law and Vohralik [276] with total emissions of 448.9 Tg CH_4 yr^{-1}.

and each model was run with alternative parameterisations of convection. One case used ECMWF analysed wind-fields. Although these studies were mainly designed to explore inversion techniques, having a suite of results is very useful in identifying which features of the inversion are insensitive to the details of the model. Overall, no long-term trend was found in the sources, implying that the slowdown in rate of growth represents a relaxation towards an equilibrium concentration. However, the trends were found to be sensitive to the number of stations in the network and whether the inversion used data from a fixed number of stations or used data from all available stations. The calculations produced an apparent additional CH_4 source in the period after the Pinatubo eruption. This is almost certainly a reduced sink, arising from changes in concentration of OH due to a decrease in intensity of ultra-violet radiation that the inversion omitted by using OH fields with no interannual variation.

The differences among the various calculations make it difficult to determine how much our understanding has been improved by spatial inversions of CH_4. It is encouraging that smaller-than-expected emissions for rice growing are estimated when global budgeting is augmented either by spatial distributions or by pre-industrial constraints.

Figure 15.2 shows two estimates of the latitudinal distribution of emissions of methane that were obtained using different techniques [165, 276]. The source in high northern latitudes in the synthesis inversion is from the clathrate component.

The analysis of the methane budget from atmospheric observations is hampered by the co-location of many of the important sources and the spread of isotopic composition within many of the classes of source. The main scope for clarification lies in the use of additional constraints. These include the use of pre-industrial concentrations to estimate emissions from natural wetlands and thus distinguish them from emissions from agricultural sources, e.g. [217].

Analysis of the reasons for the post-1985 slowing of the rate of growth is limited by the small amount of data from the time before this slowdown started. It is likely that further analysis can add little to the initial result [448, 112] that the slowdown commenced at northern latitudes. Although low-resolution networks can provide information about CFCs, CH_4 sources involve more processes with greater uncertainties. One aspect of the data that could repay more detailed analysis as more data become available is the spatial distribution of $^{13}CH_4$.

Further reading

■ The biogeochemistry of CH_4 was reviewed by Cicerone and Oremland [71]. A special edition of *Chemosphere* [257] contains proceedings from a comprehensive NATO workshop.

■ IPCC reports, especially the Radiative Forcing Report [369] and the Third Assessment Report [370], have reviewed the scientific understanding of the biogeochemistry of methane.

■ Warneck's book [498] has a useful chapter on methane and Wuebbles and Hayhoe [520] review current changes.

■ Additional material on chemistry, particularly for the stratosphere, is included in the ozone assessments listed at the end of the following chapter.

■ As with CO_2, *EOS* commonly has overview news accounts reviewing scientific issues involving the biogeochemistry of methane.

Chapter 16

Halocarbons and other global-scale studies

I have only knocked at the door of chemistry, and I see how much remains to be said.

Johannes Kepler: *The Six-cornered Snowflake.* (1611) (OUP, 1966)

16.1 Issues

The halocarbons are a group of compounds with halogen atoms (and possibly hydrogen) attached to a carbon skeleton. Most of the halocarbons are purely synthetic. However, there are significant natural sources of the methyl halides (CH_3Cl, CH_3Br and CH_3I), chloroform ($CHCl_3$) and, to a lesser extent, dichloromethane (CH_2Cl_2). The halocarbons can be regarded as derived from the alkane C_nH_{2n+2} structure by substituting hydrogen atoms with halogens. An outline of the relevant chemical nomenclature is given in Box 16.1.

As polyatomic species, all these compounds will be likely to affect the infra-red radiative balance of the earth. However, the main concern about these gases has arisen from their role in the depletion of stratospheric ozone (see Box 16.2). The extent of depletion of ozone depends on the amount of chlorine and bromine released when halocarbon compounds are photolysed in the stratosphere and on the altitude at which the photolysis occurs. Chlorine and bromine participate in catalytic cycles that destroy ozone while leaving the halogens free to participate in further cycles of destruction of ozone.

Most of the compounds that are not fully halogenated, i.e. those with one or more hydrogen atoms remaining, react primarily with tropospheric OH. The products are

Box 16.1. Nomenclature of halocarbons

The halocarbons can be regarded as derived from the alkane C_nH_{2n+2} structure by replacing hydrogen atoms with halogens. Note that, for $n \geq 4$, isomers (i.e. the same chemical formula) with different carbon skeletons can occur and, for $n \geq 2$, isomers with different combinations of halogen substitution can occur.

There are several naming systems applied to these compounds:

(i) in a number of cases there is a common name such as methyl chloroform;

(ii) there is an IUPAC systematic name that includes numerical indices to identify the halogen substitutions; and

(iii) there is a simplified numerical designation that indicates the substitutions, applied as a suffix to a series name (or abbreviation) such as chlorofluorocarbons (CFC), halons (H), etc.

The simplified scheme identifies CFCs, HFCs and HCFCs through a three- (or two-) digit index of which the first is one less than the number of carbon atoms (and is omitted if it is zero), the second is one greater than the number of hydrogen atoms and the third is the number of fluorine atoms. The number of chlorine atoms is determined by the valence. Halons are identified by four digits giving the numbers of carbon, fluorine, chlorine and bromine atoms. Isomers are identified by trailing letters. Examples are given below.

Formula	Common name	Abbreviation	IUPAC name
CH_3Cl	Methyl chloride		Chloromethane
CH_3Br	Methyl bromide		Bromomethane
CCl_4	Carbon tetrachloride	CFC-10	Tetrachloromethane
CF_4	Carbon tetrafluoride	CFC-14	Tetrafluoromethane
$CHCl_3$	Chloroform		Trichloromethane
CH_3CCl_3	Methyl chloroform		1,1,1-Trichloroethane
CCl_3F		CFC-11	Trichlorofluoromethane
CCl_2F_2		CFC-12	Dichlorodifluoromethane
CF_2HCl		HCFC-22	Chlorodifluoromethane
CHF_2CHF_2		HFC-134	1,1,2,2-Tetrafluoroethane
CH_2FCF_3		HFC-134a	1,1,1,2-Tetrafluoroethane
CF_4	Carbon tetrafluoride		Tetrafluoromethane
$CBrF_3$		H-1301	Bromotrifluoromethane
$CBrClF_2$		H-1211	Bromochlorodifluoromethane
$CBrF_2CBrF_2$		H-2402	1,2-Dibromotetrafluoroethane

Box 16.2. Destruction of ozone

Ozone is produced in the stratosphere by the reaction

$$O + O_2 + M \rightarrow O_3 + M \tag{1}$$

where the extra molecule, M, is required for conservation of momentum.

The free oxygen atoms come from photolysis of O_2:

$$O_2 + h\nu \rightarrow O + O \tag{2}$$

The gas-phase destruction of ozone occurs through the reactions

$$O_3 + h\nu \rightarrow O + O_2 \tag{3a}$$
$$O_3 + O \rightarrow 2O_2 \tag{3b}$$

and the catalytic cycles

$$NO + O_3 \rightarrow NO_2 + O_2 \tag{4a}$$
$$O_3 + h\nu \rightarrow O + O_2 \tag{4b}$$
$$NO_2 + O \rightarrow NO + O_2 \tag{4c}$$

and

$$Cl + O_3 \rightarrow ClO + O_2 \tag{5a}$$
$$O_3 + h\nu \rightarrow O + O_2 \tag{5b}$$
$$ClO + O \rightarrow Cl + O_2 \tag{5c}$$

Each of these sequences has the net effect

$$2O_3 + h\nu \rightarrow 3O_2 \tag{6a}$$

There is a similar cycle involving bromine and several cycles involving hydrogen oxides. These cycles continue until minor reactions remove the chlorine, nitrogen or bromine.

At short wavelengths ($\lambda < 315$ nm) photolysis of ozone leads to excited states:

$$O_3 + h\nu \rightarrow O(^1D) + O_2(^1\Delta) \tag{6b}$$

where the $O(^1D)$ state is responsible for oxidising halocarbons to produce ClO (leading to cycle (5a)–(5c)) and for oxidising N_2O to NO (leading to cycle (4a)–(4c)).

In the Antarctic night, chlorine is produced by the reaction

$$HCl + ClONO_2 \rightarrow Cl_2 + HNO_3 \tag{7a}$$

which occurs as a surface reaction in polar stratospheric clouds. In spring, the early sunlight photolyses the Cl_2 to 2Cl, which then destroys ozone through reaction (5a). There is not enough free oxygen to regenerate the Cl through reaction (5c), but the catalytic cycle is closed through the reaction

$$2ClO \rightarrow 2Cl + O_2 \tag{7b}$$

giving rapid depletion of ozone, creating the ozone hole.

Box 16.3. The Montreal Protocol

In 1985, the *Vienna Convention for the Protection of the Ozone Layer* established a global commitment to address the problems of ozone-depleting chemicals and specified a process to implement this.

The first step, in 1987, was the *Montreal Protocol on Substances that Deplete the Ozone Layer*. This specified decreases in the production of CFCs.

The Montreal Protocol and its various amendments involve a number of key definitions:

- a distinction between developed and developing nations;
- a specification of a base level of 'consumption' defined for a particular past year, with different base years for the different gases and developing countries having base years later than those for developed countries;
- a specified year for phase-out, again later for developing countries; and
- specified decreases (as percentages of production during the base year) for specific years prior to phase-out.

The Montreal Protocol included (for developed countries) a freeze in production of CFCs followed by decreases to 80% in 1994 and 50% in 1999.

The *London amendments* in 1990 specified (for developed countries) a phase-out of CFCs, carbon tetrachloride and halons by 2000 and methyl chloroform by 2005, with developing countries meeting the same requirements 10 years later.

The *Copenhagen amendments* in 1992 specified (for developed countries) a phase-out of CFCs, carbon tetrachloride and methyl chloroform by 1996, halons by 1994 and HCFCs by 2030, with developing countries meeting the same requirements 10 years later.

The *Vienna adjustments* in 1995 applied additional restrictions on production of HCFCs, including a freeze by developing nations in 2016.

The *Montreal amendments* in 1997 included a phase-out of methyl bromide by 2005 (2015 for developing nations).

One of the provisions of the Montreal Protocol was for the establishment of assessment panels to provide scientific information for the parties to the protocol (see references in the 'further reading' at the end of this chapter).

removed by rainout and so relatively little of the chlorine or bromine reaches the stratosphere.

The main classes of compounds of concern are the following.

CFCs. The *chlorofluorocarbons* are fully substituted (i.e. there is no remaining hydrogen) with chlorine and fluorine on an alkane carbon skeleton. The lack of hydrogen means that there is essentially no tropospheric sink and the compounds enter the stratosphere where photolytic destruction releases free chlorine, leading to destruction of ozone.

HCFCs. The *hydrochlorofluorocarbons* are only partly substituted with chlorine and fluorine, retaining some hydrogen. This allows reaction with tropospheric OH, so that the input of chlorine to the stratosphere is reduced. These are treated under the Montreal Protocol as 'interim replacements' for the CFCs, with caps on their emission introduced by the Copenhagen amendments and a phase-out specified by the Montreal amendments.

HFCs. The *hydrofluorocarbon* compounds are also subject to breakdown in the troposphere because of the presence of hydrogen. Since they lack chlorine, the small proportion that does enter the stratosphere does not lead to destruction of ozone. The Montreal Protocol (as amended) identifies these compounds as long-term replacements for CFCs [319].

PFCs. The *perfluorcarbons* are fully substituted with fluorine. Since they contain neither chlorine nor bromine, they are not involved in destruction of ozone. However, they are radiatively active and have extremely long lifetimes and so they are of concern as greenhouse gases.

Halons. These are the halocarbons containing both bromine and fluorine (and possibly chlorine). These compounds are of great concern regarding depletion of ozone because bromine is more active in catalytic destruction of ozone than is chlorine.

The chlorofluorocarbons have become very widely used because of their low chemical reactivity and consequent low toxicity. It is only in the stratosphere that they can be broken down through photo-dissociation by ultra-violet radiation. It is this stratospheric breakdown that leads to catalytic destruction of ozone.

Some of the key points in the history have been the following.

1928. Invention of synthetic CFCs.

1950s. Introduction of CFCs as refrigerants.

1970s. Widespread use of CFCs as aerosol propellants.

1974. Measurement of very low concentrations of trace gases becomes possible with the development of the electron-capture detector [291].

1973–74. Recognition of the potential of destruction of stratospheric ozone by catalytic cycles involving nitrogen [86] and chlorine from CFCs [321].

1985. Discovery of the ozone hole [152] and its cause [442, 440].

1985. The *Vienna Convention for the Protection of the Ozone Layer* was agreed.

1987. The Montreal protocol (under the Vienna Convention) prescribed restrictions on production of CFCs (for developed countries). This was subject to the London amendments (1990) which specified a phase-out of CFCs and extended the restrictions (with a later date of application) to developing countries. The Copenhagen amendments (1992) extended the coverage and accelerated the phase-out. The Vienna amendments (1995) imposed additional restrictions on production of methyl bromide, which was later subject to phase-out under the Montreal amendments (1997). Further details of the Montreal Protocol and its amendments are given in Box 16.3.

A number of observational programmes have been established, including those of the University of California, ALE/GAGE/AGAGE and NOAA/CMDL. Brief summaries are given in Box 5.1 and further details can be found in the ozone assessments listed at the end of this chapter. The AGAGE review paper [382] identified the objectives of the halocarbon-measurement programme as

- determination of the lifetimes (with respect to stratospheric destruction) for the long-lived halocarbons – see Section 16.3;
- analysis of the biologically important gases N_2O and CH_4 – see Section 16.4 and Chapter 15;
- estimation of the tropospheric sink due to oxidation by the OH radical; see Section 16.3;
- estimation of halocarbon sources on global and regional scales – see Section 16.3; and
- determining surface concentrations as a boundary condition for analysis of halocarbon distributions in the oceans.

Halocarbon studies have played an important part in the development of atmospheric trace-gas inversions, particularly the time-stepping-synthesis approach described in Chapter 11.

16.2   Estimation of sources

Initial studies of halocarbons in the atmosphere concentrated on estimating atmospheric lifetimes. Halocarbon production was largely confined to a small number of producers in western nations and quite comprehensive reporting procedures were implemented. However, for applications such as refrigeration and insulation, there are considerable time delays between production and emission, leading to a degree of uncertainty in estimates of emission. In later times, production became more widespread and the reporting less comprehensive. The issue of estimating sources became more important. Decreases in the calibration uncertainties made such estimation feasible.

Most of the studies estimating fluxes of halocarbons have used some form of recursive estimation based on a state-space formalism described in Section 4.4. The basis of state-space modelling is the definition of the state, $\mathbf{x}$, and a (prior) model of $\mathbf{x}(t)$, described by matrices $\mathbf{F}$ and $\mathbf{Q}$ that specify the evolution of the state. Many of these studies have aimed to estimate mean sources. The sources to be estimated are set as the components of the state-vector $\mathbf{x} = [s_1, \ldots, s_M]$. These sources, being defined as means, are regarded as constant. Therefore, the evolution of $\mathbf{x}$ is represented by $\mathbf{F} = \mathbf{I}$ and $\mathbf{Q} = \mathbf{0}$. The matrix $\mathbf{H}$ relating the state to observations is defined by a set of Green-function elements $G_{j\mu} = \partial c_j / \partial s_\mu$. This is an example of using the Kalman-filter formalism as a recursive implementation of linear regression. The basic estimation equations have been given as (11.2.3a)–(11.2.3c) (see also Box 4.4). This approach has been used in a number of studies using the ALE/GAGE/AGAGE 12-box model (see Box 7.3).

The same estimation procedure was used with a three-dimensional model by Hartley and Prinn [197]. The estimation procedure was a slight variation on the standard Kalman-filter approach: the Kalman gain matrix was defined by the special case (11.2.3c). However, the update to the state was calculated as

$$\hat{x}(n) = \hat{x}(n-1) + K(n)[c_{obs}(n) - c_{model}(n)] \tag{16.2.1a}$$

rather than using the linear model to give the standard form

$$\hat{x}(n) = \hat{x}(n-1) + K(n)[c_{obs}(n) - H(n)\hat{x}(n)] \tag{16.2.1b}$$

The 'projections' H, which are still required by the update equations, were defined as averages of $\partial c / \partial s_\mu$.

Mahowald et al. [298] used a three-dimensional model and daily data to estimate means of five regional sources, again using the Kalman filter with the $F = I, Q = 0$ model.

Using $F = I$ and $Q = 0$ in the underlying statistical model means assuming constant sources. Therefore the Kalman-filter formalism is not suitable for analysing time-varying sources unless the model is changed. Mulquiney et al. [332] used a random-walk model as their prior model for time-evolutions of varying CFC sources. They used the Kalman-filter estimation technique in a highly innovative way, including the influence of past sources by using a recursive representation of the Green function as described by Mulquiney and Norton [330] and outlined, omitting some details of the original, in Section 11.2.

A way of estimating time-varying sources, while formally retaining the $F = I$ statistical model, was to use a generalisation of the Kalman filter known as adaptive filtering. This uses departures from the theoretical model for the residuals to adapt the uncertainties used in the model, relaxing restrictions that prevent the estimates from tracking the data. This was used with a three-dimensional model by Haas-Laursen et al. [187] for estimating emissions of CFCs and more recently by Huang [218] for estimating sources of HCFC-142b and HCFC-141b.

An exception to the use of state-space modelling in halocarbon studies was a synthesis inversion using annual pulses in the three-dimensional model TM2 [34]. The estimation was a generalised least-squares estimation with the covariance matrix, X, including off-diagonal elements representing an autoregressive error structure. The synthesis inversion was embedded in an iterative outer minimisation loop that adjusted the atmospheric lifetime to find an overall best fit to the data by adjusting annual sources and scaling the sink.

16.3   Tropospheric and stratospheric sinks

16.3.1   Lifetimes

An initial objective of the observational programmes was to establish the atmospheric lifetimes of the various compounds. Cunnold et al. [89] defined the lifetime, $(\lambda_\eta)^{-1}$, on the basis of the equation

$$\frac{d}{dt}c_\eta = s_\eta(t) - \lambda_\eta c_\eta(t) \tag{16.3.1a}$$

or equivalently

$$\lambda_\eta = \frac{s_\eta(t) - (d/dt)c_\eta}{c_\eta(t)} \tag{16.3.1b}$$

If the emissions and concentrations are related by a response-function relation

$$c_\eta(t) = \int_{t_0}^t \rho_\eta(t - t')s_\eta(t')\,dt' \tag{16.3.2a}$$

then this leads to

$$\lambda_\eta = \frac{\int \dot\rho_\eta(t - t')s_\eta(t')\,dt'}{c_\eta(t)} \tag{16.3.2b}$$

so that λ_η, as defined by (16.3.1a) and (16.3.1b) will be constant only for the special cases when the sources grow exponentially, $s_\eta(t) = Ae^{\gamma t}$, or when $\rho_\eta(t) \propto e^{-\lambda_\eta t}$.

Cunnold et al. proposed two ways of estimating λ_η, first by iteratively fitting the relation

$$c_\eta(t) = \int_{t_0}^t s_\eta(t')e^{-\lambda_\eta(t-t')}\,dt' \tag{16.3.3a}$$

and secondly by transforming (16.3.1b) to

$$\lambda_\eta = \frac{s_\eta(t)}{\int_{t_0}^t s_\eta(t')\exp[-\lambda_\eta(t-t')]\,dt'} - \frac{1}{c_\eta(t)}\frac{d}{dt}c_\eta(t) \tag{16.3.3b}$$

This was solved iteratively for λ.

Cunnold et al. noted that (16.3.2b) had the advantage of not relying on the absolute accuracy of the calibration of the measurements. However, a comprehensive comparison of the two approaches would need to be based on treating them as alternative estimators of λ and considering the sampling distributions for various assumptions about the measurements, $c(t)$.

Later studies used estimated lifetimes, $\hat\lambda_\eta^{-1}$, and inverted the concentration data in order to estimate the sources, specifically sources not covered by industry reporting (mainly from eastern Europe and the USSR). The assumption is that such unreported sources were small over the period for which the data were used to estimate λ_η. AGAGE data have recently confirmed that emissions of CFC-11 and CFC-12 are decreasing [91].

Box 16.4. Ozone-depletion potentials

Ozone-depletion potentials (ODPs) are defined for the various halocarbons as the amount of destruction of ozone caused by release of unit mass of that compound, integrated over its atmospheric presence [519]. They are determined by model calculations. The ODPs are defined as ratios of destruction of ozone relative to that arising from CCl_3F (CFC-11), thus removing much of the uncertainty associated with differences in model transport and (to a lesser extent) chemistry. The ODP concept is the forerunner of the global-warming potentials used to compare greenhouse gases (see Box 6.2).

Within the Montreal Protocol ODPs acquired a 'legal role'. Production limits are expressed as percentages of base-year levels, in terms of the ODP-weighted contribution of each class of gas. With total phase-out, this role disappears.

Solomon and Albritton [441] have extended the ODP concept to include a time horizon in order to assess shorter-term effects. They proposed a semi-empirical approximation:

$$\text{ODP}_\eta(t) = \alpha_\eta \frac{F_\eta}{F_{\text{CFC-11}}} \frac{\mu_{\text{CFC-11}}}{\mu_\eta} \frac{n_\eta}{3} \frac{\int_{t_s}^t \exp[-\lambda_\eta(t - t_s)]\,dt}{\int_{t_s}^t \exp[-\lambda_{\text{CFC-11}}(t - t_s)]\,dt} \qquad (1)$$

where t_s is the time of reaching the stratosphere and, for compound η, μ_η is the relative molecular mass, $1/\lambda_\eta$ is the atmospheric lifetime, n_η is the number of chlorine or bromine atoms and α_η is a correction factor for compounds containing bromine, being 40 if bromine is present and 1 otherwise. F_η is the proportion of the compound that is dissociated in the stratosphere, i.e. an inverse measure of the stratospheric lifetime.

Bloomfield *et al.* [34] reported three types of calculation.

- A Kalman-filter analysis with the AGAGE 12-box model (see Box 7.3), where the elements of the unknown state-vector are a set of inverse lifetimes.

- A least-squares analysis with the TM2 model, where the least-squares fit to (10.3.2) is embedded within an outer loop that varies the atmospheric lifetime in order to fit both the source distribution and the lifetime.

- A least-squares analysis with the GISS model that used a range of basis components in order to estimate the lifetime.

The problem of using surface concentrations to estimate sources and sinks in the free atmosphere was noted in Chapter 8. The toy-model study by Enting and Newsam [138] suggested that what could be deduced was vertical averages of the net source minus sink and that the greater the horizontal variability the more the vertical averaging would emphasise the lower levels of the atmosphere.

Table 16.1. *Estimates of global mean concentrations of OH and trends derived from CH_3CCl_3 data with ± 1-SD ranges (in [379], (a) assumes that there is no ocean sink and (b) assumes a $1/85$ yr^{-1} ocean sink).*

Period	OH (10^5 cm^{-3})	CH_3CCl_3 lifetime (years)	Reference
1978–85	7.4 ± 1.4	5.4–7.5	[377]
1978–90	8.7 ± 1.0	5.1–6.4	[379] (a)
1978–90	8.1 ± 0.9	5.1–6.4	[379] (b)
1978	10.0 ± 1.0	4.7	[266]
1993	10.7 ± 1.3	4.5	[266]
1998–99	11.0 ± 2.0	$5.2^{+0.2}_{-0.3}$	[323]
Period	(d/dt)OH (% yr^{-1})		Reference
1978–90	1.0 ± 0.8		[379]
1978–99	-0.5 ± 1.0		[218]

16.3.2 Tropospheric hydroxyl

Apart from the atmospheric-lifetime studies, the main applications of inversions to estimate source–sink processes in the free atmosphere have been concerned with the role of hydroxyl radicals as a sink for trace gases in the troposphere. The main species used to investigate this has been methyl chloroform (CH_3CCl_3). Notionally, such calculations estimate a mean concentration of OH. In practice it is a vertically weighted average, due to the nature of the inverse problem. It is also a rate-weighted average, since what is really estimated is the rate of loss of CH_3CCl_3.

The main reason for estimating the OH sink of CH_3CCl_3 is to apply the results to other constituents. A particularly important case is the tropospheric sink of methane. The validity of using such OH estimates derived from one compound in order to deduce rates of loss of a second compound depends on the extent to which the spatial patterns of the sinks for the two compounds are proportional. For CH_4 and CH_3CCl_3, the dependences of the reaction rates on temperature are quite similar and the spatial distribution of each compound is close to uniform because of the long lifetimes. Therefore results of studies of CH_3CCl_3 can usefully be applied to CH_4. Table 16.1 summarises a number of estimates based on CH_3CCl_3 data, including estimates of a secular trend.

Some of the early work deducing concentrations of OH from CH_3CCl_3 data became the subject of dispute. The main steps were as follows.

■ Prinn *et al.* [377] produced a set of estimates using various aspects of an 'optimal-estimation' (i.e. Kalman-filtering) technique, based on the nine-box model of Box 7.3, applied to data from five ALE/GAGE sites, obtaining inverse lifetimes of 0.145 ± 0.021 yr^{-1} from a 'trend method' fitting logarithms of concentrations (without assumptions about the absolute calibration), $0.167 \pm 0.025 \ \text{yr}^{-1}$ from fitting concentrations and $0.167 \pm 0.031 \ \text{yr}^{-1}$ from fitting spatial gradients.

■ Spivakovsky et al. [445] performed a three-dimensional modelling study. On comparing their results with those of Prinn et al. [377], they claimed that the process of fitting logarithms of concentrations in order to deduce both a lifetime and a measurement calibration was extremely ill-conditioned. They also claimed that the latitudinal gradients gave a measure of transport (as described by Plumb and MacConalogue [359], see Section 10.2) rather than of the OH sink.

■ Hartley and Prinn responded [196] with a comment that queried the extent to which the three-dimensional model of Spivakovsky et al. accurately reproduced the data and suggested that the defects crippled the ability of the model to be used for the inverse problem of deducing sources.

■ In reply, Spivakovsky et al. [446] noted that deduction of sources was not one of their objectives and disputed the statistical analysis presented by Hartley and Prinn [196]. In particular, they noted the statistical issues involved in taking model results based on repeating a single year of transport and comparing them with multi-year averages from observations.

■ A comment by Cunnold and Prinn [88] addressed the concerns of ill-conditioning raised by Spivakovsky et al. [445]. They claimed that the trend method, based on fitting logarithms of concentrations, was not ill-conditioned and that, with a proper account of the initial conditions, the counter-example given by Spivakovsky et al. was a very poor fit to the data. They noted that their use of latitudinal gradients was based on gradients normalised with respect to the mean concentration.

■ In reply, Spivakovsky [444] noted that, with the more detailed information about the initial conditions, it was still possible to construct a counter-example (one with a >10-year lifetime) only slightly different from the original counter-example [445], which did give a good fit to the observations. She suggested that this indicated a great sensitivity to initial conditions that implied an unrecognised uncertainty in the calculations of Prinn et al. [377]. The use of normalised gradients was claimed to make the method essentially equivalent to the trend method.

In order to clarify this, we consider the analysis in terms of the transport times defined by Plumb and MacConalogue [359] (see Section 10.2). These reduce to

$$\tau(r) = \frac{c(r_0) - c(r)}{\beta' + \lambda c(r_0)} \tag{16.3.4}$$

where β' is the growth rate ($\beta c(r_0)$ in the notation of Section 10.2) and λ is the inverse lifetime.

Figure 16.1 shows a set of relative transport times calculated in this way. Since the concentrations are not measured at the 'effective location' of the source, each set is subject to an unknown offset. If the curves have the same shape, expression (16.3.4)

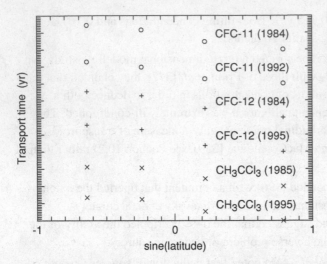

Figure 16.1 Empirical transport times for several halocarbons during various periods, as defined by (16.3.4). The curves are offset for clarity since the true origin, defined by $c(r_0)$, is unknown. Major tick marks on the time axis indicate years.

implies that the profiles are providing information about $\beta + \lambda$. The comparisons for CCl_3F, CCl_2F_2 and CH_3CCl_3 use lifetimes of 52, 185 and 4.8 years and mean concentrations and trend values from Tables 3, 4 and 5 of Prinn *et al.* [382]. It will be seen that these empirically estimated distributions of 'transport times' remain the same even with great changes in the source strengths and consequent rates of growth, even extending to the case of negative rates of growth for the concentration of CH_3CCl_3 around 1995. CCl_3F and CCl_2F_2 have similar profiles, indicating that they have similar source distributions, whereas the difference for CH_3CCl_3 can be attributed to there being a different pattern of sources and sinks.

Krol *et al.* [266] revisited the problem using a Monte Carlo implementation of Bayesian estimation to determine the mean OH field and a linear trend. The OH field was specified as a multiple, α, of the OH field produced by a chemical transport model. The parameters estimated were $x_1 = \bar{\alpha}$, $x_2 = \dot{\bar{\alpha}}$ and a nuisance parameter, x_3, characterising the error in the initial CH_3CCl_3 concentration fields. The data that were fitted were the coefficients obtained by fitting the time series of CH_3CCl_3 at the five AGAGE sites in terms of Legendre polynomials. The parameter estimates were

$$\hat{x}_i = \frac{\sum_m p(\mathbf{c}|\mathbf{x}) x_i}{\sum_m p(\mathbf{c}|\mathbf{x})} \qquad (16.3.5)$$

where the sum is over all Monte Carlo realisations of $\mathbf{x}$ that are generated according to the prior distribution. The 'measurement' error was treated as a product of independent normal distributions with variances estimated from the data fitting. (Fitting the data in terms of Legendre polynomials rather than as single-time values should greatly reduce the correlations among the items of data.) The concentration of OH was estimated as changing from $1.00^{+0.09}_{-0.15} \times 10^6$ molecules cm^{-3} in 1978 to $1.07^{+0.09}_{-0.17} \times 10^6$ molecules cm^{-3} in 1993. The reasons for the differences from the results of Prinn *et al.* regarding the trend were not identified, but possibilities such as the resolution of the model, the estimation procedure and the treatment of the stratospheric sink were considered.

Table 16.2. *Estimates of the inverse lifetime (yr^{-1}) for*
CH_3CCl_3 from the ALE/GAGE/AGAGE programme with
1-SD uncertainties and including uncertainties from
emissions and calibration.

Method	[377]	[379]	[380]
Trend	0.145 ± 0.021	$0.210^{+0.037}_{-0.022}$	0.217 ± 0.025
Content	0.167 ± 0.025	0.164 ± 0.031	0.222 ± 0.017
Gradient	0.167 ± 0.031	0.166 ± 0.030	0.208 ± 0.034

To investigate the issues of ill-conditioning in more detail, we look at how the estimates from 'optimal estimation' have evolved over time. The results for the three main methods are summarised in Table 16.2. The final column from Prinn *et al.* [380] in 1995 incorporates a number of differences:

- the possibility of actual changes in the lifetime due to changes in concentration of OH;
- a great decrease in calibration uncertainty; and
- an expectation that the degree of ill-conditioning will be less because of the turn-around both in the emissions and in the concentration of CH_3CCl_3.

Given the last point, what is surprising about the results in Table 16.2 is that the change in the pattern of emissions does not lead to a reduction of uncertainty. More specifically, the reduction in uncertainty appears in the 'content' method which, being a zero-dimensional case of the linear Green-function method, would be expected to have uncertainties that depend on the observational uncertainties but not on the observational values.

The continued phase-out of CH_3CCl_3 has led to a situation in which emissions are close to zero and the concentration has a near exponential decay. This makes it possible to estimate the lifetime with little sensitivity to calibration errors. Montzka *et al.* [323] have analysed the period 1998–99 and, as well as estimating the lifetime (and concentration of OH), they were able to identify southern-hemisphere levels of OH as being (15 ± 10) % higher than those in the northern hemisphere.

Several other compounds have been used to estimate abundances of OH. The estimated global means are $11.0^{+5.0}_{-3.6} \times 10^5$ cm^{-3}, using CHClF$_2$ [320]; and $9.4^{+2.7}_{-1.7} \times 10^5$ cm^{-3}, using a combination of CH_3CCl_3, CHClF$_2$, CH$_2$FCF$_3$, CH$_3$CCl$_2$F and CH$_3$CClF$_2$ [218]. ^{14}CO from the stratosphere has also been used, e.g. [52, 389].

16.3.3   A linearised sensitivity study

Some indication of the relation between uncertainties can be obtained by considering the linearised equations for the single-box atmosphere. We write the CH_3CCl_3

budget as

$$\dot{c} = (1 + \gamma)E - (\lambda + \beta t)c \tag{16.3.6a}$$

where λ is the inverse lifetime, β is the trend in the inverse lifetime and γ characterises the calibration uncertainty. This transforms to the regression equation.

$$\gamma E(t) - \lambda c(t) - \beta t c(t) = \dot{c}(t) - E(t) \tag{16.3.6b}$$

Equation (16.3.6b) can be solved recursively for γ, λ and β, using a state-space representation with

$$\mathbf{x} = [\gamma, \lambda, \beta]^T \tag{16.3.7a}$$

$$\mathbf{F} = \mathbf{I} \tag{16.3.7b}$$

$$\mathbf{H}(t) = [E(t), -c(t), -t\,c(t)] \tag{16.3.7c}$$

$$z(t) = E(t) - \dot{c}(t) \tag{16.3.7d}$$

The conditions $\mathbf{F} = \mathbf{I}$ and $\mathbf{Q} = \mathbf{0}$ reduce the Kalman-filter equations to the simple 'recursive-regression' form given in equations (11.2.3a)–(11.2.3c).

The estimation is initiated with $\mathbf{x} = \mathbf{0}$ and with $\mathbf{P}$ diagonal. In the standard case we make the diagonal elements of $\mathbf{P}(0)$ large so that the regression tries to estimate all three state variables. Various restricted cases in which one of the state variables, $x(j)$, is fixed can be simulated by making $P_{jj}(0)$ small.

The linearisation reduces the estimation problem to that of a regression fit to three functions. The degree of ill-conditioning depends on the extent to which the functions are orthogonal. Figure 16.2 shows the functions, scaled for the purposes of comparison. In the early part of the record it will be seen that concentrations and emission grow in a comparable way.

The calculations have been performed using the recursive-regression algorithm (Box 4.4) with initial values and standard deviations $\hat{\gamma} = 0.0 \pm 0.2$, $\hat{\lambda} = 0.2 \pm 0.1$ and $\hat{\beta} = 0.0 \pm 10^{-2}$ and an uncertainty of ± 2 ppb in the concentration data.

Figure 16.3 shows how $[P_{kk}(n)]^{1/2}$ (for $k = 1, 2$ and 3), the variances of $\hat{\gamma}$ (dotted line), $\hat{\lambda}$ (solid line) and $\hat{\beta}$ (dashed line) decrease as the record extends in time. The curves are normalised by their respective prior variances. The dashed line shows a

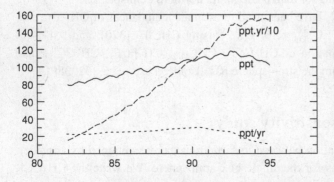

Figure 16.2 The three 'basis functions' used in the regression estimate based on linearising the global CH_3CCl_3-budget equations: $E(t)$ (in ppt yr^{-1}), $c(t)$ (in ppt) and $tc(t)$ (in ppt yr, scaled by a factor of ten) (with coefficients γ, β and λ, respectively).

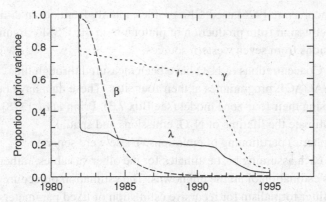

Figure 16.3 The variances of the estimates of λ (solid line), γ (dotted line) and β (dashed line), normalised with respect to the respective prior variances. This plot is based on a linearised model, (16.3.6a), with parameters estimated by fitting the linearised budget equation (16.3.6b) by recursive regression fits to the basis functions shown in Figure 16.2.

steady reduction in the uncertainty of any trend coefficient, β, in the decay rate. However, by allowing a calibration uncertainty γ, the uncertainties in the loss rate decrease relatively slowly. This changes around 1992. It is at this point that the behaviours of the three basis functions become more distinct. Therefore the observational data can identify the relative contributions with significantly less ambiguity. However, an inspection of the full covariance matrix shows that, even at the end of the record, the estimates of γ and λ are still very highly correlated. This emphasises the importance of the spatial distribution. As described above, the normalised spatial distribution gives information about the growth rate plus loss rate, independently of any knowledge of the source strength.

16.4   Other surface fluxes

Inversion studies of trace atmospheric constituents have concentrated on those gases with the greatest projected impacts: CO_2 and CH_4 among the greenhouse gases and CFC-11 and CFC-12 among the ozone-depleting substances. In many cases, these studies have been augmented by including additional species for diagnostic purposes – most notably the inclusion of isotopic species in CO_2 inversions. However, there is potential for extending inversion studies to other trace constituents. Those of global significance, in terms both of distribution and of impact, are the other greenhouse gases: N_2O and SF_6, HFCs and PFCs and other ozone-depleting substances. In addition, gases with indirect effects are important, the main examples being those that affect the concentration of methane via changes in atmospheric chemistry: CO and non-methane hydrocarbons.

At the time of writing, there have been only a few such studies. Some examples are the following.

Krypton-85. Jacob *et al.* [226] used ^{85}Kr profiles, mainly from ocean cruises, to estimate the emission from production of plutonium in the USSR, assuming known emissions from seven western sources.

Nitrous oxide. Concentrations of N_2O have been measured through the ALE/GAGE/AGAGE programmes at their four sites. These data have been interpreted using their four-zone model (see Box 7.3). Prinn *et al.* [378] used the data to estimate the lifetime of N_2O, emissions and spatial (semi-hemispheric) partitioning of emissions. These were separate calculations, each assuming best estimates for the other variables, rather than a simultaneous estimation of all the unknowns. The estimation procedure was the Kalman-filter formalism for recursive estimation of fixed parameters. Even though isotopic data provide additional constraints on the global scale [391], the N_2O budget is still rather uncertain [264]. A new type of isotopic information comes from isoptomer differences, regarding which Yoshida and Toyoda [524] have reported finding spatial differences in the extent to which ^{15}N occurs at the central location of the $^{15}N^{14}NO$ molecule, i.e. $^{15}N^{14}NO$ versus $^{14}N^{15}NO$.

Carbon monoxide. Bayesian synthesis inversions of concentrations of CO have been undertaken by Bergamaschi and co-workers. The first stage was the inversion of the CO mixing ratios from 31 sites [30]. The second stage added data on the $^{13}C/^{12}C$ and $^{18}O/^{16}O$ ratios in CO at five of these sites [31]. They found that the inclusion of isotopic data significantly improved the inversions. They also found that their results, especially for technological sources, were quite sensitive to the spatial distribution of emissions of terpene. An additional result was the estimated range, 0.80–0.88, for the yield of CO from the destruction of methane (i.e. the parameter denoted ζ in the globally lumped model of Box 1.5).

As noted above, ^{14}CO data have been used to estimate OH sinks. Conny [77] gives an overview of isotopic characterisation of CO.

Methyl chloride. Khalil and Rasmussen [256] used a low-resolution model to invert CH_3Cl data and estimate that 85% of the source was from the region 30 °S to 30 °N.

Further reading

■ The Montreal Protocol established assessment panels to provide scientific information. The main scientific assessments [513–516] have been supplemented by a number of special-purpose reports. These and other publications are listed in [489].

■ The volume edited by Brasseur *et al.* [51] provides a comprehensive coverage of atmospheric chemistry in the context of earth-system science and global change.

■ The report [64] on 'Atmospheric Chemistry Research Entering the Twenty-First Century' gives a broad perspective on current knowledge and future research needs.

■ The book by Wayne [501] describes the chemistry of the earth's atmosphere and relates this to the atmospheres of other planets and their satellites.

■ Poisson *et al.* [362] have recently reviewed tropospheric chemistry, concentrating on the role of non-methane hydrocarbons. The account of updated estimates of global OH fields by Spivakovsky *et al.* [447] is based on a comprehensive review of the relevant tropospheric chemistry.

Chapter 17

Regional inversions

Valéry has called philosophy 'an attempt to transmute all we know into what we should like to know'.

Aldous Huxley: *Beyond the Mexique Bay*.

17.1 General issues

The previous chapters of this book have described the progress in tracer inversions applied to global-scale problems. In parallel with these developments, there has been increasing interest in problems on smaller scales. Motivations for such work include applications such as monitoring of emissions and also improving our understanding of the observational data used in global inversions.

In this section we consider problems defined with a spatial resolution of order 100 km defined over continental-scale regions. This is a rapidly developing field, with, as yet, relatively little sign of any consolidation of techniques.

Some of the new features that can be expected in going from the global to regional scale are the following.

■ Data will be needed on synoptic time-scales, rather than monthly means: at Cape Grim, the operational conditions defined as 'baseline' apply for less than 50% of the time (see Section 5.2 above). For the remainder of the time, the concentrations of CO_2 reflect more local influences. Model studies indicate that, for as much as 70% of the time, air at Cape Grim has been influenced by Tasmania, south-eastern Australia (as illustrated in Figure 9.1), or more westerly parts of the Australian continent). This raises the possibility of using these 'non-baseline' data to interpret the local sources and sinks [393]. Similar possibilities arise for other sites on the coasts of continents or on nearby islands.

■ The nature of the inverse problem is likely to change since the transport characteristics are dominated by advection rather than appearing diffusive.

■ It appears likely that new forms of statistical characterisation will be needed for time series that exhibit intermittent peaks (see, for example, Figure 5.2). In particular, modelling observations as a stationary time series may become inadequate.

In considering regional scales, a useful starting point is to consider what can be learned from the experience with global-scale problems. Some of the possibilities are as follows.

■ Studies of the use of observational data on synoptic time-scales. For global models based on GCM wind-fields, such studies would need to be in the form of 'identical-twin' experiments (see Box 13.2).

■ Comparisons of global models at various spatial resolutions. This makes it possible to characterise the degree of spatial and temporal variability associated with different length-scales in the transport processes.

■ Use of adjoint models. This allows the exploration of the extent to which the discretisation of sources and the spatial statistics of sources become important, extending the analysis of discretisation effects [241] (c.f. Section 8.3) to specific cases.

■ Determining the 'boundary conditions' for regional modelling. One important issue is the degree of small-scale variability generated by atmospheric transport acting on concentration distributions from large-scale flux distributions.

Some relevant modelling studies have already occurred. A 1989 study by Heimann *et al.* [205: Figure 26] showed that even at 8° by 10° resolution, a transport model using analysed winds exhibited some skill at calculating synoptic-scale variations in concentration of CO_2. Using a model with 2.5° by 2.5° resolution, Ramonet and Monfray [393] found the representation of synoptic-scale features sufficiently good that they explicitly suggested the scope for making inferences on the regional scale. Monfray *et al.* [322] have used longitudinal gradients to estimate air–sea fluxes for the Southern Ocean. Brost and Heimann [55] have investigated the extent to which regional distributions of concentrations are affected by global-scale sources with a simple exponential decay rate λ. They found, as expected, little distant influence for $\lambda^{-1} \ll 5$ days; and for $\lambda^{-1} \gg 5$ days the contributions from distant sources had little variation. At around $\lambda^{-1} = 5$ days, the percentage influence tended to increase with height. Enting and Trudinger [142] performed related calculations, assessing the extent to which large-scale flux distributions would lead to small-scale variability at Cape Grim. They found that the low-resolution (8°-by-10°) model with only large-scale fluxes (zonal means from a two-dimensional inversion) produced a degree of synoptic-scale variability comparable to that observed at Syowa on the Antarctic coast (c.f. [23]). For Cape Grim the observed variability was much greater than was modelled, confirming the influence of regional fluxes.

There are also global-scale synthesis-inversion studies that have used low resolution over most of the world while attempting to resolve small-scale detail of one particular region [214, 149]. This approach is likely to be very sensitive to the distribution of the network and hence the results of such studies need to be treated with caution. Although the 'truncation error' from distant sources is likely to be low, it is also likely to be very highly correlated at sites whose separation is small relative to the distance from the source. Neglect of such correlated error can seriously bias estimates of local fluxes.

Observational studies can also contribute to the development of techniques for regional inversion. In particular, aircraft-based measurement can be used to determine the relation between spatial and temporal variabilities. The phase-out of CFCs in industrial nations means that measurements of concentrations of CFCs at locations such as Cape Grim can also be used to assess the degree of synoptic-scale variation generated by distant sources.

As noted above, regional inversions have used a range of estimation techniques. These are summarised in the following section and, in keeping with one of the major themes of this book, the underlying statistical assumptions are noted. This statistical emphasis provides a basis for some preliminary proposals on how a more integrated framework might be established.

At distances smaller than 100 km, we come to the scales applicable to urban-pollution problems. Some of the aspects in which these studies differ from global- and regional-scale studies are

- that there is a much greater importance of reactive chemistry in problems of practical interest;
- co-location of sources for different processes;
- that there is a requirement for meteorological information on space- and time-scales less than those provided by standard forecasting models;
- that there is frequently a requirement for forecasting particular events such as episodes of pollution.

These issues go beyond the scope of the present book, although a few urban-scale studies are cited below as examples. As well as developing regional inversion techniques by 'down-scaling' techniques that have successfully been used in global inversions, 'up-scaling' of techniques used in urban-scale studies is important. The studies based on the 'tracer-ratio' method are a prime example of this.

17.2   Types of inversion

17.2.1   Tracer ratios

The 'tracer-ratio' technique described in Section 12.3 has commonly been applied and is used in many of the applications cited in the following section. The two main reasons for this are (a) the method does not require the use of a transport model and (b) estimates can be obtained with data from a single site.

Notionally, the estimates take the form of the source, s_η, of gas η being expressed as a multiple of a reference source strength s_{ref}:

$$\hat{s}_\eta = \hat{\alpha} s_{ref} \tag{17.2.1a}$$

where the multiple α is estimated from sets of simultaneous observations of gas η and the reference gas:

$$\hat{\alpha} = \frac{\langle c_\eta - c_{\eta,0} \rangle}{\langle c_{ref} - c_{ref,0} \rangle} = \frac{\langle c'_\eta \rangle}{\langle c'_{ref} \rangle} \tag{17.2.1b}$$

where $c_{\eta,0}$ denotes a background concentration, c'_η denotes perturbations from $c_{\eta,0}$ and $\langle . \rangle$ denotes some form of averaging over conditions, events or cases.

The discussion in Section 12.3 indicates the need for a more general form based on the statistical model underlying such estimates:

$$c'_{xn} = G_n s_x + \epsilon_{xn} \tag{17.2.1c}$$

with the scale factors G_n unknown. For a pair of tracers, eliminating G_n leads to the form

$$c'_{xn} - \epsilon_{xn} = \frac{s_x}{s_{ref}}(c'_{ref,m} - \epsilon_{ref,m}) = \alpha(c'_{ref,m} - \epsilon_{ref,m}) \tag{17.2.1d}$$

The regression solution, corresponding to ignoring errors in c_{ref}, is a special case of (17.2.1b) with $\langle . \rangle$ taken as the sum over observations, weighted by $c'_{ref} - \bar{c}'_{ref}$, with $\bar{c}'$ denoting $\langle c' \rangle$:

$$\hat{\alpha}_{REGRESS} = \frac{C_{12}}{C_{22}} = \frac{\sum(c'_{ref} - \bar{c}'_{ref})(c'_\eta - \bar{c}'_\eta)}{\sum(c'_{ref} - \bar{c}'_{ref})^2} \tag{17.2.2}$$

Accounting for errors in both data sets leads to the maximum-likelihood estimate, $\hat{\alpha}_{ML}$, given by the solution of

$$\hat{\alpha}^2 C_{12} + \hat{\alpha}(\xi C_{11} - C_{22}) - \xi C_{12} = 0 \tag{17.2.3}$$

where ξ is the ratio of the variances of the errors in ϵ_η and ϵ_{ref} [254: Chapter 29; and Section 12.3]. The requirements for using the technique are as follows:

- one needs measurements of concentrations of multiple species, including a reference species whose source, s_{ref}, is known;
- frequent measurements must be obtained, so that the peaks can be measured;
- periods of background conditions must exist in order to determine the background levels, $\bar{c}_{ref}$ and $\bar{c}_\eta$; and
- as discussed in Section 12.3, one needs a statistical model that specifies ξ, the relative variances of the $\epsilon_{\eta,m}$ and $\epsilon_{ref,m}$.

Such 'tracer-ratio' methods have commonly been applied in urban-pollution studies. However, the combination of requirements has meant that most of the applications

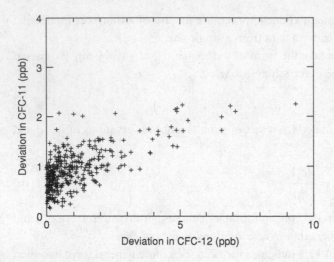

Figure 17.1 An illustration of tracer-ratio relations for CCl_3F and CCl_2F_2 at Cape Grim, using AGAGE data for the first 3 months of 1997, excluding data below 263 and 529 ppb, respectively.

on a regional scale have been based on a small number of sites. In particular, the following two.

Cape Grim, north-west Tasmania. Cape Grim is located at the north-west of the island state of Tasmania. Concentrations of many tracer species are measured in the Cape Grim laboratory. In addition, samples are collected for analysis off-site. The emphasis is on measurements made under defined 'baseline' conditions (see Section 5.2), which occur a little less than 50% of the time. These require winds from the 190°–280° sector. However, some instruments at the Cape Grim laboratory operate in a continuous mode and so record concentrations from non-baseline conditions. The land influences on Cape Grim are (a) from agricultural and forest areas of Tasmania, (b) from the agricultural areas on King Island to the north-west and (c) from the Australian mainland. As noted above, model studies suggest that there is a land influence on about 70% of the data.

 Figure 17.1 shows illustrative data: concentrations of CCl_3F and CCl_2F_2 measured at Cape Grim, showing individual samples selected from the first 3 months of 1997. This is a relatively weak 'signal' since these gases have been phased out under the Montreal Protocol.

Adrigole/Mace Head, western Ireland. As with Cape Grim, these sites on the west coast of Ireland were chosen because they experience a high proportion of baseline conditions – in this case westerly winds from the North Atlantic. However, there are long periods during which the weather systems bring air-masses from Europe over the North Atlantic.

Figure 17.2 shows the locations of Cape Grim and Mace Head, relative to the regions of major sources. In each case there are large ocean regions nearby, from which air will have 'baseline' concentrations of many tracers. In other directions,

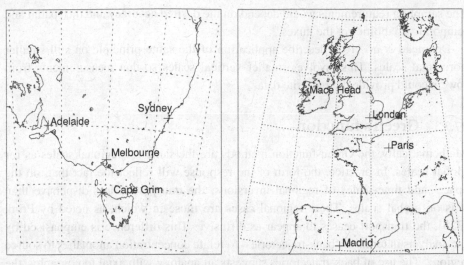

Figure 17.2 Locations, *, of observing sites at Cape Grim and Mace Head, relative to continents with major source regions (on a 5° grid).

there are agricultural and industrial areas. The location of the observing sites makes it possible, in many cases, to distinguish between these two types of air-mass.

17.2.2 Boundary integration

The boundary-integral (sometimes called mass-balance) method aims to determine the integrated surface flux over a region by balancing the total changes above the region against the flux across the boundaries, i.e. the vertical integral of $c\mathbf{v} \cdot \mathbf{n}$ integrated around the boundary, where $\mathbf{n}$ is a unit-length vector normal to the boundary. If $\mathbf{v}$ is steady and uniform then only the upwind and downwind boundaries need be considered. Conservation of mass expresses the estimate, $\hat{\bar{s}}$, of the mean flux, $\bar{s}$, as

$$A_S \hat{\bar{s}} = \int s\, dS = \frac{d}{dt} \int c\, dV + \int_{p_{surf}}^{p_{min}} \int c\mathbf{v} \cdot \mathbf{n}\, dL\, dp \qquad (17.2.4)$$

where the surface integral is over the region, with area A_S, bounded by the line integral $\int dL$ and $\mathbf{n}$ is the unit vector normal to the surface defined by extending the boundary vertically. An additional term integrating over the upper boundary will be needed in conditions of subsidence or convergence of the wind flux. Often the actual changes in concentration are small so that the volume-integral term can be neglected. The statistics of the estimate will be determined by the measurement and sampling statistics of the observations on the right-hand side.

This method has proved practical on small scales, e.g. a subtropical island of order 20 km × 20 km [523] using a succession of flights in a rectangular path around the island at various heights. In a similar study, Wratt *et al.* [518] estimated agricultural emissions of CH_4 from upwind–downwind differences in aircraft profiles 20–40 km apart. On larger scales, some form of selective sampling would be needed.

The sampling requirements would depend on the desired accuracy and the spatial and temporal variabilities of the fluxes.

Denmead *et al.* [103] describe application of the same principle on still smaller horizontal scales. This involves smaller vertical scales so that measurements from towers could provide the requisite data.

17.2.3 Green's function

Mathematically, the Green-function methods are the same for regional scales as for global scales. In practice, the form of the response will reflect the fact that, on the space- and time-scales of regional inversions, the transport is less dispersive than it is on global scales. These regional cases are those in which, as noted by Prinn [375], the transport ceases to appear as diffusive. This difference is emphasised by the analysis in terms of 'back-trajectories' to relate concentration anomalies to source regions. The use of back-trajectories suggests an analogy with axial tomography (the basis of CAT-scans) and in particular the 'back-projection' algorithm for the tomographic inverse problem (see Box 17.1). As with global-scale studies, the adjoint formalism is computationally efficient for the case limited to sets of observations and a multiplicity of possible source components. Several theoretical studies of such adjoint techniques have been undertaken [385, 494] for determining sensitivity to sources.

Box 17.1. Axial tomography

Tomography is the process of reconstructing spatial distributions from a set of projections. Our interest is the analogy with interpreting concentrations that reflect the integral of fluxes along a back-trajectory.

Axial tomography aims to recover a two-dimensional distribution expressed in polar coordinates as $f(r, \theta)$ from a set of projections onto a line (with coordinate x), at an angle of $\theta' + \pi/2$ to the direction $\theta = 0$. These are

$$g(x, \theta') = \int_{-\infty}^{\infty} f\left[\sqrt{x^2 + y^2}, \theta' + \tan^{-1}\left(\frac{x}{y}\right) \right] \mathrm{d}y \tag{1}$$

The most widely known case is the CAT-scan whereby a three-dimensional patient is analysed through a sequence of two-dimensional slices, each derived from X-rays taken at a range of angles.

The analytical inversion of (1) was derived by Radon [390] as

$$f(r, \theta) = \frac{1}{2\pi^2} \int_{-\pi/2}^{\pi/2} \int_{-\infty}^{\infty} \frac{\partial g(x, \theta')}{\partial x} \frac{1}{r \sin(\theta - \theta') - x} \, \mathrm{d}x \, \mathrm{d}\theta' \tag{2}$$

The review by Gordon and Herman [182] lists four classes of techniques for estimating $f(r, \theta)$, given noisy discrete data.

> **Summing over projections.** This is a relatively crude type of approximation but can be improved by additional processing.
>
> **Approximations to analytical solution.** This is numerical integration of Radon's formula, (2).
>
> **Fourier-transform techniques.** These techniques can exploit the special property that the Fourier transform of the projection g is a one-dimensional slice of the two-dimensional Fourier transform of f [182].
>
> **Expansion in terms of basis functions.** In this class of methods, which are essentially synthesis inversions, the full set of projections is calculated for each basis function and the coefficients of the basis functions are determined by a best fit to the measured projections. A basis of discrete cells provides direct pixel-based imaging.
>
> As with other inverse problems, the best form of approximation and the appropriate regularisation depend on the statistics of the observations and the signal that is sought.

In the literature of urban and regional pollution, these Green-function techniques have sometimes been termed 'source-receptor' techniques, but this term covers a range of estimation techniques [209].

17.2.4   Fitting variability

The Green-function method relies on the accuracy of the transport model that is used to calculate the responses. In contrast, the tracer-ratio method dispenses with the transport model altogether. One intermediate case, between a reliable model and no usable model, is a model that is correct only in a statistical sense. The obvious example is the use of transport fields from a GCM. Global-scale inversions have commonly used GCM-derived wind-fields and addressed the issue of lack of fidelity on synoptic time-scales by fitting monthly mean data. An alternative possibility is to fit the variability. The estimation equations for the single-component case are presented in Section 12.3. A possible generalisation is to estimate model parameters by fitting the power spectrum of the concentrations [Y.-P. Wang, personal communication, 2000].

17.2.5   Combining the approaches

The methods described in previous subsections for using remote measurements of trace-gas concentrations are

- tracer-ratio methods, involving equation (17.2.1a) or some generalisation;
- boundary integration to apply mass-balance over a volume of the atmosphere – equation (17.2.4); and
- Green-function-type methods.

Information from these techniques can be combined with other information from

- process models, as described in Chapter 6;
- local flux measurements from techniques such as those involving chambers or micrometeorological relations;
- remote sensing; and
- boundary-layer accumulation (Box 17.2).

In combining information from various techniques, it is important to avoid double-counting the observations. This is particularly important for the three atmospheric techniques.

The Bayesian formalism provides a means for combining information, but the requirement is that the information provided by the observations is independent of the prior information. This may be applicable when the prior information comes from such sources as satellite data or process modelling. However, if we wish to combine the atmospheric approaches, we need to do this through a single combined estimation procedure.

For the atmospheric estimates, the way to combine them is to develop a statistical model for the observational and model errors, capturing the various types of error and then using this model to define the form of estimates, using some principle such as maximum likelihood.

A possible direction in which to generalise the statistical model to incorporate Green-function and tracer-ratio methods is to use the multiplicative form

$$c_{j\eta} = G_j s_\eta \epsilon_{j\eta} \tag{17.2.5}$$

The Green-function method is applicable to the case in which ϵ_j has a mean of 1 and a small variance. The tracer-ratio method corresponds to the case in which the ϵ_j has a large variance, but, at a single place and time, the ϵ_j for different species are highly correlated. Fitting the variability corresponds to the case in which the variance of the ϵ_j is large but its mean is near 1. Quantitative analysis of the intermediate cases of this type of combined model remains to be done. However, one group of techniques that reflect a Green-function approach, while acknowledging uncertainty regarding transport, consists of those based on trajectory climatologies, e.g. [384, 62]. The European Tracer Experiment (ETEX) [146] provided an opportunity for testing methodologies for estimating sources [408, 326].

17.3 Some applications

This section and the following section on carbon dioxide review some of the examples of tracer inversions on regional scales.

Urban pollution. Although these scales are outside the scope of this book, such studies should be noted for the approaches that may contribute to studies on larger scales. Many urban-scale analyses have been based on some form of the 'tracer-ratio' method. An example of a Green-function analysis on an urban scale is the estimation of emissions of CO in the Los Angeles area by Mulholland and Seinfeld [329].

17.3.1 Tracer-ratio estimates of sources

Nitrous oxide. Several studies have used the 'tracer-ratio' method to estimate fluxes of N_2O. Prather [365] used ratios of excess CCl_3F, CCl_2F_2, CCl_4, CH_3CCl_3 and N_2O in identified 'pollution events' at Adrigole (Ireland) to estimate the relative source strengths of these gases. The underlying assumption is that, for western Europe, the sources of N_2O and CCl_3F were relatively dispersed. In analysing events of elevated concentrations of N_2O at Cape Grim (Tasmania), Wilson *et al.* [509] noted that elevated concentrations of CCl_3F would be mainly from the city of Melbourne. As a 'reference' tracer, with a distribution of emissions comparable to that of N_2O, they used radon. They estimated fluxes of 160 ± 45 kg N km^{-2} yr^{-1} for Tasmania, 1300 ± 500 kg N km^{-2} yr^{-1} for King Island, 130 ± 30 kg N km^{-2} yr^{-1} for the south-east of mainland Australia and 115 ± 40 kg N km^{-2} yr^{-1} for this region if Melbourne were excluded.

Halocarbons. In contrast to the study by Wilson *et al.* [509], in which the city of Melbourne was a complicating factor in the choice of reference tracer, Dunse [115] analysed Cape Grim data for pollution events associated with air that had passed over Melbourne (see Figure 17.1). She used CO as the reference gas, with emissions estimated by the state's Environmental Protection Authority. From this, estimates of emission were obtained for the halocarbons CCl_4, CCl_3F, CCl_2F_2, CCl_2FCClF_2 and $CHCl_3$, as well as for CH_4, H_2 and N_2O.

Methane, from Rn ratios. Levin *et al.* [284] used ratios of CH_4 to radon to estimate a mean methane source of 0.24 ± 0.04 g CH_4 km^{-2} s^{-1} for 1996 in the region around Heidelberg in the Rhine valley. The CH_4 : Rn relation was based on the slope of the concentration plots, avoiding the need to define a specific reference level of CH_4 perturbations. However, the trend in differences in concentration of CH_4 between Heidelberg and the Izaña station in the Canary Islands was used to infer a long-term decline in the emissions around Heidelberg. Isotopic data indicated that there had been a change in the mean composition of the methane source.

Greenhouse gases from Europe. Biraud *et al.* [33] presented estimates of emissions of CO_2, CH_4, N_2O and CFCs over western Europe, using data from Mace Head with ^{222}Rn as a reference. They also presented estimates in which the presence of ^{220}Rn (or more specifically its decay product ^{212}Pb with a half-life of 10.6 h) was taken as an indication of local sources so that these cases were excluded from the analyses.

17.3.2 Tracer-ratio estimates of sinks

Hydroxyl. Derwent *et al.* [108] reported a study of a single pollution event during a special measurement campaign at Mace Head. An air-mass that had been advected over the UK and north-east Europe remained stationary for several days, as indicated by stable concentrations of long-lived constituents. From the decay of trichloroethylene ($CHCl=CCl_2$), which has a lifetime of the order of days, a mean concentration of $2.8 \pm 0.3 \times 10^6$ OH molecules cm^{-3} was estimated. Cárdenas *et al.* [63] used measurements of air from north-western Europe, measured on the Norfolk coast of the UK, to estimate daytime OH levels of 3×10^5 cm^{-3}. Carbon monoxide was used as the reference, since its lifetime of months is long relative to the lifetimes of the other constituents (NO_x, O_3 and NMHCs) measured.

17.3.3 Green-function/synthesis estimates of sources

Emissions from Chernobyl. Maryon and Best [308] analysed the problem of estimating emissions from the Chernobyl nuclear accident as a function of time and 'injection height'. This was based on a Green-function approach. Their study was primarily a sensitivity analysis. The actual analysis of data was limited by a lack of data from eastern Europe.

Sulfur. Seibert [430] estimated mean emissions of sulfur from Europe in 1992, using daily mean sulfate-aerosol measurements from 13 sites. The calculations used a Green-function approach on a 25×25 grid ($3°$ longitude $\times$ $1.5°$ latitude), constrained by penalty functions both on magnitude and on variability, but without using any prior information about sources. The correlation between estimated sources and emission inventories was 0.52. The discrepancy was attributed to the effects of the regularisation constraints and to the limitations of the simplified Lagrangian model that was used to determine the responses. The transport was in terms of detailed statistics of trajectories [454] using iterative refinement of solutions obtained by an earlier technique [431] using trajectories in a manner analogous to the back-projection solution in axial tomography (see Box 17.1).

Prahm *et al.* [364] estimated sulfur sources for Europe, expressed in terms of 23 regions of various sizes, with the smallest regions corresponding to the greatest density of data. The sources were estimated through least-squares fitting of model responses obtained by combining trajectory incidence with calculated chemical transformation.

Methane. A study by Stijnen *et al.* [450] used a Kalman-smoother approach to refine estimates of CH_4 sources in north-eastern Europe. Initial 'bottom-up' emission estimates were refined by a 'correction' that was constant in time (i.e. essentially the Kalman-filter implementation of regression analysis) and smooth in space. A reduced-rank approximation was used in order to make the updating of the covariance matrices computationally feasible. A similar study [412] includes a comparison with other techniques for verifying national emission inventories.

Derwent *et al.* [107] estimated the source strength of methane, as well as sources for N_2O and CO, from Europe by fitting the amplitude of pollution events. They used two fitting methods: (i) fitting a climatological model of trajectory statistics; and (ii) a detailed Lagrangian dispersion model. In those years for which both models were used, the more detailed model gave worse agreement with independent inventory estimates.

Jagovkina *et al.* [229] analysed concentrations of CH_4 in western Siberia and concluded that leakage from gas fields gave about 20% of the emissions (in summer) from that region.

Emissions of greenhouse gases from Europe. Ryall *et al.* [417] analysed data from Mace Head, aiming to avoid the assumption of identical source distributions (which is required by the ratio method). They found that a direct Green-function approach (with least-squares fitting) was unstable. However, they obtained better results with an averaging technique that was similar to the trajectory averaging (and thus analogous to back-projection in tomography) that had initially been used by Seibert *et al.* [431]. The comparison with the study by Seibert [430] suggests that a Green-function technique with regularisation could give further improvement.

17.3.4 Other estimates of sources

Methane from boundary integration. Estimates of average emissions of methane along a series of near-parallel wind trajectories were obtained from measurements made on a sampling flight around the north of the UK [32]. A diffusive-box model was used to convert measurements that had been obtained mainly at a fixed altitude into vertical inventories. As noted above, Wratt *et al.* [518] estimated emissions of CH_4 from an agricultural region in the south of the North Island of New Zealand, using upwind–downwind differences in aircraft profiles.

17.4 Carbon dioxide

Yamamoto *et al.* [523] estimated day-time fluxes of CO_2 from the island of Iriomote, Japan, using data from aircraft transects, in an application of the boundary-integration method. However, in most situations, the wide variety of processes contributing to the carbon cycle makes the problem of estimating regional fluxes of carbon too complex for any single technique. This is particularly true when, as is often the case, one wishes to estimate the small residual net fluxes in the presence of large diurnal and seasonal fluxes from natural systems.

Owing to the complexity of the carbon cycle, a wide range of techniques has been proposed, including measurements from aircraft, towers and satellites [464, 416, 267]. The various programmes listed below usually involve several, if not all, of the following components, with the expectation that successful analyses will involve the integration of multiple approaches:

- flux towers;
- aircraft profiles;
- linking biology and hydrology;
- integration with global-scale CO_2-observing networks; and
- the use of satellite data to address the issues of spatial and temporal heterogeneity in the terrestrial biota.

Authors of a number of studies have investigated fluxes of carbon in an ecosystem or on regional scales, either through intensive campaigns or through long-term studies. This is a rapidly evolving area and in some cases the only published information is (at the time of writing) in electronic form.

These programmes include the following.

LBA. The Large-scale Biosphere–Atmosphere Experiment in Amazonia.

BOREAS. The Boreal Ecosystem Atmosphere Study in central Canada. The 1999 overview [189] serves as the introduction to a special issue of *J. Geophys. Res.* (See also http://boreas.gsfc.nasa.gov/BOREAS/Papers.html.)

CarboEurope. CarboEurope is a 'cluster' of projects aimed at quantifying the carbon balance of Europe. These include

- AEROCARB, consisting of surface observations of CO_2 and biweekly aircraft transects;
- CARBOAGE addresses the age-related dynamics of European forests;
- CARBOEUROFLUX is based on a network of flux towers;
- EUROSIBERIAN-CARBONFLUX is a composite activity comprising measurements from flux towers, aircraft measurements and surface observations of CO_2 and related tracers;
- LBA-CARBONSINK covers the European contribution to LBA;
- CARBODATA for data support of continental-scale projects, RECAB for regional assessments of carbon balance and FORCAST for forest carbon–nitrogen trajectories.

Several other continental-scale programmes of measurement and modelling have been proposed (and in some cases established), in order to quantify regional emissions of CO_2.

Tans *et al.* [470] have proposed a sampling programme based on recording vertical profiles at 40–50 sites in North America [470]. Beswick *et al.* [32] report studies on emissions of CH_4 in Finland, using a range of scales; although they experienced

---

Box 17.2. Boundary-layer accumulation

Scales in between those addressed by local flux measurements and the regional scales that are the subject of this chapter can be addressed by techniques that treat the convective boundary layer as an effective chamber [396, 102, 103].

The requisite data are

- concentrations in the boundary-layer as a function of time;
- the concentrations in the air above the boundary-layer;
- the height of the boundary-layer as a function of time; and
- the advective flow through the top of the boundary-layer, most commonly reflecting subsidence.

The method produces a time-integrated flux, representative of fetches of 10–100 km. However, the assumptions limit the techniques to conditions of steady wind over homogeneous terrain. In addition, high precision is required in the concentration measurements. Cleugh and Grimmond [72] have described some of the difficulties of the technique, particularly in heterogeneous landscapes.

As a variation, Denmead *et al.* [103] describe the use of measurements made in the nocturnal boundary-layer. This is much shallower and so a greater accumulation occurs, but it represents much smaller horizontal scales.

---

considerable difficulty, they concluded that balloon-based profiles could lead to reliable flux estimates.

Some results from various flux programmes are reviewed by Buchmann and Schulze [59]. It is hoped to integrate all the long-term flux programmes into a global programme termed FLUXNET [416] and, as noted above, to integrate these data with other information. A recent review of such an integrating framework is given by Canadell *et al.* [61].

Further reading

Many aspects of the regional-inversion problem are simply down-scaling of the global problem and therefore much of the relevant further reading is the same as for global inversions. Areas of more specific application to regional problems include the following.

- Mesoscale meteorological modelling is a specialised field. A key reference is [357].
- There is also an extensive literature on analysis of urban pollution that will, in some cases, overlap with the problems considered in this chapter.

Chapter 18

Constraining atmospheric transport

The wind goeth towards the south, and turneth about unto the north; it whirleth continually, and the wind returneth again, according to his circuits.

Ecclesiastes: 1:6.

18.1 Principles

The use of tracer distributions to determine atmospheric transport plays a relatively small role in studies of atmospheric dynamics. This is in considerable contrast to the situation in the oceans, where the classic ocean inverse problem is the inversion of indirect data, generally temperature and salinity, to determine circulation [521]. Of course, in the ocean, temperature and salinity are not *passive* tracers – through their effect on density, they cause the large-scale circulation. The fact that the foci of inverse problems in oceanography and meteorology are different mainly reflects the atmosphere being better understood than the oceans.

First, there is an extensive atmospheric measurement network, primarily for prediction of weather, with observations at several thousand locations several times a day, which is supplemented by satellite measurements and regular vertical profiles from radiosondes. The closest analogue of the ocean tracer-inversion problem is that of assimilation of meteorological data (see Section 7.3), but the analysis of the atmosphere differs from that of the ocean by virtue of the much greater frequencies both of observations and of analyses. A second difference between the atmosphere and the oceans is that it is easier to model global circulation in the atmosphere than it is for the oceans. The smaller scale of oceanic eddies means that only very recently has it been computationally feasible to model the global ocean with sufficiently fine grids to resolve these eddies. However, in spite of the better knowledge of atmospheric circulation, results of a number of studies such as the TRANSCOM inter-comparison (see Box 9.1) have revealed significant uncertainties in our quantitative knowledge of atmospheric transport. These uncertainties in transport mean that tracer inversions have corresponding uncertainties, which are, at present, poorly quantified.

In many global-scale modelling studies, investigations of tracer distributions have been undertaken as part of the model-validation process. However, this chapter concentrates on the more specific question of using trace-constituent data to *refine* the model representation of atmospheric transport. The difficulty is that there are few tracers for which there is good knowledge both of sources and of concentrations on a global scale.

The first CO_2 inversion by Bolin and Keeling [37] was actually directed towards estimating atmospheric transport. This was parameterised in terms of a single latitudinal diffusion coefficient. The source estimates were an incidental by-product. In order to deduce both transport and fluxes, Bolin and Keeling used symmetry relations. Bolin and Keeling also noted that the details of the source estimates would be unreliable, but some of their caution would seem to be due to their analysis, which overestimated the degree of ill-conditioning in the inverse problem (see Section 8.2). Even earlier, some of the first studies of distributions of ozone aimed to use these to help understand atmospheric circulation [113, 337, 210, 237].

Enting [123] suggested that a comparable inverse problem, requiring deduction both of sources and of transport, occurs in seismology. In such studies, the aim is to estimate transport of seismic waves generated by earthquakes whose amplitude, location and time of occurrence are unknown [350]. This seismological inverse problem is feasible because of the special source characteristics (sources are highly localised in space and time) combined with fixed transport characteristics (making it possible to combine data from multiple earthquakes). Transferring these requirements to the atmosphere would imply the need for considering time-scales on which we have at least statistical uniformity of transport and special information about source distributions.

A method that effectively 'factors out' the role of transport is the 'tracer-ratio' method described in Section 12.3 (with some recent applications described in Section 17.3). These techniques use observations, c_{ref}, of concentrations of a reference tracer with a known source, s_{ref}, to 'calibrate' concentrations, c_η, of tracer η whose source, s_η, is initially unknown but is estimated from an averaging relation $\hat{s}_\eta = s_{\mathrm{ref}} \langle c_\eta / c_{\mathrm{ref}} \rangle$. This technique is generally applied on small spatial scales, often with the concentrations c_η and c_{ref} being perturbations from a background level. The contrast between the Green-function methods (which assume that details of transport are known as accurately as required) and tracer-ratio methods (which assume no knowledge of transport) suggests that, in multi-tracer problems, the optimal form of estimation is likely to combine elements of both techniques. However, development of such an approach remains a task for the future.

There are a few general principles that shape the ways in which tracer distributions can be used to refine the model representation of atmospheric transport.

Estimating transport fields. The more complete the model the harder it is to estimate what is needed. At the simplest extreme, the study by Bolin and Keeling [37] represented atmospheric transport using a single diffusion parameter for (vertically averaged) latitudinal transport. This made it possible to use CO_2 data to estimate this number. In contrast, the transport fields in a full three-dimensional model involve thousands of numbers at each time-step. The study by Plumb and Mahlman [360] described below shows the requirements on data for estimating transport fields for a two-dimensional model.

Estimating corrections. A less demanding approach to using tracer data to estimate transport fields in three-dimensional models is the estimation of 'correction' parameters to allow for presumed model error. Often the correction takes the form of a diffusion coefficient, whereby the diffusion represents the effect of transport operating on scales smaller than the model grid.

Estimating indices. Another variation is that, rather than estimating parameters in transport equations, tracer data are used to estimate integrated (transport-related) properties of the model. A common form of transport 'inversion' is the estimation of a single characteristic 'index' of transport, such as the interhemispheric exchange time. As illustrated below, the analysis by Plumb and MacConalogue [359; see also Section 10.2 and Figure 18.1] shows that such indices will depend on the source distribution.

This book has presented two alternative ways of representing the transport of trace constituents in the atmosphere for the purpose of tracer inversions: the differential form and the integral form. This dichotomy extends to studies that aim to use tracer distributions to refine our knowledge of atmospheric transport.

Some preliminary studies of the problem of deducing wind-fields from concentrations have been presented by Daley [95]. He considered the one-dimensional advection equation, generally with time-varying winds with convergence and divergence. A series of 'perfect-model' experiments, as well as 'perfect-model-plus-noise' cases, tested the ability of the extended Kalman filter to deduce the advection from concentration data. The requirements for successful estimation were identified as (i) sufficient structure in the concentration fields and (ii) data closely spaced in time and position with few gaps, especially in regions of convergence.

In order to appreciate what would be involved in estimating a more complete set of transport coefficients from observational data, we consider the study by Plumb and Mahlman [360]. They derived transport coefficients for a two-dimensional model from modelled distributions of hypothetical tracers in a three-dimensional model. This calculation involved two simplifications. First, the desired coefficients were for two-dimensional, not three-dimensional, transport. Secondly, the three-dimensional model could produce more detail about the hypothetical tracers than could plausibly be observed for real tracers.

The two-dimensional representation used a flux form,

$$\frac{\partial c}{\partial t} = -\frac{\partial \Phi_x}{\partial x} - \frac{\partial \Phi_z}{\partial z} \tag{18.1.1}$$

with

$$\begin{bmatrix} \Phi_x \\ \Phi_z \end{bmatrix} = \begin{bmatrix} \kappa_{xx} & \kappa_{xz} - \chi \\ \kappa_{zx} + \chi & \kappa_{zz} \end{bmatrix} \begin{bmatrix} \partial c/\partial x \\ \partial c/\partial z \end{bmatrix} \tag{18.1.2}$$

where the elements κ_{xz} etc. define a symmetric diffusion tensor and χ is a stream function defining the advective component. This study is of theoretical interest since the transport fields are derived from tracer distributions and so reflect Lagrangian averaging. In particular, the stream functions, χ, define a single cell in each hemisphere, circulating about four times the hemispheric air-mass per year equatorwards in the lower troposphere with a return poleward flux at higher altitudes.

Plumb and Mahlman ran their three-dimensional model with two idealised tracers, one with only latitudinal variation and one with only variation in the vertical. Over each month, model statistics for the concentration gradients and the calculated fluxes were obtained and the averages used in (18.1.2) to determine monthly averaged two-dimensional transport coefficients. Even with the longitudinal averaging and the coarse time resolution, further smoothing was required in order to produce transport fields that could be used in numerical modelling. The small diffusive contributions in the stratosphere reflected the validity of effective circulations as specified by the residual circulation (see Section 2.2).

This suggests that the approach of estimating a complete specification of transport from tracer data is impractical, although the approach of estimating 'corrections' with a simple structure is sometimes feasible. The alternative to estimating the parameters in the transport operator, $\mathcal{L}$, is to estimate parameters that describe the Green function, $G(\mathbf{r}, t, \mathbf{r}', t')$, i.e. the third approach listed above. This has the advantage that the quantities that are estimated are close to what is required for analysing trace-gas distributions. The following sections review some such applications, treating the stratosphere and troposphere separately because of the differences in their structure and circulation.

18.2 Stratospheric transport

The circulation pattern in the stratosphere means that inputs of trace gases from the troposphere are highly localised, occurring mainly at the tropical tropopause. This leads to a simplified representation for stratospheric tracers with tropospheric sources. A Green-function representation would express concentrations as

$$c(r, t) = \int G(r, t, r_0, t') c(r_0, t') \, dt' \tag{18.2.1}$$

where r_0 represents the tropical tropopause region and the Green function is calculated with boundary conditions of specified concentration.

On time-scales long enough to allow one to ignore seasonal variations in circulation, the Green function $G(r, t, r_0, t')$ will depend on time only through $t-t'$ and so (18.2.1) can be written using a position-dependent response, $\rho_r(t)$, as

$$c(r, t) = \int \rho_r(t - t')c(r_0, t')\, dt' \qquad (18.2.2)$$

This approximation was discussed by Hall and Plumb [190] and, with three-dimensional model calculations, by Hall and Waugh [191]. They termed $\rho_r(.)$ the 'age spectrum' for point r and used it to clarify relations among various types of observations.

For conserved tracers with a linear growth, $c(r_0, t) = c_0 + \beta t$, (18.2.2) becomes

$$c(r, t) = c_0 + \beta(t - \tau_r) \qquad (18.2.3a)$$

where

$$\tau_r = \int t\rho_r(t)\, dt \qquad (18.2.3b)$$

In other words the 'time delay' for a location r is given by the mean of the age spectrum for location r. The point emphasised by Hall and Waugh is that this simple relation does not hold when $c(r_0, t)$ is not changing linearly. For cyclic variations in $c(r_0, t)$, the phase delay is generally closer to the mode of the age spectrum than to its mean. These concepts were used by Boering et al. [36] in relating CO_2 and N_2O data to stratospheric transport.

A later study by Hall and Waugh [192] extended the Green-function concept to cases with chemical destruction. The Green function for a non-conserved tracer was written as the product of the age spectrum, as defined for conserved tracers, and a 'loss spectrum' whose value increased with time. They found that, even for compounds with very long lifetimes (e.g. SF_6 with a lifetime of over 100 years) there was a significant bias in interpreting differences $[c(r, t) - c(r_0, t)]/\beta$ as the mean of the age spectrum. For sufficiently long lifetimes, the relations established by Plumb and Mac-Conalogue expressed the dependence in the form of $\tau = [c(r, t) - c(r_0, t)]/(\beta + \lambda)$ (see Section 10.2). Stratospheric data presented by Fabian et al. [148] generally reflect this pattern of dependence both on lifetime and on rate of growth, although the data are too scattered to provide convincing evidence of a universal 'transport-time' distribution.

As well as the direct use of tracers to estimate ages of stratospheric air, such data can be used indirectly so that the slopes of correlation plots give, to first order, the ratios of stratospheric lifetimes [358]. Another form of indirect use is to use one tracer to 'date' the air and then use this to estimate $\hat{c}_\eta(r_0, t) = c_\eta(r, t - \tau_r)$ for a second tracer (such as CF_4 or C_2F_6 [195]) whose tropospheric history is unknown. The analysis by Hall and Waugh shows that this sort of 'dating' requires both tracers to have a near-linear increase in concentration.

Early tracer studies with H_2O and helium [53] and ozone [113] suggested that the overall circulation of the stratosphere was dominated by global-scale circulatory cells. More direct observations tended to confuse the issue until the importance of the distinction between Eulerian and Lagrangian means was appreciated. Multi-tracer studies based on satellite data have allowed the determination of the diabatic circulation from tracer observations [443]. This describes an effective circulation transporting heat with specified heat sources and sinks from radiative balance. Later applications of this approach are reviewed by Holton *et al.* [213].

The dynamics of the upper troposphere and the mechanisms of cross-tropopause transport have been reviewed by Mahlman [296]. Volk *et al.* [493] describe the use of a range of tracers to quantify both the injection of air from the tropical tropopause into the stratosphere and the entrainment of extra-tropical air. (A schematic model of this is the 'leaky-pipe' model of Neu and Plumb [336]). Rind *et al.* [407] describe the use of SF_6 as a tracer for transport from the troposphere to the stratosphere.

Another way of using tracers to study stratospheric transport is through compounds deliberately released for this purpose. This approach is common in studies of small-scale pollution and is used occasionally in regional studies such as the ETEX experiment [146]. However, its application on large scales is possible only under very special conditions such as the use of $^{12}CD_4$ and $^{13}CD_4$ to study transport in the Antarctic polar vortex [328]. It required a confined circulation, a tracer with near-zero background levels and very sensitive detectors.

18.3   Tropospheric transport

In discussing the use of tracer distributions to deduce aspects of the transport, we consider the two main approaches identified above: estimating corrections and estimating integral characteristics. We also consider some specific aspects of tropospheric transport: the so-called rectifier effect, ENSO-related changes and transport from the stratosphere into the troposphere.

18.3.1   Model tuning

As noted above, one of the simplest forms of estimating transport parameters is the estimation of single parameters as 'corrections' to an existing set of model transport parameters. There have been many studies of this type. Some examples are the following.

∎ Prather *et al.* [368] included an additional diffusion in the three-dimensional GISS model. This was scaled in proportion to the amount of deep convection in the tropics and was regarded as representing the sub-grid-scale horizontal mixing associated with convective overturning. The diffusion was expressed in terms of a single horizontal length-scale, of order 200 km, for the processes. The length-scale was tuned so that the model gave the observed

interhemispheric gradient in concentration of CCl_3F. The estimated
length-scale corresponded to an average diffusion of about 10^6 m^2 s^{-1}. The
interhemispheric transport of the tuned model was then checked against
observations of ^{85}Kr.

■ Levin and Hesshaimer [282] used a two-dimensional model with transport
obtained by zonal averaging of fields from a three-dimensional model.
Additional diffusion in the horizontal and vertical directions was estimated as a
scaling factor multiplying the variances of the original horizontal and vertical
winds. The two scale factors were estimated by fitting the distribution
of ^{85}Kr.

■ Several studies have attempted to use tracer data, especially as vertical profiles,
to refine the modelling of vertical transport due to small-scale convective
overturning. This has included the use of the annual cycle in CO_2 [465] and
studies using radon [174, 407]. Radon profiles were also a key test in the
inter-comparison described by Jacob et al. [227]. (This study also tested models
for synoptic-scale variation such as that seen at Amsterdam Island [363]).

18.3.2 Time-scales for tropospheric transport

As an alternative to estimating transport parameters, we can estimate integrated rates of
transport, generally expressed in the form of characteristic time-scales. It is only at the
very broadest level that we can compare various estimates. Plumb and MacConalogue
[359] give examples pointing out the ambiguities that arise because such time-scales
depend on the source distribution (see also Section 10.2). Table 18.1 lists a number of
estimates of τ_{NS}, the time-scale for interhemispheric transport. The normalisation is
such that an interhemispheric gradient will decay as $\exp(-2t/\tau_{NS})$.

Junge [234] used the north–south ratio of the seasonal cycle of CO_2 to estimate
an exchange time $\tau_{NS} > 400$ days. Newell et al. [338] estimated 320 days, mainly on
the basis of meteorological data. The main mechanism for interhemispheric transport
was seasonal movement of the Hadley cell (the process represented in toy model C.1,
Box 1.3) with an additional contribution from eddy diffusion, mainly in the upper
troposphere.

Table 18.1. *Estimated time-scales of interhemispheric transport,*
defined so that perturbations decay as $exp(-2t/\tau_{NS})$.

Reference	τ_{NS} (yr)	Data/model
[37]	0.5 ± 0.1	CO_2, one-dimensional diffusion
[234]	>1.1	CO_2 cycle, two-box
[338]	0.88	Observed winds
[204]	<1.4	CO_2 cycle, one-dimensional diffusion
[299]	1.7 ± 0.2	SF_6-concentration gradient

Heimann and Keeling [204] analysed the propagation of the seasonal cycle of CO_2, across the latitudes 14.5° N to 14.5° S, in terms of a one-dimensional diffusion model. The estimated diffusion coefficient was converted to give an upper limit of 1.4 years on the interhemispheric exchange time. Even though only surface data were used, vertical representativeness can be justified due to the small seasonality of surface sources. However, the use of this one-dimensional model to deduce mean fluxes from mean horizontal gradients [245] is less tenable because of the vertical gradients.

Maiss *et al.* [299] used observations of interhemispheric gradients in concentration of SF_6 to estimate an exchange time of 1.7 ± 0.2 years. However, they noted that alternative ways of defining the hemispheric means could lead to values as low as 1.3 or as high as 2 years. SF_6 was used as the reference tracer in the TRANSCOM-2 inter-comparison [106].

To illustrate this we extend the use of the formalism based on the work of Plumb and MacConalogue [359] (see Section 10.2). Following Figure 16.1 we plot transport times defined by

$$\tau(r) = \frac{c(r_0) - c(r)}{\beta' + \lambda c(r_0)} \tag{18.3.1}$$

where again β' is the rate of growth ($\beta c(r_0)$ in the notation of Section 10.2) and λ is the inverse lifetime. Figure 18.1 shows, as x, CFCs 11 and 12 (as in Figure 16.1), where the profile reflects a common pattern of release. In contrast, CH_4 and N_2O, shown as +, have quite different profiles, reflecting patterns of release that differ from the

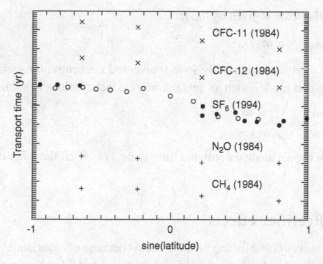

Figure 18.1 Transport times, relative to an unknown source, calculated according to (18.3.1). Major tick marks on time axis represent 1 year – the displacement of cases is for visual convenience. Absolute time origins are not determined, but the minima in each case are expected to be $\ll 1$ year. x denotes CFCs 11 and 12. + denotes CH_4 and N_2O. For SF_6 filled circles denoted fixed sites (marine boundary-layer only) and open circles are extrapolations for Atlantic cruises. SF_6 data are from [299] as used in TRANSCOM-2 [106]. Other data are from [382].

'industrial' pattern of release of the CFCs. More surprisingly, SF_6, shown as circles, exhibits a different profile from that of the CFCs, with southern-hemisphere transport times of 2 years (or more precisely 2 years greater than the transport time for Mace Head) rather than 1.5.

The ambiguities noted by Plumb and MacConalogue [359] and illustrated in Figure 18.1 raise the question of whether the time-scales that are estimated are meaningful other than as orders of magnitude. A possible answer is in terms of the 'transport modes' noted by Prather [367] in his study of chemical modes.

For the transport equation

$$\frac{\partial}{\partial t} m(r, t) = T[\mathbf{m}] \tag{18.3.2a}$$

we seek solutions of the form

$$T[\mathbf{m}_q] = -\lambda_q \mathbf{m}_q \tag{18.3.2b}$$

or more precisely, with the property

$$\langle T[\mathbf{m}_q] \rangle = -\lambda_q \langle \mathbf{m}_q \rangle \tag{18.3.2c}$$

where $\langle . \rangle$ denotes an annual average.

Here λ_q is an eigenvalue, not a chemical loss rate. Equation (18.3.2a) shows that $(\lambda_q)^{-1}$ is the time-scale of decay of this mode of atmospheric variation.

Steady-state distributions will be obtained from source distributions specified by

$$s_q(\mathbf{r}, t) = T[\mathbf{m}_q] \tag{18.3.3a}$$

and the time-scale is given by the ratio

$$(\lambda_q)^{-1} = \mathbf{m}_q / s_q \tag{18.3.3b}$$

If we can expand actual steady-state source and concentration vectors as sums over these transport modes, then an actual ratio of concentrations to sources is

$$\frac{\mathbf{m}}{\mathbf{s}} = \frac{\sum_q \alpha_q \mathbf{m}_q}{\sum_q \alpha_q \lambda_q s_q} \tag{18.3.4}$$

which can approximate a dominant time-scale $1/\lambda_1$ if all the other $\alpha_q \lambda_q \ll \alpha_1 \lambda_1$ for $q > 1$.

18.3.3 Rectifier effects

The possibility of establishing mean gradients because of covariance between transport and concentrations has been appreciated for some considerable time. The issue came to prominence in studies of CO_2 through the calculations of the TRANSCOM inter-comparison [398, 277] and the more detailed analysis by Denning et al. [105]. For CO_2 inversions, as currently practised, the important issue is not the extent to which the surface mean gradient is due to covariance effects. What matters is the extent to which these covariances extend to the sampling sites, most of which are in the marine boundary-layer.

The TRANSCOM study found that models differed considerably in the mean inter-hemispheric gradient arising from a specified annual cycle of CO_2 fluxes. In most cases, the large gradients were calculated by those models that had the more detailed models of the boundary-layer. Denning *et al.* [105] considered the issue in detail with calculations of mean gradients in concentration of CO_2 established both by seasonal and by diurnal covariance effects.

Stephens *et al.* [449] describe a proposed measurement programme for helping resolve the issues of the rectifier effect. The aim is to distinguish terrestrial, oceanic and fossil fluxes of CO_2, using CO as a marker for fossil CO_2 and oxygen ratios as a constraint (corrected for the fossil contribution) to distinguish terrestrial from oceanic contributions. The proposal combines aircraft and tower measurements with boundary-layer budgeting.

18.3.4   ENSO-related changes

The large-scale circulation changes associated with the phenomenon of El Niño/Southern Oscillation will clearly affect some aspects of tracer transport. The effect is particularly apparent in tracer concentrations at Samoa. ENSO variations determine the extent to which air-masses at Samoa are typical of northern-hemisphere versus southern-hemisphere air. The EOF analysis presented in Section 5.4 identified an anomalous autocorrelation in the CO_2 record from Samoa once coherent global-scale variations had been subtracted from the record.

Prinn *et al.* [379] report an ENSO modulation of CH_3CCl_3 data from Samoa, which they interpreted as a decrease in cross-equatorial transport during warm phases, leading to concentrations of CH_3CCl_3 at Samoa and Cape Grim at these times being very similar.

18.3.5   Input of stratospheric air into the troposphere

As well as transport within the troposphere, we need to consider the input of strato-spheric air. This is discussed in this section because most of the relevant observations are made in the troposphere. As with transport within the troposphere, a first-order characterisation of exchange between the stratosphere and troposphere can be given in terms of characteristic time-scales. The most commonly used time-scale is the strato-spheric turnover time, defined as the ratio of the mass exchange rate to stratospheric mass.

As with horizontal exchange, it is difficult to find an unambiguous relation between observations and characteristic time-scales. The appropriate 'transport modes' would seem the most plausible basis for making quantitative analyses.

Several tracers have proved useful for estimating stratosphere–troposphere exchange.

Ozone. Distributions of ozone have had a long history of application to interpreting atmospheric transport, particularly in the stratosphere, as described in the previous section. Junge [235] found that tropospheric ozone data implied a stratospheric residence time in the range 1.1–5.2 years.

CFCs. Holton *et al.* [213] review a number of applications of concentration data, including CCl_3F, in the estimation of stratosphere–troposphere exchange.

^{14}CO. More recently, ^{14}CO has been used as an indication of stratospheric air entering the troposphere [407]. Although the excess ^{14}C from nuclear testing is close to equilibrium throughout the atmosphere, the natural production of ^{14}C, primarily in the stratosphere, creates a difference between stratospheric and tropospheric air. The ^{14}C atoms that are produced are mainly oxidised to ^{14}CO in the stratosphere. ^{14}CO has also been used to estimate tropospheric concentrations of OH, e.g. [389].

Other natural radionuclides. Naturally produced radionuclides such as ^7Be and ^{32}P have also proved useful. Junge [235] used ^7Be data to suggest a stratospheric residence time of 3.2 years, thereby refining his analysis of the tropospheric ozone budget. However Benitez-Nelson and Buessler [29] found that significant fractionation between species limited the applicability of using deposition ratios to infer rates of transfer from the stratosphere. Sanak *et al.* [419] suggest that the simpler meteorology over the Antarctic may allow realistic estimates of the mass flux from the stratosphere.

Anthropogenic radionuclides. The major periods of nuclear testing produced large amounts of radionuclides, especially in the stratosphere. Early work is reviewed by Reiter [405]. This area of research has declined over recent decades as the signal has decayed and dissipated.

In conclusion, it would appear that the complexity of atmospheric transport means that available trace-gas-concentration data fall dramatically short of what is required to estimate atmospheric transport. Tracer data can be used for validating transport models and perhaps for tuning parameterisations. The scope for going beyond these limitations lies largely in the realm of assimilation of satellite data.

Further reading

The aspects of estimating transport covered in this chapter have involved the disparate topics of (i) indices of transport, (ii) tuning of transport models and (iii) assimilation of chemical data.

■ The work on transport indices is primarily of historical interest. The transport modes noted by Prather may be a basis for a revival of this type of approach, but this is as yet undeveloped.

- Tuning of transport models is more of an art than a science and, without a more formal characterisation of model error, is likely to remain so.

- Assimilation of satellite data into chemical transport models is a rapidly evolving area. The report from the SODA [439] workshop is a recent example, as is Ménard's review in the Crete volume [316].

- To the extent that the ocean analogy is relevant in estimating atmospheric transport, Wunsch's book [521] is a key reference.

Chapter 19

Conclusions

The end of our Foundation is the knowledge of causes, and the secret motions of things, and the enlarging of the bounds of human empire, to the effecting of all things possible.

Sir Francis Bacon: *New Atlantis* (1627).

Inverse modelling of the atmospheric transport of trace constituents is a rapidly growing field of science. In a little over a decade it has developed from two-dimensional inversions of CO_2 sources and sinks to encompass the range of activity described in the preceding chapters. Among the important developments has been the recognition of the ill-conditioned nature of most tracer-inversion problems, particularly in the case of estimating surface fluxes from surface concentrations. This recognition of the ill-conditioning, with associated amplification of errors, emphasises the need for a statistical approach that treats the deduction of trace-gas fluxes and other biogeochemical quantities as problems in statistical estimation.

The main applications of tracer inversion have been for the long-lived constituents, particularly CO_2 and CH_4. As a consequence, most development of inversion techniques has been of methods that are suitable for such cases and that are oriented to the use of monthly mean data from flask-sampling networks.

The two main types of inversion have been identified as 'mass-balance' and 'synthesis'. The choice between these involves a number of trade-offs such as ease of inversion versus ease of analysis of uncertainty. However, the most important difference is the choice between *either* assumptions about the degree of spatial variation in the concentration fields (in the case of mass-balance inversions) *or* assumptions about spatial variation in the source fields (in synthesis inversions). The hybrid (or time-stepping-synthesis) methods described in Sections 11.2 and 11.3 may, to some extent, provide a compromise between the advantages and limitations of mass-balance and synthesis. However, in terms of the type of interpolation, the hybrid techniques are firmly aligned to the synthesis approach: the interpolation takes place in the description of the sources rather than the description of the concentrations. The practice of mass-balance and synthesis inversions has also involved a split between the approaches to analysis of uncertainty. The synthesis methods (including time-stepping synthesis) allow a formal propagation of uncertainty from observations to estimates and this is the approach that has generally been adopted. In contrast, such direct uncertainty analysis in mass-balance techniques is computationally much less feasible. In mass-balance inversions, stochastic techniques, mainly the bootstrap method, have been used. This has required greater computational demands, but less reliance on a specific statistical model, than the direct error propagation used in synthesis techniques.

Even though a large majority of global-scale inversion studies analysed CO_2, there is still considerable uncertainty concerning the global carbon budget. A major split is between calculations that imply a low uptake by the ocean (<1 Gt C yr^{-1}) and those that imply a high uptake by the ocean (≈ 2 Gt C yr^{-1}). This split runs through the inversion calculations as well as dividing other techniques for estimating global carbon budgets. In many cases, the global budget calculated by an inversion is determined mainly by the same information as that which is used in globally aggregated studies. For example, our own CO_2 inversions [143, 145] using ^{13}C have the global partitioning determined by the same information as that in the global CO_2–^{13}C budget analysis by Tans et al. [469].

For other globally distributed tracers, the generally greater complexity of the processes and the limited amount of auxiliary data result in wide ranges of uncertainty rather than apparent discrepancies.

The prospect of performing tracer inversions on a continental scale, resolving details within such regions, is highly attractive. If such inversions could be shown to be widely applicable, they could play a very important role in verifying the extent to which emission targets under the Kyoto Protocol were being met. However, initial studies, reviewed in Chapter 17, suggest that such regional inversions are going to prove difficult. The scope for obtaining a useful estimate of regional emissions from atmospheric data alone would seem to be limited to very special cases. The real potential for exploiting the integrating role of the atmosphere will need to come from analyses that combine several types of data.

Inversions to deduce atmospheric transport from tracer distributions suffer from the severely underdetermined nature of the problem. At present there is no systematic approach for using constraints from tracer transport to improve the realism of transport in general circulation models. Chapter 18 presents some preliminary ideas.

In presenting atmospheric tracer inversions, this book has emphasised the need for a statistical approach and has treated inversion problems as problems of statistical estimation. More particularly, I have adopted a Bayesian approach. I have given my reasons for adopting this approach in Chapters 1 and 3, so I will not repeat them in full. However the key points are the following.

- The quantification of uncertainty needs to be expressed in statistical terms.
- Any statistical analysis will be based (either explicitly or implicitly) on some statistical model. It is important to make the underlying statistical model explicit so that its assumptions can be checked.
- A Bayesian statistical approach is essential because we are never working in a context that is independent of prior knowledge and experience.
- Many important types of analyses (such as the SVD) are undefined (in the sense that they depend on arbitrary normalisations) except in a Bayesian context.

The statistical analysis has been based on a distinction between a *system*, which we may model in either deterministic or stochastic terms, and *our knowledge about the system*, which, being incomplete, always needs to be expressed in statistical terms.

A concise summary of what needs to be done in order to improve atmospheric tracer inversions is 'both to characterise and to reduce model error'. However, as a summary, it is true only because of the multiple meanings of the word 'model'. Improvements are needed both in the transport models and in the statistical models that are used to define the context in which the transport models are used. The TRANSCOM inter-comparison study emphasised the differences among members of a set of transport models, which were mainly used for tropospheric studies. The inter-comparison showed that many, if not all, of them are in need of refinement. For transport models, the characterisation and reduction of model error seems likely to be a slow process. To some extent progress will come from the improvements in resolution arising from increases in available computing power and the rate of such improvement will be one of the constraints on the rate of development of transport models. In contrast, the statistical characterisation of observational data is an area with considerable potential for further refinement with relatively small computational requirements. Improvements in the statistical characterisation of the observations have the potential to help identify anomalies due to unrecognised model error.

The statistical approach, in particular the use of Bayesian statistical analysis, is by no means universal among those working in the field. In describing other studies I have used the Bayesian framework as a common basis for description without, I hope, distorting the presentation through this reformulation. However, I see the complete Bayesian analysis as an ideal for which we should strive, but also as a goal that we will probably never fully achieve. In particular, the full incorporation of 'model error' into the Bayesian formalism (or any other operational formalism) remains a distant objective.

Chapter 12 presents some idealised examples that may form the basis of more general non-linear inversion techniques. This is very much an exploratory chapter – many of the ideas discussed there remain to be tested in practice.

Looking forwards, it is possible to make some guesses about which areas will define the cutting edge of the science of atmospheric tracer inversions. Some 'growth areas' that I would identify are the following.

Regional inversions. As noted above, regional inversions seem most likely to be useful as part of a synthesis combining atmospheric-concentration data with additional information. In particular, much of the impetus for process inversions may come from regional problems.

Process inversions. Inversions aimed at characterising surface processes more directly than through flux estimates seem likely to become more important because they address the problems of most direct scientific interest. It is this type of method that would allow inverse calculations to become key diagnostic approaches in the development of earth-system models.

Inversions of chemistry. The use of tracer inversions to constrain chemical processes in the atmosphere is of great importance but also of formidable difficulty. The highly non-linear nature of atmospheric chemistry greatly complicates the analyses that seek to determine which processes (or combinations of processes) can be constrained (or, in the terminology of Section 8.3, 'resolved') by measurements of concentrations of trace gases. The 'chemical-data-assimilation' studies described in Section 7.4 represent only a first step.

Earth-system science. The last two cases may appear as part of a more general framework of 'earth-system science' that seeks to understand the interactions among the atmosphere, oceans, lithosphere and cryosphere in terms of physical, chemical and biological processes. In the development of 'earth-system models' designed to quantify our understanding of the system, it is likely that diagnostic modelling in general (and inverse modelling in particular) will play an essential role in validation of models.

Regardless of the accuracy of these predictions, the present pace of activity in the field of atmospheric tracer inversions shows no sign of slowing down – these studies will continue to be important. This implies that it has been worthwhile to write this book, even with the risk that at least the sections on applications will be superseded by new developments. I look forward to joining with my readers in this endeavour.

Appendix A

Notation

Where feasible, the notation listed below follows the proposal for a unified notation in data assimilation [222]. **Boldface** is used for matrices (upper case) and vectors (lower case), sans serif is used for matrix transpose and subscripts N and S for hemispheres. $\mathcal{C}$aligraphic letters denote operators (and vector spaces: Box 10.3 only). This list excludes generic mathematical variables (e.g. matrices $\mathbf{A}$, $\mathbf{B}$, $\mathbf{C}$, $\mathbf{Y}$ and $\mathbf{Z}$) with no fixed physical meaning, which are used in problems and some of the breakout boxes.

$\mathbf{A}^{-1}$	Inverse of matrix $\mathbf{A}$.
$\mathbf{A}^{\mathsf{T}}$	Transpose of matrix $\mathbf{A}$.
$\mathbf{A}^{\#}$	Pseudo-inverse of matrix $\mathbf{A}$.
$\mathcal{A}^{\dagger}$	Adjoint of operator $\mathcal{A}$.
a_j	General coefficient in expansion, such as regression analysis, time-series model etc.
a_η	Radiative absorption strength of species η.
A	Subscript denoting atmosphere.
A_e	Radius of the earth.
$\mathbf{A}$	Expanded 'model' matrix in Bayesian estimation.
B	Subscript denoting terrestrial biota.
$\mathbf{B}$	Matrix defining autoregressive model of source components.
$\mathcal{B}$	Backwards shift operator for time series.
c_j	Generic item of observational data. Specific cases: $c_1, c_2, \ldots, c_N$.
c_η	Concentration of tracer η. Specific cases c_{CH_4} etc.
$\mathbf{c}$	Vector of observations, elements c_j.
$\mathbf{c}'(n)$	Sub-block of $\mathbf{c}$, consisting of observations at nth time point.
$\mathbf{C}$	Covariance matrix, various contexts.

d	Combined observation/source error vector in Bayesian estimation.
D	Interpolation/smoothing operator, used to define surface concentration fields for input to mass-balance inversions.
$\mathcal{D}$	Operator defining boundary conditions.
e	Base of natural logarithms.
e(n)	Error vector for observations at time-point n in state-space model.
$E[.]$	Statistical expectation, i.e. mean.
$\hat{f}$	Estimate of f. Subscripted forms such as $\hat{f}_{\text{OLS}}$ identify the type of estimate.
$\bar{f}$	Average (generally a time-average) of f.
$\dot{f}$	Derivative of f, especially with respect to time.
f_X^+, f_X^-	Isotopic fractionation factors for fluxes Φ_X^+ and Φ_X^-.
$f(.)$	Arbitrary function of a continuous variable, especially of time.
$f'(.)$	Arbitrary function of a discrete variable, especially of time in units of Δt.
$f_{\text{reg}}(\lambda)$	Function specifying regularisation in terms of SVD.
f	Vector of differences between sources and prior estimates.
F	Subscript denoting fossil carbon.
F	Evolution operator in state-space model. **F**(n) in cases in which **F** is not necessarily constant.
g(n)	State-vector in state-space model.
$\hat{\mathbf{g}}(n)$	One-sided estimate of **g**(n), i.e. using observations **c**(n') for $n' \leq n$.
$\tilde{\mathbf{g}}(n)$	One-step prediction of **g**(n), i.e. using observations **c**(n') for $n' \leq n - 1$.
$G(.,.)$	Green's function. The kernel of the integral operator, $\mathcal{G}$, defining the solution of inhomogeneous ODEs or PDEs.
$\mathcal{G}$	The Green-function operator that defines the 'solution' of a set of inhomogeneous ODEs or PDEs. The main applications in this book involve mapping source distributions onto concentration distributions.
G	Generic Green-function matrix, a discretisation of $G(.,.)$, with elements $G_{j\mu}$.
$\tilde{\mathbf{G}}$	Green-function matrix, **G**, defined in terms of a basis in which **X** and **W** are identity matrices.
$h'(\theta)$	Power spectrum of time series in terms of dimensionless frequency.
$h(\omega)$	Power-spectral density for a process continuous in time.
$h\nu$	Denotes quantum of radiative energy (Planck's constant, h, times frequency, ν) in photo-chemical reactions.
H	Approximate inverse of **G** (Chapters 3, 8 and 10).
H(n)	Matrix specifying projection from modelled state, **g**(n), onto observation set, **z**(n), in state-space modelling.
I	Identity matrix.
$I(\theta)$	Periodogram.
i	$\sqrt{-1}$.
i	Generic integer index.

j	Generic index of observational data, or other integer index.
$J(.)$	Objective function. (Also called cost function, penalty function etc.)
k	General inverse length-scale, especially with descriptive subscript. General integer index, particularly for coefficients of digital filters.
$\mathbf{k}$	Wave-vector.
$\mathbf{K}_{\mathrm{diff}}$	Diffusion tensor (Chapter 2 only).
$\mathbf{K}(n)$	In general, the gain matrix used to define 'correction' from predicted state to estimated state, especially the optimal gain matrix.
ℓ_1	Vector norm defined by mean absolute size of components (see Box 12.2).
ℓ_2	Vector norm defined by root-mean-square size of components.
ℓ_∞	Vector norm defined by maximum size of components.
L	Boundary length (with differential element $\mathrm{d}L$)
$\mathbf{L}$	Factorisation of Green's function for transport model, introduced by Mulquiney *et al.* [330, 332], see Section 11.3.
$\mathcal{L}$	Generic differential operator.
$m(\mathbf{r}, t)$	Modelled value of trace-gas concentration.
$\mathbf{m}$	Vector of modelled concentrations (or other modelled variables), a subset, or other linear combination, of $m(\mathbf{r}, t)$.
M	Number of parameters in an estimation problem.
M_{atmos}	Mass of the atmosphere (kg).
M_X, $M_{X:\mathrm{C}}$	Mass of carbon in reservoir X.
M_η, $M_{\mathrm{A}:\eta}$	Amount of species η in atmosphere.
M_X^{13}	Mass of ^{13}C in reservoir X.
n	Time index in units of Δt.
N	Number of items of data or length of time series.
N_{atmos}	Size of the atmosphere (moles).
N	(as subscript) North, northern hemisphere.
NS	(as subscript) North–south difference.
O	Subscript denoting ocean.
p	Pressure coordinate.
p	Index of singular values (or eigenvalues), in decreasing order.
$p(.)$	Probability density function.
$p(\mathbf{x})$	Unless otherwise indicated, this is an abbreviation for $p_{\mathrm{prior}}(\mathbf{x})$.
$P(.)$	Cumulative probability distribution.
P_n	Legendre polynomial.
P_n^m	Associate Legendre function, used in spherical harmonics $P_n^m(\cos\theta)\mathrm{e}^{\mathrm{i}m\phi}$.
$p_{\mathrm{prior}}(\mathbf{x})$	Prior probability density function for model parameters, i.e. before incorporating information from observations.
$p_{\mathrm{inferred}}(\mathbf{x}|\mathbf{c})$	Inferred probability density function for model parameters, after incorporating information from observations.
$\hat{\mathbf{P}}(n)$	Covariance matrix for state estimate, $\hat{\mathbf{g}}(n)$, in Kalman-filter formalism.

$\tilde{\mathbf{P}}(n)$	Covariance matrix for one-step prediction, $\tilde{\mathbf{g}}(n)$, in Kalman-filter formalism.
$\mathbf{P}_{\mathrm{I}}(n)$	Covariance matrix for innovations, $\mathbf{z}(n + 1) - \mathbf{H}(n + 1)\mathbf{F}(n)\hat{\mathbf{g}}(n)$, in Kalman-filter formalism.
q	Index of singular values (or eigenvalues), in decreasing order.
$\mathbf{q}$	Deviation from prior prediction of observations, i.e. $\mathbf{c} - \mathbf{m}(\mathbf{z})$ or $\mathbf{c} - \mathbf{Gz}$ for linear models.
$\mathbf{Q}$	Covariance matrix for $\mathbf{w}(n)$ in state-space modelling. $\mathbf{Q}(n)$ in cases in which it depends on time.
Q	Variance of random forcing in ARIMA models (a scalar analogue of $\mathbf{Q}$).
$\mathbf{r}$	Position vector, $[x, y, z]$ in locally Euclidean coordinates.
r_{s}	Standard ratio used in definition of $\delta^{13}\mathrm{C}$.
r	Radial coordinate in spherical polar system.
$R'(n)$	Autocovariance function for discrete time-steps: $R'(n) = R(n\,\Delta t)$.
$R(t)$	Autocovariance as function of continuous time.
$\mathbf{R}$	Covariance matrix for observational error, $\mathbf{e}(n)$, in state-space modelling. $\mathbf{R}(n)$ in cases in which the error covariance varies over time.
R_{A}	Isotopic ratio ($^{13}\mathrm{C} : \mathrm{C}$ unless specified otherwise) in atmosphere.
R_{e}	Radius of the earth.
R_{r}	Reference $^{13}\mathrm{C} : \mathrm{C}$ ratio used to define isotopic anomalies.
R_X	Isotopic ratio ($^{13}\mathrm{C} : \mathrm{C}$ unless specified otherwise) in reservoir X.
R_X^+	Isotopic ratio ($^{13}\mathrm{C} : \mathrm{C}$ unless specified otherwise) of flux from reservoir X to atmosphere.
R_X^-	Isotopic ratio ($^{13}\mathrm{C} : \mathrm{C}$ unless specified otherwise) of flux to reservoir X from atmosphere.
$s(\mathbf{r}, t)$	Source strength as function of space and time.
s_μ	Generic coefficient in source discretisation: $s(\mathbf{r}, t) = \sum_\mu s_\mu \sigma_\mu(\mathbf{r}, t)$.
$\mathbf{s}$	Vector of source strengths, with components s_μ.
S	Surface area (with differential element $\mathrm{d}S$).
S	(as subscript) South, southern hemisphere.
$\hat{\mathbf{s}}$	Estimate of $\mathbf{s}$.
t	Time. In applications and examples, units are years unless indicated otherwise.
T	Absolute temperature.
T_{cycle}	Period of cycle; a year in most contexts.
$\mathcal{T}[.]$	Transport operator.
$\mathbf{T}$	Matrix representation of transport operator.
$\mathbf{T}^*$	Matrix representation of mapping of sources onto concentrations over a single time-step. (Chapter 11 only.)
u	East–west component of $\mathbf{v}$.
$\mathbf{u}(n)$	Deterministic forcing at time n in state-space model.
$\mathbf{U}$	Matrix of singular vectors, elements U_{jp}.

v	North–south component of $\mathbf{v}$.
$\mathbf{v}$	Advective velocity of air-mass, with components $[u, v, w]$ in locally Euclidean coordinates.
$\mathbf{V}$	Matrix of singular vectors, elements, $V_{\mu q}$.
w	Vertical component of $\mathbf{v}$.
$\mathbf{w}(n)$	Stochastic forcing in state-space evolution.
$\mathbf{W}$	Inverse covariance matrix for prior estimates of parameters z_μ and z_v, elements $W_{\mu v}$.
x	Euclidean spatial coordinate.
x	Position index for observational data.
$\mathbf{x}$	Vector of model parameters.
X	Generic reservoir index.
$\mathbf{X}$	Inverse covariance matrix for observational errors. Elements X_{jk} are covariances of ϵ_j and ϵ_k.
y	Euclidean spatial coordinate.
$\mathbf{y}$	Extended 'data vector', combining observations, $\mathbf{c}$, and priors, $\mathbf{z}$, fitted in Bayesian estimation.
$\mathbf{y}$	General coordinate vector in description of Green's functions (Section 10.1).
Y	Generic index for flux or process.
$\mathbf{Y}$	Combined *a priori* covariance matrix for sources and observations in Bayesian estimation.
z	Euclidean spatial coordinate.
$\mathbf{z}$	Vector of prior estimates of $\mathbf{x}$.
$\mathbf{z}(n)$	Vector of observations at time n in state-space modelling.
z_μ	Prior estimate of s_μ, element of $\mathbf{z}$.
α	Generic numerical coefficient.
β	Generic numerical coefficient, especially one describing a rate of change.
β_η	Global rate of increase in concentration of species η.
γ_k	Ratio of source strength at wavenumber k to concentration at wavenumber k.
δ_X	Normalised ^{13}C : C anomaly in flux or process X. Approximately equal to anomaly measure δ^{13}C, with appropriate choice of normalisation.
δ_{ij}	Kronecker delta (i, j integer): $= 1$ if $i = j$; $= 0$ otherwise.
$\delta(.)$	Dirac delta-function.
Δm	Difference in tracer mass between regions, especially northern and southern hemispheres.
Δt	Time-step, in time-series model or numerical integration.
$\in$	Set inclusion (also $\notin$ as set exclusion).
$\epsilon(n)$	Stochastic forcing in time-series models.
ϵ_j	Error in c_j. Also stochastic forcing in time-series models.
ζ	Partitioning between reactions: specifically proportion of CH_4 oxidised

	to CO.
η	Species index (for observational data, response functions etc.).
θ	Coordinate in spherical polar system, the colatitude in radians.
θ_L	Latitude (negative for southern hemisphere).
θ	Dimensionless frequency, $= 2\pi\nu\,\Delta t = \omega\,\Delta t$, range $[-\pi, \pi]$.
Θ	Potential temperature.
$\Theta(.)$	Unit step function.
κ	Diffusion coefficient, subscripted by coordinate symbols as required.
λ_η	Loss rate (inverse lifetime) for constituent η.
λ, λ_q	Eigenvalue or singular value.
Λ, Λ_q	Eigenvalue or singular value in contexts in which confusion with loss rates, λ_η, is possible.
Λ	Lagrange multiplier, generally for specifying relative weights in an objective function.
μ	Generic component index for sources or other parameters.
μ_η	Relative molecular mass of species η.
ν	Frequency, in cycles per unit time, range $[-\frac{1}{2}\,\Delta t, \frac{1}{2}\,\Delta t]$.
ξ	Integrating factor (Box 1.1 only).
ξ	Variance ratio in two-variable regression.
ξ	Unit conversion in toy model of CH_4–CO–OH.
Ξ	Basis function for expanding space–time distribution of tracer or flux.
ρ	Density.
$\rho(.)$	Generic response function.
$\rho(.)^*$	Kernel in formal deconvolution relation, $\rho_\eta^*(.)$ for deconvolution of global budget of gas η.
$\rho_\eta(t)$	Response function giving the proportion of a pulse of constituent η remaining in the atmosphere after time t.
$\sigma_\mu(r, t)$	The μth specified source distribution in synthesis inversion.
τ	Transport time in the formalism of Plumb and MacConalogue [359].
ϕ	Coordinate in spherical polar system.
ϕ	Phase of a cycle, especially $\phi(t)$ for phase of slowly varying cycle.
Φ_x, Φ_y	Trace-gas fluxes (mass flow per unit area) in atmosphere (Chapter 18 only).
Φ_X	Net flux (of carbon unless otherwise specified) to atmosphere from reservoir X, with 'flux' in this and following cases meaning the rate of exchange of mass integrated over a finite area. Specific cases are Φ_F, Φ_B and Φ_O for fossil, biospheric and ocean fluxes, respectively,
Φ_X^+	Gross flux (of carbon unless otherwise specified) to atmosphere from reservoir X.
Φ_X^-	Gross flux (of carbon unless otherwise specified) from atmosphere to reservoir X.
Φ_X^{13}	Net flux of ^{13}C to atmosphere from reservoir/process X.
Φ_X^*	^{13}C anomaly flux.

χ	Stream function.
$\Psi(k)$	Coefficient in digital filter. $\Psi_\gamma(k)$ where γ identifies particular filter.
$\psi(\theta)$	Frequency response of digital filter.
ω	Angular frequency $= 2\pi\nu$, range $[-\pi/\Delta t, \pi/\Delta t]$.
∇_1	One-step difference operator.
∇_{12}	12-step difference operator.
∇	Gradient operator.

Appendix B

Numerical data

B.1 Data bases

Many observational data sets for gas concentrations and a number of related data such as emission inventories are available on the internet. Since this is constantly evolving, URLs for institutions are given as well as URLs for specific data sets.

CDIAC The Carbon Dioxide Information and Analysis Center is at http://cdiac.esd. ornl.gov.

NOAA CMDL The Climate Monitoring and Diagnostics Laboratory is at http://www. cmdl.noaa.gov, from which links lead to data files. Data are also available by anonymous ftp from ftp.cmdl.noaa.gov with Globalview on path ccg/co2/ GLOBALVIEW.

WDCGG The World Data Center for Greenhouse Gases provides data to researchers via CD-ROM. The ftp data access is currently (2001) only for registered users. Details can be obtained via http://gaw.kishou.go.jp/wdcgg.html.

GEIA The Global Emissions Inventory Activity (GEIA) [183] is an activity of the IGAC which is an IGBP core project and so GEIA can be accessed via the IGAC (or less directly via the IGBP). Currently the data base can be accessed more directly at http://weather.engine.umich.edu/geia.

EDGAR The EDGAR data base [345, 7] is at RIVM (http://www.rivm.nl (or specify file 'index_en.html' for the English language). Currently the introduction to this data base is at http://www.rivm.nl:80/env/int/coredata/edgar/intro.html.

B.2   **Biogeochemical parameters**

Physical constants and properties

Description	Value	Units	Reference
Avogadros's number	6.02214×10^{26}	$kmol^{-1}$	[285]
Relative molecular mass of N_2	28.0134	—	
Relative molecular mass of O_2	31.9988	—	
Relative molecular mass of Ar	39.948	—	
Relative molecular mass of CO_2	44.00995	—	
Half-life of ^{14}C	5730	years	

Earth-system quantities

	Description	Value	Units	Reference
R_e	Radius of the earth (mean equatorial)	6378.14	km	[285]
A_e	Area of the earth	5.101×10^8	km^2	
M_{atmos}	Mass of atmosphere	5.1441×10^{18}	kg	[486]
	Dry mass of atmosphere	5.132×10^{18}	kg	[486]
N_{atmos}	Dry mass of atmosphere	1.773×10^{20}	moles	

The mass of the atmosphere has a range of 0.00193×10^{18} kg, due to seasonal variations in water content, and an uncertainty of order 0.0005×10^{18} kg [486].

B.3   **Numerical values for toy models**

A.3 Diffusive transport (Box 2.1)

	Description	Value	Units
κ_p	Vertical diffusion	$\frac{4}{9} \times 10^3$	$Pa^2 \, s^{-1}$
κ_θ	North–south diffusion	3×10^6	$m^2 \, s^{-1}$
α	East–west enhancement of diffusion	16	—
p_0	Surface pressure	100	kPa
p_1	Pressure at upper limit	20	kPa

With R_e, A_e and N_{atmos} as above giving sources in moles per unit area.

E.1 Toy chemistry (Box 1.5)

Exercises using the toy chemistry model of Box 1.5 should, unless specified otherwise, use the following representative numbers when numerical values are required.

	Description	Value	Units
λ_{CH_4}	Rate of loss of CH_4 (except for OH)	0.0091	yr^{-1}
λ_{CO}	Lifetime of CO (except for OH)	0.473	yr^{-1}
λ_{OH}	Rate of loss of OH (except CO and CH_4)	0.0	yr^{-1}
k_1	Rate constant for $CH_4 + OH$	0.123×10^{-3}	$mm^3 \, yr^{-1}$
k_2	Rate constant for $CO + OH$	5.93×10^{-3}	$mm^3 \, yr^{-1}$
ζ	Proportion of CH_4 oxidised to CO	1	—
ξ	Unit conversion	1.3×10^4	$ppm \, mm^3 \, yr \, s^{-1}$
S_{CH_4}	CH_4 source	500	$Tg \, CH_4 \, yr^{-1}$
S_{CO}	CO source	2000	$Tg \, CO \, yr^{-1}$
S_{OH}	OH source	1.5×10^3	$mm^{-3} \, s^{-1}$

Appendix C

Abbreviations and acronyms

AES   Atmospheric Environment Service (Canada).

AGAGE   Advanced Global Atmospheric Gases Experiment. Successor to GAGE
            [382].

AGU   American Geophysical Union.

AGWP   Absolute global warming potential (see Box 6.2).

ALE   Atmospheric Lifetime Experiment [382].

AMIP   Atmospheric Model Intercomparison Project [170, 171].

AMS   Accelerator mass spectrometry.

AR   Autoregressive. Time-series model (see Chapter 4).

ARIMA   Autoregressive integrated moving average. Time-series model.

ARMA   Autoregressive moving average. Time-series model.

AVHRR   Advanced Very-High-Resolution Radiometer. (A satellite instrument).

BAPMoN   Background Air Pollution Monitoring Network (WMO).

BLS   Bayesian least-squares (estimator).

CAR   CSIRO Atmospheric Research (formerly Division of Atmospheric
            Research).

CBM   Carbon-bond mechanism. A chemical-reaction scheme (see
            Section 15.3).

CCM   Community Climate Model (from the NCAR).

CDIAC   Carbon Dioxide Information Analysis Center (Oak Ridge National
            Laboratory, Tennessee, USA).

CFC   Chlorofluorocarbon: CFC-11 is CCl_3F and CFC-12 is CCl_2F_2 (see
            Box 16.1 for more details of nomenclature).

CFR   Centre des Faibles Radioactivités (France).

CMDL   Climate Monitoring and Diagnostics Laboratory (NOAA).

CoP	Conference of Parties (i.e. parties to the FCCC). Denoted CoP1, CoP2, etc. for the years 1995, 1996, ...
CRCSHM	Cooperative Research Centre for Southern Hemisphere Meteorology.
CSIRO	Commonwealth Scientific and Industrial Research Organisation (Australia).
DAR	Division of Atmospheric Research (CSIRO).
DARLAM	Division of Atmospheric Research Limited Area Model.
DE	Differential equation (see also ODE, PDE).
ECMWF	European Centre for Medium-range Weather Forecasting.
EDGAR	Emissions Database for Global Atmospheric Research [345].
ENSO	El Niño/Southern Oscillation (phenomenon) [6].
EOF	Empirical orthogonal functions. Analysis technique based on the SVD (see example in Section 5.4).
ETEX	European Tracer Experiment [146]. A controlled release of PFCs.
FCCC	Framework Convention on Climate Change (see Box 6.1).
FGGE	First Global Geophysical Experiment.
FTIR	Fourier-transform infra-red (spectroscopic measurement technique).
GAGE	Global Atmospheric Gases Experiment, successor to ALE [382].
GAIM	Global Analysis, Interpretation and Modelling. A project of the IGBP.
GAW	Global Atmospheric Watch. WMO programme.
GC	Gas chromatograph.
GCM	General Circulation Model.
GCMS	Gas-chromatograph mass spectrometer.
GEIA	Global Emissions Inventory Activity [183]. An activity of the IGAC IGBP project.
GFDL	Geophysical Fluid Dynamics Laboratory.
GISS	NASA Goddard Space Flight Center, Institute for Space Studies (also used to identify transport model developed at GISS).
GMCC	Geophysical Monitoring for Climatic Change (NOAA). Activities now at NOAA CMDL.
GPP	Gross Primary Production (of biological systems).
GWP	Global warming potential (see Box 6.2).
HCFC	Hydrochlorofluorocarbon.
HFC	Hydrofluorocarbon.
IGAC	International Global Atmospheric Chemistry Project (IGBP).
IGBP	International Geosphere–Biosphere Program.
IMAGE	Integrated Model to Assess the Greenhouse Effect (RIVM) [4].
INM	Instituto Nacional de Meteorologia (Spain).
IPCC	Intergovernmental Panel on Climate Change.
JGOFS	Joint Global Ocean Flux Study (an IGBP project).
LBA	Large-scale Biosphere Experiment in the Amazon.
LUCF	Land-use change and forestry (IPCC emissions category).
LULUCF	Land Use, Land-Use Change and Forestry [500] (IPCC report).

MA	Moving average. Time-series model (see Chapter 4).
MATCH	Model of Atmospheric Chemistry and Transport, developed at the NCAR [395].
MATCH	Meso-scale Atmospheric Transport and Chemistry model [410] developed at the Swedish Meteorological and Hydrological Institute.
ML	Maximum likelihood (estimator) (see Chapter 3).
MPD	Mode of posterior distribution (estimator).
MPI	Max-Planck-Institut, e.g. MPI für Meteorologie.
NASA	National Aeronautics and Space Administration (USA).
NCAR	National Center for Atmospheric Research (Boulder, USA).
NDIR	Non-dispersive infra-red (type of CO_2 analyser).
NDVI	Normalised difference vegetation index. Measure of biological activity from AVHRR data.
NEP	Net ecosystem production (of biological systems).
NIRE	National Institute for Resources and Environment (Japan).
NIWA	National Institute of Water and Atmospheric Research (New Zealand).
NMHC	Non-methane hydrocarbons.
NOAA	National Oceanic and Atmospheric Administration (US Department of Commerce).
NPP	Net primary production (of biological systems).
ODE	Ordinary differential equation.
ODP	Ozone-depletion potential [519, 441, see also Box 16.4].
OLS	Ordinary least-squares (estimator).
PDB	Pee Dee belemnite. The limestone that defines the $^{13}C : {}^{12}C$ standard ratio.
PDE	Partial differential equation.
PFC	Perfluorocarbon.
RIVM	National Institute for Public Health and the Environment (Rijksinstituut voor Volkgezendheid en Milieu, The Netherlands).
SIO	Scripps Institution of Oceanography (USA).
SODA	Satellite Ozone Data Assimilation [439].
SOI	Southern Oscillation Index [6].
SRES	Special Report on Emission Scenarios [333] (IPCC report).
SVD	Singular-value decomposition (see Box 3.2).
TEM	Transformed Eulerian mean (zonal average effective circulation).
TM2	Transport model developed at the MPI für Meteorologie.
WDCGG	World Data Center for Greenhouse Gases.
WLS	Weighted least-squares (estimator).
WMO	World Meteorological Organization.

Appendix D

Glossary

(See also list of acronyms and abbreviations in Appendix C.) **Boldface** denotes another item in the glossary.

accuracy The extent to which a **bias** is absent; c.f. **precision**.

adjoint model A modification of a numerical model such that the adjoint model provides an efficient calculation of the adjoint of the Green-function matrix calculated by the original model. See Section 10.6.

adjustment time For methane, the time-scale for decay (to 1/e) of a methane perturbation due to a pulse input. Differs from the **lifetime** since the additional methane from the pulse decreases the concentration of OH and changes the rate of loss for all atmospheric methane.

alkanes Hydrocarbons with the general formula C_nH_{2n+2}. The basic structure is a linear chain of carbons with attached hydrogen, but for $n \geq 4$ **isomers** with branched carbon chains are possible.

AMIP An inter-comparison of atmospheric GCMs. Of particular interest is the case in which the forcing included long-term observational records of sea-surface temperatures [170].

a priori Sometimes used for the Bayesian **prior** values. The use of the term '*a priori*' is avoided in this book because it is often used to mean information existing prior to experience.

atmospheric transport model A numerical model describing the time-evolution of minor atmospheric constituents in response to atmospheric motions.

back-trajectory The path along which an air-mass has travelled to reach a specified point at a specified time.

baseline Conceptually, conditions under which the concentrations of trace atmospheric constituents are representative of large-scale air-masses. In practice, site-specific operational definitions specify which data are selected as baseline.

Bayesian estimation A statistical estimation procedure that produces estimates by combining new information (possibly indirect) with existing information, by use of Bayes' theorem.

bias A systematic offset between the true value of a quantity and the value that is measured or estimated.

C3 Photosynthetic pathway (or plant using such a mechanism) with 3 carbon atoms in the initial photosynthetic product. Significant in biogeochemical studies due to differences in isotopic fractionation between the C3 and C4 processes.

C4 Photosynthetic pathway (or plant using such a mechanism, most notably grasses) with four carbon atoms in the initial photosynthetic product.

CFCs = chlorofluorocarbons Compounds with **alkane** structure, in which all the hydrogen atoms have been replaced by chlorine or fluorine.

compact operator Compactness refers to the degree of continuity between the inputs and outputs of operators acting on functions. The ubiquity of ill-conditioned inverse problems comes from the fact that a compact operator cannot have a bounded inverse.

concentration Strictly, the amount of constituent per unit volume. Often (including in this book) used less specifically to include **mixing ratios**.

consistent A property of estimators, requiring that the estimate converge to the true value as the size of the sample increases.

convolution Integral relation of the form $\int_0^t f_1(t - t') f_2(t') \, dt'$.

cost function See **objective function**.

cyclo-stationary Constant apart from a seasonal cycle. Often, 'cyclo-stationary' refers to the sources, in which case concentrations will have a fixed, globally constant, trend added to cyclo-stationary variations in concentration.

deconvolution The processes of inverting a relation involving a **convolution** integral.

earth-system science The science of the lithosphere, hydrosphere, cryosphere, biosphere, atmosphere and their interactions considered as a whole.

El Niño A period of anomalously warm sea-surface temperatures in the east Pacific. Associated with the **southern oscillation** in the **ENSO** phenomenon [6].

ENSO El Niño/Southern Oscillation phenomenon. The combination of El Niño and the **southern oscillation** [6].

estimator A statistical procedure or function for obtaining statistical estimates.

expectation The mean, or average, over a probability distribution.

feedback A closed loop in a causal chain.

filter An operation that transforms one time series into another, by linear combination of terms in the input series. Often restricted to transformations that are invariant in time. The **Kalman filter** is a special usage of the term 'filter' in that it is by definition one-sided and explicitly allows time-evolution of the filter.

fingerprint method An inversion technique based on using ratios of concentrations of trace gases (a chemical fingerprint) to infer information about sources [258, 5].

flux Literally, a rate of exchange of mass per unit area; it is used with this meaning in the context of partial differential equations where the notation $s(\mathbf{r}, t)$ is used. Also used in this book for integrated exchanges of mass over finite areas, often denoted Φ, with an appropriate descriptive subscript.

forward problem For the purposes of this book, a calculation in which the sequence of mathematical calculation is in the same direction as the chain of real-world causality (c.f. **inverse problem**).

Gaia The hypothesis [292] that the biota acts like a single living organism, manipulating the physical environment in such a way as to maintain the earth in conditions suitable for life.

gain matrix Matrix used in the Kalman-filtering formalism. Projects a vector of differences between observations and predictions onto a correction term for the estimated state-vector.

greenhouse gas A gas that affects the radiative balance of the earth by absorption of infra-red radiation. Such gases have three or more atoms since infra-red energies correspond to the energies of bending modes of many molecules. (Rotational modes spread these energy levels into absorption bands.)

Green's function Inhomogeneous differential equations can be solved by applying, to the forcing term, an integral operator, which is the inverse (subject to specified boundary conditions) of the differential operator and whose kernel is known as Green's function (see Section 10.1).

Green-function matrix A discretisation of **Green's function**.

GWP = global-warming potential A measure of the relative radiative effects of different greenhouse gases, taking into account both the difference in **radiative forcing** and the different time-scales on which the greenhouse gases are removed from the atmosphere. See Box 6.2.

halons Halocarbons that contain bromine and fluorine. They may contain chlorine. The term is usually applied only to compounds with no hydrogen, but in some cases the usage is extended to compounds that include hydrogen.

ill-conditioned A calculation that is subject to arbitrarily large amplification of errors in the inputs.

inter-comparison A comparison between models. In contrast to comparisons with real-world observations, inter-comparisons can explore details of mechanisms, by adding additional diagnostics to the model and by running multiple experiments.

inverse problem For the purposes of this book, a calculation in which the sequence of mathematical calculation is in the opposite direction to the chain of real-world causality (c.f. **forward problem**). Since many real-world processes are dissipative, many such inverse problems are subject to **ill-conditioning**. A common alternative terminology is to use such ill-conditioning as the defining characteristic of an inverse problem.

isoflux The component of a two-way exchange of mass that represents a change in isotopic composition through isotopic equilibration, independent of net mass flux.

isomers Two or more compounds with the same stoichiometry but different structures are termed isomers.

Jacobian Matrix of derivatives of a set of functions with respect to a set of common arguments.

Kalman filter A recursive estimation technique applied to a time series of observations that are used to derive time series of estimates of an unknown state-vector. The estimates at time n are based on data for time $n' \leq n$. See Section 4.4.

Kalman smoother An extension to the **Kalman filter** whereby estimates of the state-vector for time n include data for times $n' > n$ as well as $n' \leq n$. There are several different forms of Kalman smoothing. See Section 4.4.

kernel A representation, in a particular basis, of an operator. Generally, in this book, the Green function is the kernel of the operator that inverts a specified differential operator.

La Niña The opposite of **El Niño**: a period of anomalously cold sea-surface temperatures in the east Pacific.

Least-squares An estimation technique in which the estimates are those which minimise an **objective function** defined as the sums of squares of **residuals**.

lifetime A characteristic time-scale of a system. There are several usages; it is sometimes used as a synonym for **turnover time**. In all but the simplest systems, any use of the term 'lifetime' needs to be carefully qualified.

likelihood In our notation, $p(\mathbf{c}|\mathbf{x})$, the probability density for obtaining a given set of observations, $\mathbf{c}$, from a parameter set $\mathbf{x}$.

mass-balance An inversion technique, deriving trace-gas fluxes from concentration data by ensuring conservation of mass at each time-step in the model, by assigning fluxes to account for the difference between the observations and calculations of rates of change of concentration. See Section 11.1. The term 'mass-balance' is also applied to what this book terms the 'boundary-integral technique'.

maximum likelihood An estimation technique, in which the estimates $\hat{\mathbf{x}}$ are the values that maximise the likelihood, $p(\mathbf{c}|\mathbf{x})$.

mixing ratio Amount of tracer as a molar fraction of a total sample of air. This book uses the older form of units such as ppm (parts per million) rather than the IUPAC standard of μmol mol^{-1}.

National inventories Emission inventories of greenhouse gases, produced as a requirement of the FCCC.

null space Set of inputs that have zero effect on the outputs, e.g. the set of source distributions that cause no net changes in concentration at the observing sites. See Box 3.1.

objective function A function that is to be minimised in a particular mathematical calculation (also called cost function or penalty function).

ozone-depletion potential A comparative measure for ozone-depleting substances, in terms of the relative amount of destruction of ozone over the entire time that the gas is in the atmosphere [519, 441; see also Box 16.4].

parameter A numerical quantity in a mathematical relation; in this book generally a numerical quantity in a mathematical expression of a model.

PDB Notional standard material (a particular limestone sample, no longer available) for ^{13}C : ^{12}C ratio.

penalty function Often used as a synonym for **objective function**. Alternatively, a function that is one term in an objective function, designed to favour a particular aspect of the solution when one is minimising the objective function.

PFC = perfluorocarbon Compounds with the alkane structure with all hydrogen replaced by fluorine.

precision The degree of spread of an estimate or measurement, often characterised by the variance of the value concerned; c.f. **accuracy**.

prior (value/estimate/distribution) The information available before making a particular set of measurements.

radiative forcing A measure of the radiative effect of one or more greenhouse gases. Notionally it is the instantaneous effect, but the actual definition is in terms of radiation balance at the tropopause, after allowing the stratosphere to achieve radiative balance, a process that requires time-scales of months.

rectifier effect A time-average gradient established by a flux whose time-average is zero, due to temporal correlation between the flux and transport.

residual The difference between an observation and the model calculation for the observation (or similar difference).

sampling distribution If observations are treated as random variables, then estimates derived from such observations will also be random variables. For a particular estimator (i.e. technique of estimation) the estimates will have a probability distribution, called the sampling distribution, due to the expected distribution of observations.

scenario An internally consistent set of estimated future changes. The most important cases for greenhouse-gas studies are emission scenarios.

sigma-coordinates Vertical coordinate in the atmosphere, expressed as the ratio of pressure to the surface pressure.

Southern Oscillation (Index) The Southern Oscillation is a large-scale fluctuation of atmospheric circulation with a characteristic time-scale of 2–5 years. It is quantified by the Southern Oscillation Index (SOI), which is defined in terms of monthly mean differences in pressure between the east and west Pacific, variously Tahiti – Darwin, Easter Island – Djakarta etc.

state-vector A set of mathematical quantities representing some aspect of a system, whose single-step time-evolution can be described by a specified stochastic/deterministic model. The purpose of the **Kalman filter** and **Kalman smoother** is to provide estimates of such state-vectors.

stationary (time series) A time series (or similar sequence) in which the statistical properties are invariant under shifts in time.

statistical model A mathematical model in which some of the quantities are treated as random variables.

Suess effect The change in carbon-isotope ratios in the atmosphere due to input of fossil carbon that is isotopically depleted.

synthesis inversion An inversion technique based on fitting a set of pre-specified basis functions. In transport modelling, synthesis inversion is based on a discretised representation of Green's function.

time series A set of numbers with a one-dimensional ordering, generally representing time. Generally used for cases in which the values are treated as random variables. The term is often restricted to cases in which the time-steps between successive values are equal.

tracer-ratio technique The use of observations of multiple tracers to infer ratios of sources or sinks from ratios of concentrations.

turnover time A measure of the rate of a flux, particularly that due to a loss process. The turnover time (for a particular constituent) from reservoir X due to process Y is $\tau_Y = M_X/F_{X,Y}$, where M_X is the amount of the constituent in reservoir X and $F_{X,Y}$ is the rate of loss from reservoir X due to process Y.

unbiased Estimator whose expectation is equal to the unknown true value of the quantity being estimated.

validation The process of ensuring that the (symbolic) model corresponds to the relevant aspects of the real world [423].

verification The process of ensuring that the computer implementation corresponds to the symbolic model [423].

Appendix E

Data-source acknowledgements

Many of the figures are based on synthetic values chosen to illustrate the discussion. When results of actual cases are used, most of the model results come from my own work or from published sources that are referenced in the figure captions. Results from previously unpublished CSIRO research are used with permission.

However, in a number of cases colleagues have kindly supplied unpublished results or supplied numerical values of information previously published as graphs.

- ■ Rachel Law supplied
  - the full latitudinal dependence of CO_2 fluxes used in Figure 14.5;
  - the full latitudinal dependence of CH_4 fluxes used in Figure 15.2;
  - the MATCH model results plotted in Figure 9.1.
- ■ Peter Rayner supplied the estimates of CO_2 flux plotted in Figure 14.9 and other data from calculations reported in [402].
- ■ Ying-Ping Wang supplied the DARLAM model results plotted in Figure 9.1.
- ■ Figure 1.2 is based on GISS model responses calculated by Cathy Trudinger.

The sources for the observational data used in examples are generally cited in the figure captions. Additional details are as follows.

- ■ The Cape Grim CO_2 data in Figure 5.2 were supplied by L. P. Steele, supported by staff in GASLAB and at the Cape Grim Baseline Atmospheric Pollution Station.
- ■ The NOAA CO_2 data used for Figure 5.4 are described by Conway et al. [78] and are available by ftp from NOAA/CMDL.

■ The AGAGE data in Figures 16.1, 17.1 and 18.1 were taken from the WDCGG CD-ROM as described in the WDCGG report [383; and other sections by these authors in the same report].

In addition, the latitudinal distribution of methane sources in Figure 15.2 was constructed from source inventories described by Fung and co-workers [165; and references therein] accessed via http://www.giss.nasa.gov/data/ch4fung/index.html.

Some of the model results are based on GlobalView data sets [179; see also 309 and Box 5.2].

For the author's synthesis results used in illustrations, response functions were calculated by Cathy Trudinger using the GISS tracer-transport model kindly provided by Inez Fung. The inversions used the 'biome-oriented' basis described in Box 10.1. Experience suggests that this may be excessively vulnerable to the sampling-bias error described in Section 8.3, so the results should be regarded as illustrative only. The illustrations use several different cases.

■ Figure 8.2 uses a 20-site inversion of CO_2 data.

■ Figure 13.1 shows various combinations of data for the rate of growth in concentration of CO_2 and rate of growth of $\delta^{13}C$ for a 14-site spatial distribution, each representing 1985–95. As well as changing the time period for the observations, relative to earlier work [143, 145], the prior estimates of the isofluxes have been revised to reflect more recent information cited by Trudinger [487].

■ Figure 14.7 represents the 1980s and is the author's contribution to the Third Assessment Report of the IPCC [371]. As required for that report, it uses CO_2 only. The estimates for the three specified regions (bounded by 30° S and 30° N) were obtained by taking area-weighted combinations of the components specified in Box 10.1.

The calculations in this book do not make direct use of the TRANSCOM data sets but many of the concepts presented in this book have been refined through discussions at TRANSCOM workshops and the practice of performing and documenting the TRANSCOM inter-comparisons. TRANSCOM and the TRANSCOM workshops have been supported strongly by the IGBP/GAIM project, largely with funding from the US National Science Foundation.

Solutions to exercises

As well as the problems set at the end of the chapters, the reader is referred to problems in some of the suggested reading. In particular, the AGU volume from the Heraklion conference contains a set of problems [403] including computer routines on CD and also instructions for remote access to the adjoint and tangent linear compilers. Many of the problems are based on the two-reservoir model (Box 2.1). Tarantola [471] also includes a number of problems on various aspects of inversions. Some of the problems given by James [230] (especially in the early chapters) address issues of atmospheric dynamics that are relevant to this book.

The solutions

Chapter 1

1. *Calculate change in atmospheric mass from burning 6 Gt of carbon, if half the CO_2 goes into the oceans.*

   The input of 6 Gt of carbon is combined with $6 \times (32/12)$ Gt of oxygen (from the atmosphere!) to give 22 Gt of CO_2. If half of this is taken up by the oceans then the mass of the atmosphere will have a net *decrease* of 5 Gt.

2. *Calculate the variance of the slope estimated by linear regressions.*

   Use $\sum_{m=1}^{n} m = n(n+1)/2$ and $\sum_{m=1}^{n} m^2 = n(n+1)(2n+1)/6$, $E[\sum n^a \epsilon_n] = 0$, $E[\sum n^a \epsilon_n \sum n^b \epsilon_n] = R \sum n^{a+b}$. From (4b) in Box 1.4,

$$(\hat{\beta} - \beta)^2 [N^2(N+1)(N-1)/12)]^2 = \left(\sum n \sum \epsilon_n - N \sum (n\epsilon_n)\right)^2$$

$$= \left(\sum n \sum \epsilon_n\right)^2 + N^2 \left(\sum n \sum \epsilon_n\right)^2$$

$$- 2N \left(\sum n\right)\left(\sum \epsilon_n\right)\left(\sum n\epsilon_n\right)$$

$$= N^2 R \left(\sum n^2\right) - N R \left(\sum n\right)^2$$

whence

$$E[(\hat{\beta} - \beta)^2] = R \Big/ \left[N^2 \left(\sum n^2\right) - N \left(\sum n\right)^2\right] \sim \frac{12R}{N^3}$$

For $E[\epsilon_n \epsilon_{n+1}] = R/2$, neglecting 'end effects', each non-zero term (proportional to R) in the sums above has two corresponding extra terms proportional to $R/2$, so the variances of the estimates are doubled.

Chapter 2

1. *Show that, for steady flow, $m(x, y) = f(\chi(x, y))$ is a steady-state for arbitrary $f(.)$.*
   For $\alpha = 0$ and $m(x, y) = f(\chi(x, y))$, the PDE is

$$\frac{\partial}{\partial t} m(x, y, t) = \frac{\partial}{\partial x}\left(\chi \frac{\partial f}{\partial \chi} \frac{\partial \chi}{\partial y}\right) - \frac{\partial}{\partial y}\left(\chi \frac{\partial f}{\partial \chi} \frac{\partial \chi}{\partial x}\right)$$

$$= f'(\chi_x \chi_y + \chi \chi_{xy}) + f'' \chi \chi_y \chi_x$$

$$- f'(\chi_y \chi_x + \chi \chi_{xy}) - f'' \chi \chi_y \chi_x = 0$$

where $f' = \partial f / \partial \chi$ and $f'' = \partial^2 f / \partial \chi^2$.

2. *Determine transport modes of a diffusive model.*
   For time-invariant transport, the transport modes are defined by

$$\frac{\partial}{\partial t} \mathbf{c} = -\lambda \mathbf{c} = T[\mathbf{c}]$$

Expand as

$$c = \sum_{j,n} a_{jn} P_n(\cos \theta) \cos[j\pi (p - p_1)/(p_0 - p_1)] e^{-\lambda t}$$

to satisfy the boundary conditions. The PDE for the zonally averaged diffusive model gives

$$-\lambda_{jn} = -\frac{\kappa_p j^2}{(p_0 - p_1)^2} - \frac{\kappa_\theta n(n+1)}{R_e^2}$$

These solutions are independent combinations of relaxations of horizontal and vertical gradients. The slowest-decaying modes (i.e. the smallest λ) will be for

$[n, j] = [0, 1]$ or $[1, 0]$. With the values from Appendix B, each of these has λ of order 10^{-7} s^{-1}, i.e. decay times of rather less than a year.

3. *Write and test numerical integration of one-dimensional advection.*
   Numerical exercise; no solution given, but see Rood [413].

Chapter 3

1. *Prove the matrix-inversion lemma.*
   Multiplying both sides on the right by $\mathbf{A} + \mathbf{BC}^{-1}\mathbf{B}^{\mathsf{T}}$ gives $\mathbf{I}$ on the left and

$$\mathbf{I} + \mathbf{A}^{-1}\mathbf{BC}^{-1}\mathbf{B}^{\mathsf{T}} - \mathbf{A}^{-1}\mathbf{B}[\mathbf{B}^{\mathsf{T}}\mathbf{A}^{-1}\mathbf{B} + \mathbf{C}]^{-1}\mathbf{B}^{\mathsf{T}}$$
$$- \mathbf{A}^{-1}\mathbf{B}[\mathbf{B}^{\mathsf{T}}\mathbf{A}^{-1}\mathbf{B} + \mathbf{C}]^{-1}\mathbf{B}^{\mathsf{T}}\mathbf{A}^{-1}\mathbf{BC}^{-1}\mathbf{B}^{\mathsf{T}}$$
$$= \mathbf{I} + \mathbf{A}^{-1}\mathbf{BC}^{-1}\mathbf{B}^{\mathsf{T}} - \mathbf{A}^{-1}\mathbf{B}[\mathbf{B}^{\mathsf{T}}\mathbf{A}^{-1}\mathbf{B} + \mathbf{C}]^{-1}[\mathbf{B}^{\mathsf{T}}\mathbf{A}^{-1}\mathbf{BC}^{-1} + \mathbf{I}]\mathbf{B}^{\mathsf{T}}$$
$$= \mathbf{I}$$

as required.

2. *Prove the perturbation form of Bayesian estimation from (3.3.13c).*
   Minimising (3.3.13c) gives

$$\mathbf{f} = [\mathbf{G}^{\mathsf{T}}\mathbf{x}\mathbf{G} + \mathbf{W}]^{-1}\mathbf{G}^{\mathsf{T}}\mathbf{X}\mathbf{q}$$

where $\mathbf{f} = \mathbf{x} - \mathbf{z}$ and $\mathbf{q} = \mathbf{c} - \mathbf{Gz}$. Substituting these definitions into the solution above gives

$$\mathbf{x} = \mathbf{z} + [\mathbf{G}^{\mathsf{T}}\mathbf{x}\mathbf{G} + \mathbf{W}]^{-1}[\mathbf{G}^{\mathsf{T}}\mathbf{X}\mathbf{c} - \mathbf{G}^{\mathsf{T}}\mathbf{X}\mathbf{Gz}]$$
$$= [\mathbf{G}^{\mathsf{T}}\mathbf{x}\mathbf{G} + \mathbf{W}]^{-1}[\mathbf{G}^{\mathsf{T}}\mathbf{X}\mathbf{c} + \mathbf{Wz}]$$

as given by (3.3.12c).

3. *Prove that a Bayesian estimate using priors from an earlier estimate is equivalent to a combined Bayesian estimate.*
   The independence condition is expressed as $p(\mathbf{c}_1, \mathbf{c}_2 | \mathbf{x}) = p(\mathbf{c}_1 | \mathbf{x}) p(\mathbf{c}_2 | \mathbf{x})$. Therefore Bayesian estimation on the combined data set using a prior $p_{\text{prior}}(\mathbf{x})$ gives

$$p_{\text{inferred}}(\mathbf{x} | \mathbf{c}_1, \mathbf{c}_2) = \frac{p(\mathbf{c}_1 | \mathbf{x}) p(\mathbf{c}_2 | \mathbf{x}) p_{\text{prior}}(\mathbf{x})}{\int p(\mathbf{c}_1 | \mathbf{x}) p(\mathbf{c}_2 | \mathbf{x}) p_{\text{prior}}(\mathbf{x}) \, d\mathbf{x}}$$

Bayesian estimation using only data $\mathbf{c}_1$ gives

$$p_{\text{inferred}}(\mathbf{x} | \mathbf{c}_1) = \frac{p(\mathbf{c}_1 | \mathbf{x}) p_{\text{prior}}(\mathbf{x})}{\int p(\mathbf{c}_1 | \mathbf{x}) p_{\text{prior}}(\mathbf{x}) \, d\mathbf{x}}$$

Bayesian estimation of $\mathbf{x}$ from data $\mathbf{c}_2$ using $p_{\text{inferred}}(\mathbf{x} | \mathbf{c}_1)$ as the prior gives

$$p_{\text{inferred}}(\mathbf{x} | \mathbf{c}_1, \mathbf{c}_2) = \frac{p(\mathbf{c}_2 | \mathbf{x}) p_{\text{inferred}}(\mathbf{x} | \mathbf{c}_1)}{\int p(\mathbf{c}_2 | \mathbf{x}) p_{\text{prior}}(\mathbf{x} | \mathbf{c}_1) \, d\mathbf{x}}$$
$$= \frac{p(\mathbf{c}_1 | \mathbf{x}) p(\mathbf{c}_2 | \mathbf{x}) p_{\text{prior}}(\mathbf{x})}{\int p(\mathbf{c}_1 | \mathbf{x}) p(\mathbf{c}_2 | \mathbf{x}) p_{\text{prior}}(\mathbf{x}) \, d\mathbf{x}}$$

since the denominator in $p_{\text{inferred}}(\mathbf{x}|\mathbf{c}_1)$ cancels out between numerator and denominator of $p_{\text{inferred}}(\mathbf{x}|\mathbf{c}_1, \mathbf{c}_2)$.

Chapter 4

1. *Convert AR(1) parameters to describe statistics of an even-numbered subset as AR(1).*

   The connection between the process $g(n)$, $n = 0, 1, 2, \ldots$ and its subprocess $g'(n) = g(2n)$ is established via the autocovariance. From Box 4.4
   $R_g(n) = Qa^{|n|}/(1 - a^2)$.

   Putting $R_{g'}(m) = Q'(a')^{|m|}/(1 - (a')^2) = Qa^{|2m|}/(1 - a^2)$ gives $a' = a^2$ and $Q' = Q(1 + a^2)$.

2. *Show that the spectral density of the subprocess in problem 1 is an aliased form of the original.*

   As described in Box 4.1, the quantity $\Delta t \times h'_g(2\pi \nu \, \Delta t)$ provides a comparison that is independent of series length (for estimates) and spacing of data. For $g(n)$ the spectral density is $h'_g(\theta) = Q/[2\pi(1 - 2a\cos\theta + a^2)$. Doubling the spacing of data aliases the frequencies onto $\phi \in [0, \pi]$, where the range 0 to π combines contributions from $\theta = 0$ to $\pi/2$ and $\theta = \pi$ to $\pi/2$. Using $\cos(\pi - \theta) = -\cos(\theta)$, the aliased spectral density (putting $\theta = \phi/2$) is

   $$\frac{Q}{2\pi}\left(\frac{1}{1 + a^2 - 2\cos(\phi/2)} + \frac{1}{1 + a^2 + 2\cos(\phi/2)}\right)$$

   $$= \frac{Q}{2\pi}\frac{2(1 + a^2)}{(1 + a^2)^2 - 4a^2\cos^2(\phi/2)}$$

   $$= \frac{2Q'}{2\pi}\frac{1}{1 + a^4 - 2a^2\cos\phi}$$

   $$= \frac{2Q'}{2\pi}\frac{1}{1 + (a')^2 - 2a'\cos\phi}$$

   $$= 2h'_{g'}(\phi)$$

   where the factor of 2 is the Δt required for invariance of the normalisation.

3. *Show that, if the gain $\mathbf{K}$ is optimal, the estimation covariance can be written as $\hat{\mathbf{P}} = (\mathbf{I} - \mathbf{KH})\tilde{\mathbf{P}}$.*

   The difference between the two forms of $\hat{\mathbf{P}}$ is

   $$[(\mathbf{I} - \mathbf{KH})\tilde{\mathbf{P}}(\mathbf{I} - \mathbf{KH})^{\mathsf{T}} + \mathbf{KRK}^{\mathsf{T}}] - [(\mathbf{I} - \mathbf{KH})\tilde{\mathbf{P}}]$$

   $$= (\mathbf{I} - \mathbf{KH})\tilde{\mathbf{P}}\mathbf{H}^{\mathsf{T}}\mathbf{K}^{\mathsf{T}} + \mathbf{KRK}^{\mathsf{T}} = [\tilde{\mathbf{P}}\mathbf{H}^{\mathsf{T}} - \mathbf{K}(\mathbf{H}\tilde{\mathbf{P}}\mathbf{H}^{\mathsf{T}} + \mathbf{R})]\mathbf{K}^{\mathsf{T}}$$

   $$= \mathbf{0}$$

   from the definition of $\mathbf{K}$.

4. *Use the state-space formalism to represent correlated error.*

Define the error process as $y(n) = a \times y(n-1) + \epsilon(n)$ and an extended state as $\mathbf{g}' = [\mathbf{g}, y]^T$ so that the evolution is defined by

$$\mathbf{F}' = \begin{bmatrix} \mathbf{F} & 0 \\ 0 & a \end{bmatrix}; \qquad \mathbf{Q}' = \begin{bmatrix} \mathbf{Q} & 0 \\ 0 & R \end{bmatrix}$$

with $\mathbf{H}' = [\mathbf{H}, 1]$ and $\mathbf{R}' = 0$.

To show that the case $a = 0$ is equivalent to the 'standard form', consider a state $\hat{\mathbf{g}}' = [\hat{\mathbf{g}}^T, \hat{y}]^T$ with covariance

$$\hat{\mathbf{P}}' = \begin{bmatrix} \hat{\mathbf{P}} & \hat{\mathbf{p}} \\ \hat{\mathbf{p}}^T & \hat{p} \end{bmatrix}$$

The evolution equation gives the prediction

$$\tilde{\mathbf{g}}' = [\mathbf{F}\hat{\mathbf{g}}^T, 0]^T$$

with

$$\tilde{\mathbf{P}}'(n+1) = \begin{bmatrix} \mathbf{F}(n)\hat{\mathbf{P}}(n)\mathbf{F}(n)^T + \mathbf{Q}(n) & 0 \\ 0 & R(n+1) \end{bmatrix}$$

$$= \begin{bmatrix} \tilde{\mathbf{P}}(n+1) & 0 \\ 0 & R(n+1) \end{bmatrix}$$

The innovation covariance is (a scalar) given by

$$\mathbf{P}'_I = \mathbf{H}'\tilde{\mathbf{P}}'\mathbf{H}'^T = \mathbf{H}[\mathbf{F}\hat{\mathbf{P}}\mathbf{F}^T + \mathbf{Q}]\mathbf{H}^T + R = p_I$$

which is equivalent to the standard case, whence

$$\mathbf{K}' = \begin{bmatrix} [\mathbf{F}\hat{\mathbf{P}}\mathbf{F}^T + \mathbf{Q}]\mathbf{H}^T/p_I \\ R/p_I \end{bmatrix} = \begin{bmatrix} \mathbf{K} \\ R/p_I \end{bmatrix}$$

Thus $\hat{\mathbf{g}}'(n+1) = [\hat{\mathbf{g}}(n+1)^T, \hat{y}]^T$ with $\hat{\mathbf{g}}$ equivalent to the standard case. Its variance is given (using the result from problem 4.3) by

$$\hat{\mathbf{P}}'(n+1) = \begin{bmatrix} \mathbf{I} - \mathbf{KH} & \mathbf{K} \\ RH/p_I & R/p_I \end{bmatrix} \begin{bmatrix} \tilde{\mathbf{P}} & 0 \\ 0 & R/p_I \end{bmatrix}$$

$$= \begin{bmatrix} (\mathbf{I} - \mathbf{KH})\tilde{\mathbf{P}} & KR \\ RH\tilde{\mathbf{P}}/p_I & R^2/(p_I)^2 \end{bmatrix}$$

i.e. the variance of $\hat{\mathbf{g}}$ is as given by the standard case, $\hat{\mathbf{P}} = (\mathbf{I} - \mathbf{KH})\tilde{\mathbf{P}}$ as required as a starting point for the next step in the estimation.

Chapter 5

1. *Derive the frequency response of a spline fit.*

The key is the normalisation which we can take from the definition of $R'(0)$. The first term is

$$\frac{N}{2\pi} \int_{-\pi}^{\pi} |1 - \phi(\theta)|^2 h'_c(\theta) \, d\theta$$

In the second term, taking the square of the second derivative introduces a factor $(\theta/\Delta t)^4$ and the integration introduces a scale factor of $N\,\Delta t$. Therefore, to minimise the objective function, we need to choose $\phi(\theta)$ to minimise

$$2\pi J(\theta)/N = |1 - \phi(\theta)|^2 h_c'(\theta) + \Lambda\,\Delta t(\theta/\Delta t)^4 |\phi|^2 h_c'(\theta)$$

which leads to the quoted result.

2. *Determine the autocorrelation of a sequence of averages over four successive measurements with AR(1) noise.*

For $x(n) = \sum_{j=1}^{4} y(4n + j)$ with the noise ϵ_j in $y(j)$ and $E[\epsilon_j\,\epsilon_k] = Ra^{|j-k|}$

$$R(0) = E[(x(n) - E[x(n)])^2] = \frac{r}{16}[4 + 6a + 4a^2 + 2a^3]$$

$$R(1) = E[(x(n) - E[x(n+1)])^2]$$
$$= \frac{r}{16}[a + 2a^2 + 3a^3 + 4a^4 + 3a^5 + 2a^6 + a^7]$$

and

$$R(k) = E[(x(n) - E[x(n + k + 1)])^2] = a^{4k} E[(x(n) - E[x(n + 1)])^2]$$

The plot shows the 1-month (solid line) and 2-month (dotted line) autocorrelations in the four-point means as functions of a. For $a \approx 0.7$, $R(1)/R(0) \approx 0.5$ and $R(2)/R(0) \approx 0.1$.

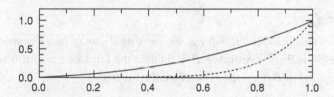

Chapter 6

1. *Derive stability limits of [186].*

For equilibrium the rates of change in (1a)–(1c) are set to zero. For the Guthrie conditions, $\lambda_{OH} = \lambda_{CO} = 0$, (1b) and (1c) can be combined to eliminate m_{CO}, giving

$$(1 + \zeta)k_1 m_{CH_4} m_{OH} = \xi S_{OH} - S_{CO}$$

Since the concentrations are required to be non-negative, we must have

$$\xi S_{OH} - S_{CO} \geq 0$$

Substituting $m_{CH_4} m_{OH}$ into (1a) gives

$$(1 + \zeta)(S_{CH_4} - \lambda_{CH_4} m_{CH_4}) = \xi S_{OH} - S_{CO}$$

whence, for non-negative m_{CH_4}, we require

$$(1 + \zeta)S_{CH_4} \geq \xi S_{OH} - S_{CO}$$

2. *Derive the deconvolution relation for response functions.*

   Using lower-case letters to denote Laplace transforms, i.e.

   $$f(p) = \int_0^\infty F(t)e^{-pt}\,dt$$

   the convolution

   $$Q(t) = \int_0^t R(t - t')S(t')\,dt'$$

   becomes $q(p) = r(p)s(p)$. $R(t) = \sum a_j \exp(-\alpha_j t)$ gives
   $r(p) = \sum a_j/(p + \alpha_j) = f_1(p)/f_2(p)$, where f_1 is a polynomial of degree N
   and f_2 is a polynomial of degree $N - 1$. Formally the inverse is

   $$s(p) = [r(p)]^{-1}q(p)$$

   This can be interpreted by writing

   $$[r(p)]^{-1} = Ap + B + g_1(p)/g_2(p)$$

   where g_1 and g_2 are of degrees $N - 2$ and $N - 1$, respectively, so that
   $g_1(p)/g_2(p)$ is the Laplace transform of a sum of $N - 1$ exponentials. Explicit
   division shows that $A = 1/\sum a_j = 1$ and $B = (\sum_j a_j\alpha_j)/(\sum a_j)^2$. Additional
   discussion of Laplace-transform analysis of CO_2 budgets is given in [135].

   # Chapter 7

1. *Derive the deconvolution relation for a response as a sum of two exponentials.*
   If $R(t) = Ae^{-\alpha t} + Be^{-\beta t}$ (with $A + B = 1$) then the Laplace transform is
   $r(p) = A/(p + \alpha) + B/(p + \beta)$. Therefore the inverse is

   $$r(p) = \frac{(A + B)p + A\beta + b\alpha}{(p + \alpha)(p + \beta)}$$

   or

   $$\frac{1}{r(p)} = \frac{p^2 + (\alpha + \beta)p + \alpha\beta}{p + \gamma} = p + (\alpha + \beta - \gamma) + \frac{\alpha\beta + \gamma^2}{p + \gamma}$$

   where $\gamma = A\beta + B\alpha$. The first term corresponds to the derivative, as required.
   The second term corresponds to $\dot{R}(0) = A\alpha + B\beta$ as required, since

   $$\alpha + \beta - \gamma = \alpha + \beta - A\beta - (1 - A)\alpha = A\alpha + (1 - A)\beta$$

2. *Show the equivalence of a recursive fit and a direct matrix inversion.*
   For the model

   $$c_n = G_n s + \epsilon_n$$

   We use $\mathbf{g} = [s]$ as the state so that $H(n) = G_n$ and denote

   $$f_n = \sum_{j=1}^n H_j X H_j + W$$

and show that, for the optimal gain,

$$\hat{P}(n) = 1/f_n$$

By definition this expression for $\hat{P}(n)$ is true if $n = 0$. The relations $F = 1$ and $Q = 0$ imply that $\tilde{P}(n+1) = \hat{P}(n)$ (see Box 4.4) and so, if $\hat{P}(m) = 1/f_m$ holds for $m < n$, then the optimal gain is

$$K_{\text{opt}}(n) = \frac{1}{f_{n-1}} H_n \left(H_n \frac{1}{f_{n-1}} H_n + \frac{1}{X} \right)^{-1} = \frac{H_n X}{f_n}$$

whence $1 - K(n)H_n = f_{n-1}/f_n$ so that the expression for

$$\hat{P}(n) = \frac{f_{n-1}}{f_n} \frac{1}{f_{n-1}} \frac{f_{n-1}}{f_n} + \left(\frac{H_n X}{f_n} \right)^2 X^{-1} = \frac{1}{f_n}$$

as required.

The source estimate follows from the recursive form

$$\begin{aligned} \hat{s}(n) &= \hat{s}(n-1) + K(n)[c_n - H_n \hat{s}(n-1)] \\ &= K(n)c_n + (1 - K(n)H_n)\hat{s}(n-1) \\ &= \frac{H_n X c_n}{f_n} + \frac{f_{n-1}\hat{s}(n-1)}{f_n} = \sum_{j=1}^{n} \frac{H_j X c_j}{f_n} \end{aligned}$$

in agreement with the result from direct matrix inversion.

In general, the Kalman-filtering formalism is requiring us to perform N matrix inversions for matrices whose dimension is the number of observations at each time-point, whereas the Green-function formalism requires inversion of a matrix whose dimension is the total number of observations.

Generalising the correspondence to the case in which X, the measurement error, depends on j is straightforward.

Chapter 8

1. *Derive properties of a pseudo-inverse.*

For the solution $\mathbf{x} = \mathbf{Hc}$ to the linear problem $\mathbf{c} = \mathbf{Ax}$, the least-squares requirement leads to $\mathbf{A}^T \mathbf{AH} = \mathbf{A}$. Expressing $\mathbf{A}$ and its pseudo-inverse $\mathbf{H}$ in terms of the singular-value decomposition, $\mathbf{A} = \mathbf{ULV}^T$ and $\mathbf{H} = \mathbf{VKU}^T$, where $\mathbf{L}$ is diagonal. The least-squares requirement is

$$\mathbf{VLU}^T \mathbf{ULV}^T \mathbf{VKU}^T = \mathbf{VLU}^T$$

which requires $\mathbf{L}^2 \mathbf{K} = \mathbf{L}$. Thus the psuedo-inverse satisifies the least-squares requirement. This implies that $\mathbf{K}$ is diagonal with elements $K_{qq} = 1/L_{qq}$ but does not place any restriction on elements K_{qq} for those q for which $L_{qq} = 0$. Putting $\mathbf{H}' = \mathbf{H} + \mathbf{VK}'\mathbf{U}^T$, where $\mathbf{K}'\mathbf{L} = \mathbf{K}'\mathbf{K} = \mathbf{0}$, it will be seen that $|\mathbf{Hc}|^2$ is minimised (with respect to elements of $\mathbf{K}'$) when $\mathbf{K}'_{qq} = 0$ for all q, i.e. when the least-squares criterion does not define a unique solution, the pseudo-inverse selects the minimum-length solution from the subspace of least-squares solutions.

2. *Derive equations for one-dimensional transport with a quadratic vertical profile.*

   (a) The quadratic profile corresponds to $c(p) = c_1 + a(p - p_1)^2$, whence $c_0 = c_1 + a(p_0 - p_1)^2$ and the surface gradient is $c_0' = 2a(p_0 - p_1)$ and the average is

   $$\bar{c} = c_1 + a(p_0 - p_1)/3 = c_0 - c_0'(p_0 - p_1)/3.$$

   (b) The vertical average of the transport equation is

   $$\frac{\partial \bar{c}}{\partial t} = \int_{p_0}^{p_1} \left[\frac{\partial}{\partial p} \left(\kappa_p \frac{\partial c}{\partial p} \right) \right] dp + \frac{1}{R_e^2 \sin \theta} \frac{\partial}{\partial \theta} \left(\kappa_\theta \sin \theta \frac{\partial \bar{c}}{\partial \theta} \right)$$

   $$= \frac{\kappa_p}{(p_0 - p_1)} \frac{\partial c}{\partial p} \bigg|_{p_0} + \frac{1}{R_e^2 \sin \theta} \frac{\partial}{\partial \theta} \left(\kappa_\theta \sin \theta \frac{\partial \bar{c}}{\partial \theta} \right)$$

   In the steady state, expanding $\bar{c}$ and s in terms of Legendre polynomials with coefficients a_n and s_n, respectively, gives the relation between components of the average (a_n) and components of the surface gradient d_n as

   $$d_n \kappa_p / (p_0 - p_1) - a_n \kappa_\theta n(n+1)/R_e^2 = 0$$

   while the quadratic expansion relates these to the surface concentration (expanded in c_n):

   $$a_n = c_n - d_n(p_0 - p_1)/3$$

   and the boundary condition relates gradients to sources as

   $$s_n = K \kappa_p d_n$$

   Eliminating d_n and a_n gives (in a notation comparable to that in Box 2.1)

   $$s_n \approx \frac{c_n \kappa_p}{p_0 - p_1} \frac{\gamma_n^2}{1 + \gamma_n^2/3}$$

   with

   $$\gamma_n^2 = \frac{1}{R_e^2} K \kappa_\theta (p_0 - p_1)^2 n(n+1)$$

   Therefore, this approximation represents the s_n-to-c_n relation for small n more accurately than does equating vertical averages to surface concentrations. It also has a form that suppresses the unbounded amplification of errors for large n [139]. However, there is no scope for adjusting the trade-off between accuracy and suppression of errors.

3. *Calculate the latitudinal resolution matrix in the diffusive model.*
   The resolution matrix is

   $$h_N(x, y) = \sum_{m=0}^{N} \frac{P_m(x) P_m(y)(2m + 1)}{2} \approx \delta(x - y)$$

The plots show $h_N(0.5, y)/N$ for $x = 0.5$ as a function of y for $N = 8, 16, 32$ and 64.

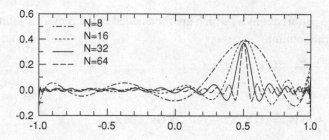

Chapter 9

1. *Prove that the data covariance is the sum of the covariances of model error and measurement error.*

   A more extensive derivation is given by Tarantola [471: p. 58]). However, the basic proposition can be proved by writing the expression of the joint distribution $p(\mathbf{c}, \mathbf{m}|\mathbf{x})$ in the form $p(\mathbf{m}|\mathbf{c}, \mathbf{x})p(\mathbf{c}|\mathbf{x})$, in which case integrating over $\mathbf{m}$ gives an integral of 1 for the factor $p(\mathbf{m}|\mathbf{c}, \mathbf{x})$ and leaves $p(\mathbf{c}|\mathbf{x})$ unchanged. The argument of the exponential in $p(\mathbf{m}, \mathbf{c}|\mathbf{x})$ is

   $$(\mathbf{c} - \mathbf{m})^\mathsf{T} \mathbf{X}(\mathbf{c} - \mathbf{m}) + [\mathbf{m} - \mathbf{g}(\mathbf{x})]^\mathsf{T} \mathbf{Y}(\mathbf{m} - \mathbf{g}(\mathbf{x}))$$

   To write this in the factorised form, the argument of the exponential needs to be written as

   $$[\mathbf{m} - \mathbf{Ac} - \mathbf{Bg}(\mathbf{x})]^\mathsf{T}[\mathbf{X} + \mathbf{Y}][\mathbf{m} - \mathbf{Ac} - \mathbf{Bg}(\mathbf{x})] + [\mathbf{c} - \mathbf{f}(\mathbf{x})]^\mathsf{T}\mathbf{Z}[\mathbf{c} - \mathbf{f}(\mathbf{x})]$$

   The terms in the first expression that are quadratic in $\mathbf{m}$ show that the inverse covariance for $p(\mathbf{m}|\mathbf{c}, \mathbf{x})$ must be $\mathbf{X} + \mathbf{Y}$. Comparing the terms that combine $\mathbf{c}$ and $\mathbf{m}$ shows that

   $$\mathbf{A}^\mathsf{T}[\mathbf{X} + \mathbf{Y}] = \mathbf{X}$$

   Terms quadratic in $\mathbf{c}$ require

   $$\mathbf{A}^\mathsf{T}[\mathbf{X} + \mathbf{Y}]\mathbf{A} + \mathbf{Z} = \mathbf{X}$$

   whence

   $$\mathbf{Z} = \mathbf{X} - \mathbf{X}[\mathbf{X} + \mathbf{Y}]\mathbf{X}$$

   so, from the matrix-inversion lemma (problem 1 of Chapter 3),

   $$\mathbf{Z}^{-1} = \mathbf{X}^{-1} - [\mathbf{X} - (\mathbf{X} + \mathbf{Y})]^{-1} = \mathbf{X}^{-1} + \mathbf{Y}^{-1}$$

   as required.

2. *Determine the limit of iterative inversion if a transport model is non-linear.*
   The iterative form involves repeatedly treating $\mathbf{c} - \mathbf{Gs}$ as the data, i.e.

minimisation of

$$J = (\mathbf{c} - \mathbf{Gs} - \mathbf{G's'})^\mathsf{T}\mathbf{X}(\mathbf{c} - \mathbf{Gs} - \mathbf{G's'}) + (\mathbf{s} + \mathbf{s'})^\mathsf{T}\mathbf{W}(\mathbf{s} + \mathbf{s'})$$

with respect to perturbations $\mathbf{s'}$, which are then added to the previous solution $\mathbf{s}$. This leads to the solution

$$-(\mathbf{G'})^\mathsf{T}\mathbf{X}(\mathbf{c} - \mathbf{Gs} - \mathbf{G's'}) + \mathbf{W}(\mathbf{s} + \mathbf{s'})$$

whence

$$[(\mathbf{G'})^\mathsf{T}\mathbf{XG'} + \mathbf{W}]\mathbf{s'} = (\mathbf{G'})^\mathsf{T}\mathbf{Xc} - (\mathbf{G'})^\mathsf{T}\mathbf{XGs} - \mathbf{Ws}$$

This converges, if at all, by having the right-hand side vanish so that

$$[(\mathbf{G'})^\mathsf{T}\mathbf{XG} + \mathbf{W}]\mathbf{s} = (\mathbf{G'})^\mathsf{T}\mathbf{Xc}$$

Chapter 10

1. *Show that the initial conditions in the integrating-factor solution of the DE* $(d/dt)m = \lambda m + s$ *is a particular case of the formalism of (10.1.6a)–(10.1.6e).* In terms of (10.1.6a)–(10.1.6e) (putting $t_0 = 0$),

$$\mathcal{L} = \frac{d}{dt} - \lambda$$

with the solution (for the boundary condition $m(0) = 0$) in terms of the Green function $G(t, t') = \xi(t)/\xi(t')$ with $\xi(t) = \exp[\int_0^t \lambda(t')\,dt']$.

For inhomogeneous boundary conditions, the simplest form of the auxiliary function $h(t)$ is $h(t) = m(0)$. Equations (10.1.6a)–(10.1.6e) define the solution as

$$m(t) = m(0) + \int_0^t \{G(t, t')[s(t') + \lambda(t')m(0)]\}dt'$$

From the definition of ξ, we have $(d/dt)\,(1/\xi) = -\lambda/\xi$ and so the terms proportional to $m(0)$ in the solution above are

$$m(0) - \xi(t)\int_0^t \frac{\lambda(t')}{\xi(t')}\,dt' = m(0)\left[1 - \xi(t)\left(\frac{1}{\xi(t)} - \frac{1}{\xi(0)}\right)\right]$$
$$= m(0)\xi(t)$$

in agreement with Box 1.1.

2. *Find the equivalent of the integrating factor for model A with boundary conditions $m(0) = m(T)$.*
   (a) From the integrating-factor solution, non-trivial solutions with $s(t) = 0$ occur if and only if $\xi(T) = \xi(0) = 1$, i.e. $\int_0^T \lambda(t')\,dt' = 0$.
   (b) In the absence of non-trivial solutions for the homogeneous equation, the boundary conditions require

$$m(0) = m(T) = m(0)\xi(T) + \xi(T)\int_0^T \frac{s(t')}{\xi(t')}\,dt'$$

whence

$$m(0) = \frac{\xi(T)}{1 - \xi(T)} \int_0^T \frac{s(t')}{\xi(t')} \, dt'$$

(c) Since the previous solution is proportional to a weighted integral of $s(t)$, the factor $m(0)\xi(t)$ in the integrating-factor solution from Box 1.1 can be combined with the Green function for initial conditions to give

$$m(t) = \int_0^T G(t, t')s(t') \, dt'$$

with

$$G(t, t') = \begin{cases} \dfrac{1}{1 - \xi(T)} \dfrac{\xi(t)}{\xi(t')} & \text{for } t' < t \\[2mm] \dfrac{\xi(T)}{1 - \xi(T)} \dfrac{\xi(t)}{\xi(t')} & \text{for } t' > t \end{cases}$$

If $s(t)$ and $\lambda(t)$ are periodic with period T, then so is $m(t)$, so long as $\xi(T) \neq 1$.

3. *Calculate integrating factors for the two-box model.*

From equations (3) and (4) of Box 2.1, the integrating-factor formalism gives

$$m_N(t) + m_S(t) = [m_N(0) + m_S(0)]e^{-\lambda t} + \int_0^t e^{-\lambda(t-t')}[s_N(t') + s_S(t')] \, dt'$$

and

$$m_N(t) - m_S(t) = [m_N(0) - m_S(0)]e^{-(2\kappa+\lambda)t} + \int_0^t e^{-(2\kappa+\lambda)(t-t')}[s_N(t') - s_S(t')] \, dt'$$

whence

$$m_N(t) = \frac{1}{2}\bigg([m_N(0) + m_S(0)]e^{-\lambda t} + \int_0^t e^{-\lambda(t-t')}[s_N(t') + s_S(t')] \, dt' \\ + [m_N(0) - m_S(0)]e^{-(2\kappa+\lambda)t} + \int_0^t e^{-(2\kappa+\lambda)(t-t')}[s_N(t') - s_S(t')] \, dt' \bigg)$$

and

$$m_S(t) = \frac{1}{2}\bigg([m_N(0) + m_S(0)]e^{-\lambda t} + \int_0^t e^{-\lambda(t-t')}[s_N(t') + s_S(t')] \, dt' \\ - [m_N(0) - m_S(0)]e^{-(2\kappa+\lambda)t} - \int_0^t e^{-(2\kappa+\lambda)(t-t')}[s_N(t') - s_S(t')] \, dt' \bigg)$$

Chapter 11

1. *Calculate the time-dependence of the Kalman-filter estimate if the mean rate of growth is β, but the model assumes that $\beta = 0$.*

This is most easily derived from the Green-function form of Box 3.3. (The equivalence to the Kalman-filter form is in problem 7.2). The source estimate after n observations is

$$\hat{s}(n) = \left(\sum_{j=1}^{n} GXc(j)\right) \bigg/ \left(W + \sum_{j} GXG\right)$$

For $c(n) = \alpha + \beta n + \epsilon(n)$ this is

$$\hat{s}(n) = \left(\alpha + \frac{\beta(n+1)}{2} + \sum_{j} \frac{\epsilon(j)}{n}\right) \bigg/ \left(1 + \frac{W}{nGXG}\right)$$

(using the relation $\sum_{j=1}^{n} j = n(n+1)/2$). Thus, although the estimates of the source change in time, the use of an excessively restrictive model means that the time-variation of the estimate does not track the true time-variation of the source.

2. *Examine the role of AR(1) noise in mass-balance inversions for the N-box model (analytically for $N = 2$, numerically for $N \geq 2$).*

   For each hemisphere, the DE is of the form

   $$s_1 = \dot{m}_1 + am_1 + bm_2$$

   Expressing the derivative as $\dot{m}(n) = [m(n+1) - m(n-1)]/(2\,\Delta t)$, in terms of autocovariance functions $R_1(.)$ and $R_2(.)$ for m_1 and m_2, the autocovariance function $R_s(.)$ of s is

   $$R_s(0) = \frac{1}{4\,\Delta t^2}[2R_1(0) - 2R_1(2)] + a^2 R_1(0) + b^2 R_2(0)$$

   $$R_s(1) = \frac{1}{4\,\Delta t^2}[R_1(1) - R_1(3)] + a^2 R_1(1) + b^2 R_2(1)$$

   and, for $n \geq 2$,

   $$R_s(n) = \frac{1}{4\,\Delta t^2}[2R_1(n) - R_1(n-2) - R_1(n+2)] + a^2 R_1(n) + b^2 R_2(n)$$

   since terms $\propto a/\Delta t$ vanish due to symmetry.

Chapter 12

1. *Prove that the median is the maximum-likelihood estimator of a quantity with errors given by the two-sided exponential distribution.*

   For a central value of a, the distribution of the c_j is

   $$p(c_j) = \frac{1}{2\gamma} \exp(-\gamma|a - c_j|)$$

For N data points,

$$p(\mathbf{c}) = \left(\frac{1}{2\gamma}\right)^N \exp\left(-\gamma \sum |a - c_j|\right)$$

On differentiating with respect to a, the likelihood is maximised when

$$N^+ - N^- = 0$$

where N^+ is the number of observations with $a - c_j > 0$ and N^- is the number of observations with $a - c_j < 0$, i.e. when a is the median (if one of the data values equals the median) or a is between two data values that immediately bracket the median.

2. *Simulate sampling distributions of the mean and median for estimating centres of Gaussian and double-exponential distributions.*

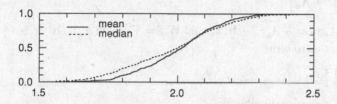

An example generated using 400 samples of 50 normally distributed values with mean $= 2$ and variance $= 1$, showing that the sampling distribution of the median is somewhat wider than that of the mean. Cases with many fewer samples exhibit variability comparable to the widths of the distributions.

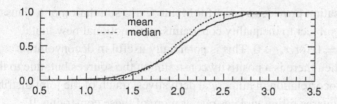

For 50 values with a double-exponential distribution the median has a narrower sampling distribution (a simulation of 400 cases).

3. *(a) Use the block-matrix inverse to factorise a multivariate Gaussian posterior distribution into a product of conditional probabilities. (b) For a set of independent items of information about each component, express the new posterior in factorised form.*
Putting

$$p_{\text{inferred}}(\mathbf{x}) \propto \exp[-(\mathbf{x} - \mathbf{z}')\mathbf{W}'(\mathbf{x} - \mathbf{z}')]$$

we put

$$\Pr(x_1, \ldots, x_n) \propto \exp[-(\mathbf{x}_n - \mathbf{z}_n)\mathbf{W}_n(\mathbf{x}_n - \mathbf{z}_n)]$$

We want to factorise this as

$$\Pr(x_1, \ldots, x_n) = \Pr(x_n | x_1, \ldots, x_{n-1}) \Pr(x_1, \ldots, x_{n-1})$$

The factorisation corresponds to the equality

$$
\begin{aligned}
(\mathbf{x}_n - \mathbf{z}_n)^{\mathsf{T}} \mathbf{W}_n (\mathbf{x}_n - \mathbf{z}_n) &= (\mathbf{x}_m - \mathbf{z}_m)^{\mathsf{T}} \mathbf{V}_m (\mathbf{x}_m - \mathbf{z}_m) + (\mathbf{x}_m - \mathbf{z}_m)^{\mathsf{T}} \mathbf{b} (x_n - z_n) \\
&\quad + (x_n - z_n) \mathbf{b}^{\mathsf{T}} (\mathbf{x}_m - \mathbf{z}_m) + (x_n - z_n) A_n (x_n - z_n) \\
&= (x_n - y_n) A_n (x_n - y_n) \\
&\quad + (\mathbf{x}_{n-1} - \mathbf{z}_{n-1})^{\mathsf{T}} \mathbf{W}_{n-1} (\mathbf{x}_{n-1} - \mathbf{z}_{n-1}) + c
\end{aligned}
$$

where $m = n - 1$ and we use the block decomposition $\mathbf{x}_n = [\mathbf{x}_m, x_n]$, $\mathbf{z}_n = [\mathbf{z}_m, z_n]$ and

$$
\mathbf{W}_n = \begin{bmatrix} \mathbf{V}_m & \mathbf{b} \\ \mathbf{b}^{\mathsf{T}} & A_n \end{bmatrix}
$$

and have $-A_n y_n = -A_n z_n - \mathbf{b}^{\mathsf{T}} \mathbf{x}_m$ and $\mathbf{W}_m = \mathbf{V}_m - \mathbf{b}(A_n)^{-1} \mathbf{b}^{\mathsf{T}}$.

This allows us to write

$$\Pr(\mathbf{x}) = \prod \Pr(x_n | x_1 \ldots x_{n-1})$$

(b) The new posterior distribution is

$$
\begin{aligned}
p_{\text{inferred}}(\mathbf{x}) &= \frac{\prod [\Pr(x_n | x_1 \ldots x_{n-1}) p_{\text{new}}(q_n | x_n)]}{\int \prod [\Pr *(x_n | x_1 \ldots x_{n-1}) p_{\text{new}}(q_n | x_n)]} \\
&= \prod \left[\frac{[\Pr(x_n | x_1 \ldots x_{n-1}) p_{\text{new}}(q_n | x_n)]}{\int [\Pr(x_n | x_1 \ldots x_{n-1}) p_{\text{new}}(q_n | x_n)]} \right]
\end{aligned}
$$

This factorisation provides a basis for Monte Carlo simulation of a Gaussian distribution subject to inequality constraints, with notional new 'data' $p_{\text{new}}(q_n | x_n) = 0$ for $x_n < 0$. This is potentially useful in deconvolution problems when there is a positivity constraint on the sources but, due to the negative autocorrelation in numerical derivatives, much of the joint distribution from least-squares fitting violates one or more of these constraints [C. Trudinger, personal communication, 2001]. However, if the inequalities greatly restrict the solution space, it is still a difficult problem to generate an unbiased sample of realisations.

Chapter 13

1. *Determine the utility of increasing the number of data by a factor of k for estimating a trend in linear regression.*

From problem 2 of Chapter 1, we have

$$E[(\hat{\beta} - \beta)^2] \sim \frac{12R}{N^3}$$

(a) Extending the range of observations by a factor of k corresponds to replacing N by kN. The variance is scaled by a factor of $1/k^3$ (i.e. reduced for $k > 1$).

(b) Changing the density of data within the same range of the independent variable replaces N by kN and β by $\beta' = \beta/k$, where β' is the gradient relative to a data-spacing of unity.

Thus

$$E[(\hat{\beta}' - \beta')^2] \sim \frac{12R}{N^3k^3}$$

whence

$$E[(\hat{\beta} - \beta)^2] \sim \frac{12R}{N^3k}$$

Case (b) is the common situation of doubling the amount of data, which halves the variance of the estimate. Case (a) represents a factor of k from a k-fold change in the amount of data and a factor of k^2 because the trend is being estimated from a longer baseline.

2. *Determine the utitility of additional observation using the recursive-regression formalism.*

For a quantity defined by $\alpha = \mathbf{a} \cdot \mathbf{x}$, after n measurements $\sigma_0 = \text{var } \alpha = \mathbf{a}^T \mathbf{P}(n)\mathbf{a}$ in state-space notation. A direct measurement of α corresponds to $\mathbf{H} = \mathbf{a}$. Using the recursive-regression relations from Box 4.4,

$$\hat{\mathbf{P}}(n+1) = \hat{\mathbf{P}}(n) - \hat{\mathbf{P}}(n)\mathbf{a}[\mathbf{a}^T\hat{\mathbf{P}}(n)\mathbf{a} + R]^{-1}\mathbf{a}^T\hat{\mathbf{P}}(n)$$

so that the revised variance is

$$\sigma = \text{var } \alpha = \mathbf{a}^T\hat{\mathbf{P}}(n+1)\mathbf{a} = \sigma_0 - \frac{(\sigma_0)^2}{\sigma_0 + R} = \frac{R\sigma_0}{\sigma_0 + R}$$

so the reduction in variance is $(\sigma_0)^2/(\sigma_0 + R)$.

References

1. Acton, F. S. (1970) *Numerical Methods That Work*. (Harper and Row: New York.)

2. Albritton, D. L., Derwent, R. G., Isaksen, I. S. A., Lal, M. and Wuebbles, D. J. (1995) Trace gas radiative forcing indices, pp. 205–231 of *Climate Change 1994: Radiative Forcing of Climate Change and an Evaluation of the IPCC IS92 Emission Scenarios*. Eds. J. T. Houghton, L. G. Meira Filho, J. Bruce, Hoesung Lee, B. A. Callander, E. Haites, N. Harris and K. Maskell. (Published for the IPCC by CUP: Cambridge.)

3. Albritton, D. L. and Meira Filho, L. G. (co-ordinating lead authors) (2001) *Technical Summary*, in *Climate Change 2001: The Scientific Basis*. Eds. J. T. Houghton, Y. Ding, D. J. Griggs, N. Noguer, P. J. van der Linden, X. Dai, K. Maskell and C. A. Johnson. (Published for the IPCC by CUP: Cambridge.)

4. Alcamo, J. (Ed.) (1994) *IMAGE 2.0: Integrated Modeling of Global Climate Change*. (Kluwer: Dordrecht). Reprinted from *Water, Air and Soil Pollution*, **76**.

5. Alcock, E., Gemmill, R. and Jones, K. C. (1999) Improvements to the UK PCDD/F and PCB atmospheric emission inventory following an emissions measurement programme. *Chemosphere*, **38**, 759–770.

6. Allan, R., Lindesay, J. and Parker, D. (1996) *El Niño Southern Oscillation and Climate Variability*. (CSIRO Publishing: Collingwood, Australia.)

7. van Amstel, A., Olivier, J. and Janssen, L. (1999) Analysis of differences between national inventories and an Emissions Database for Global Atmospheric Research (EDGAR). *Environmental Sci. Policy*, **2**, 275–293.

8. Anderssen, R. S. (1977) *Some Numerical Aspects of Improperly Posed Problems or Why Regularization Works and When Not to Use It*. Australian National University

Computer Centre. Technical Report No. 52.

9. Anderssen, R. S. and Bloomfield, P. (1974) Numerical differentiation procedures for non-exact data. *Numer. Math.*, **22**, 157–182.

10. Anderssen, R. S. and Jakeman, A. J. (1975) Abel type integral equations in stereology. II. Computational methods of solution and the random spheres approximation. *J. Micros.*, **105**, 111–126.

11. Andres, R. J., Marland, G., Fung, I. and Matthews, E. (1996) A 1° × 1° distribution of carbon dioxide emissions from fossil fuel consumption and cement manufacture, 1950–1990. *Global Biogeochemical Cycles*, **10**, 419–429.

12. Andres, R. J., Marland, G., Boden, T. and Bischof, S. (2000) Carbon dioxide emissions from fossil fuel consumption and cement manufacture, 1751–1991, and an estimate of their isotopic composition and latitudinal distribution. Chapter 3 of *The Carbon Cycle*. Ed. T. Wigley and D. Schimel. (CUP: Cambridge.)

13. Andrews, D. G. and McIntyre, M. E. (1976) Planetary waves in horizontal and vertical shear: the generalized Eliassen–Palm relation and the mean zonal acceleration. *J. Atmos. Sci.*, **33**, 2031–2048.

14. Arrhenius, S. (1896) On the influence of carbonic acid in the air upon the temperature of the ground. *Phil. Mag. S. 5.* **41**, 237–276. (Reprinted in *The Legacy of Svante Arrhenius: Understanding the Greenhouse Effect*. Eds. H. Rodhe and R. Charlson (Royal Swedish Academy of Sciences: Stockholm.)

15. Arsham, H. (1998) Techniques for Monte Carlo optimizing. *Monte Carlo Methods Applications*, **4**, 181–229.

16. Athias, V., Mazzega, P. and Jeandel, C. (2000) Nonlinear inversions of a model of the oceanic dissolved-particulate exchanges, pp. 205–222 of *Inverse Methods in Global Biogeochemical Cycles*. Eds. P. Kasibhatla, M. Heimann, P. Rayner, N. Mahowald, R. G. Prinn and D. E. Hartley. Geophysical Monograph No. 114. (American Geophysical Union: Washington.)

17. Ayers, G. P. and Gillett, R. W. (1990) Tropospheric chemical composition: Overview of experimental methods in measurement. *Rev. Geophys.*, **28**, 297–314.

18. Bacastow, R. B. (1976) Modulation of atmospheric carbon dioxide by the Southern Oscillation. *Nature*, **261**, 116–118.

19. Backus, G. and Gilbert, F. (1968) The resolving power of gross earth data. *Geophys. J. R. Astr. Soc.*, **13**, 247–276.

20. Baker, D. F. (2000) An inversion method for determining time-dependent surface CO_2 fluxes, pp. 279–293 of *Inverse Methods in Global Biogeochemical Cycles*. Eds. P. Kasibhatla, M. Heimann, P. Rayner, N. Mahowald, R. G. Prinn and D. E. Hartley. Geophysical Monograph No. 114. (American Geophysical Union: Washington.)

21. Baker, D. F. (2001) *Sources and Sinks of Atmospheric CO_2 Estimated from Batch Least-Squares Inversions of CO_2 Concentration Measurements*. Ph. D. Thesis, Princeton University.

22. Bard, Y. (1974) *Nonlinear Parameter Estimation*. (Academic: New York.)

23. Bartello, P. (2000) Using low-resolution winds to deduce fine structure in tracers. *Atmosphere–Ocean*, **38**, 303–320.

24. Barth N. and Wunsch, C. (1990) Oceanographic experiment design by simulated annealing. *J. Phys. Oceanogr.*, **20**, 1249–1263.

25. Battle, M., Bender, M. L., Tans, P. P., White, J. W. C., Ellis, J. T., Conway, T. and Francey, R. J. (2000) Global carbon sinks and their variability inferred from atmospheric O_2 and $\delta^{13}C$. *Science*, **287**, 2467–2470.

26. Beardsmore, D. J. and Pearman, G. I. (1987) Atmospheric carbon dioxide measurements in the Australian region: data from surface observations. *Tellus*, **39B**, 42–66.

27. Bender, M. L., Tans, P. P., Ellis, J. T., Orchado, J. and Habfast, K. (1994) A high precision isotope ratio mass spectrometry method for measuring O_2/N_2 ratio of air. *Geochim. Cosmochim. Acta*, **58**, 4751–4758.

28. Bengtsson, L. (1975) 4-dimensional assimilation of meteorological observations. GARP Publication Series No. 15 (ICSU/WMO.)

29. Benitez-Nelson, C. and Buessler, K. O. (1999) Phosphorus 32, phosphorus 37, beryllium 7 and lead 210: atmospheric fluxes and utility in tracing stratosphere/troposphere exchange. *J. Geophys, Res.*, **104D**, 11 745–11 754.

30. Bergamaschi, P., Hein, R., Heimann, M. and Crutzen, P. J. (2000) Inverse modeling of the global CO cycle 1. Inversion of CO mixing ratios. *J. Geophys. Res.*, **105D**, 1909–1927.

31. Bergamaschi, P., Hein, R., Brenninkmeijer, C. A. M. and Crutzen, P. J. (2000) Inverse modeling of the global CO cycle 2. Inversion of $^{13}C/^{12}C$ and $^{18}O/^{16}O$ isotope ratios. *J. Geophys. Res.*, **105D**, 1929–1945.

32. Beswick, K. M., Simpson, T. W., Fowler, D., Choularton, T. W., Gallagher, M. W., Hargreaves, K. J., Sutton, M. A. and Kaye, A. (1998) Methane emissions on large scales. *Atmos. Environ.*, **32**, 3283–3291.

33. Biraud, S., Ciais, P., Ramonet, M., Simmonds, P., Kazan, V., Monfray, P., O'Doherty, S., Spain, T. G. and Jennings, S. G. (2000) European greenhouse gas emissions estimated from continuous atmospheric measurements and radon 222 at Mace Head, Ireland. *J. Geophys. Res.*, **105D**, 1351–1366.

34. Bloomfield, P., Heimann, M., Prather, M. and Prinn, R. (1994) Inferred lifetimes. Chapter 3 of *Report on Concentrations, Lifetimes and Trends of CFCs, Halons and Related Species.* Ed. J. A. Kaye, S. A. Penkett and F. M. Ormond. NASA Reference Publication 1339. (NASA: Washington.)

35. Boer, G. (2000) Climate model intercomparisons, pp. 443–464 of *Numerical Modeling of the Global Atmosphere.* Eds. P. Mote and A. O'Neill. (Kluwer: Dordrecht)

36. Boering, K. A., Wofsy, S. C., Daube, B. C., Schneider, H. R., Loewenstein, M., Podolske, J. R. and Conway, T. J. (1996) Stratospheric mean ages and transport rates from observations of carbon dioxide and nitrous oxide. *Science*, **274**, 1340–1343.

37. Bolin, B., and Keeling, C. D. (1963) Large-scale atmospheric mixing as deduced from the seasonal and meridional variations of carbon dioxide. *J. Geophys. Res.*, **68**, 3899–3920.

38. Bolin, B., Degens, E. T., Kempe, S. and Ketner, S. (Eds.) (1979) *The Carbon Cycle.* SCOPE 13. (Wiley: Chichester.)

39. Bolin, B., Björkström, A., Holmén, K. and Moore, B. (1983) The

simultaneous use of tracers for ocean circulation studies. *Tellus*, **35B**, 206–236.

40. de Boor, C. (1978) *A Practical Guide to Splines.* (Springer-Verlag: New York).

41. Bousquet, P., Ciais, P., Peylin, P. and Monfray, P. (1999) Inverse modelling of annual atmospheric CO_2 sources and sinks 1. Method and control inversion. *J. Geophys. Res.*, **104D**, 26 161–26 178.

42. Bousquet, P., Ciais, P., Peylin, P. and Monfray, P. (1999) Inverse modelling of annual atmospheric CO_2 sources and sinks 2. Sensitivity study. *J. Geophys. Res.*, **104D**, 26 179–26 193.

43. Bouwman, A. F., Lee, D. S., Asman, W. A. H., Dentener, F. J., van der Hoek, K. W. and Olivier, J. G. J. (1997) A global high-resolution emission inventory for ammonia. *Global Biogeochemical Cycles*, **11**, 561–587.

44. Bouwman, A. F., Derwent, R. G. and Dentner, F. J. (1999) Towards reliable global bottom-up estimates of temporal and spatial patterns of emissions of trace gases and aerosols from land-use related and natural sources, pp. 3–26 of *Approaches to Scaling of Trace Gas Fluxes in Ecosystems*. Ed. A. F. Bouwman. (Elsevier: Amsterdam.)

45. Boville, B. A. (2000) Towards a complete model of the climate system, pp. 419–442 of *Numerical Modeling of the Global Atmosphere*. Eds. P. Mote and A. O'Neill. (Kluwer: Dordrecht.)

46. Box, G. E. P. and Jenkins, G. M. (1970) *Time Series Analysis: Forecasting and Control.* (Holden-Day: San Francisco.)

47. Box, G. E. P. and Tiao, G. C. (1973) *Bayesian Inference in Statistical Analysis.* (Addison-Wesley: Reading, Massachusetts.)

48. Braatz, B., Jallow, B. P., Molnar, S., Murdiyarso, D., Perdomo, M. and Fitzgerald, J. F. (1996) *Greenhouse Gas Inventories: Interim Results from the US Country Studies Program.* (Kluwer: Dordrecht.)

49. Brasseur, G. P. and Madronich, S. (1992) Chemical-transport models, pp. 491–517 of *Climate System Modeling*. Ed. K. E. Trenberth. (CUP: Cambridge.)

50. Brasseur, G., Khattatov, B. and Walters, S. (1999) Modeling. Chapter 12 of *Atmospheric Chemistry and Global Change*. Eds. G. P. Brasseur, J. J. Orlando and G. S. Tyndall. (OUP: Oxford.)

51. Brasseur, G. P., Orlando, J. J. and Tyndall, G. S. (Eds.) (1999) *Atmospheric Chemistry and Global Change.* (OUP: Oxford.)

52. Brenninkmeijer, C. A. M., Manning, M. R., Lowe, D. C., Wallace, G., Sparks, R. J. and Volz-Thomas, A. (1992) Interhemispheric asymmetry in OH abundance inferred from measurements of atmospheric ^{14}CO. *Nature*, **356**, 50–52.

53. Brewer, A. W. (1949) Evidence for a world circulation provided by the measurements of helium and water vapour distribution in the stratosphere. *Q. J. Roy. Meteorol. Soc.*, **75**, 351–363.

54. Broecker, W. S. and Peng, T.-H. (1993) *Greenhouse Puzzles* (three parts). (Eldigo Press: Palisades, New York.)

55. Brost, R. A. and Heimann, M. (1991) The effect of the global background on a synoptic-scale simulation of tracer concentration. *J. Geophys. Res.*, **96D**, 15 415–15 425.

56. Brown, M. (1993) Deduction of emissions of source gases using an objective inversion algorithm and a chemical transport model. *J. Geophys. Res.*, **98D**, 12 639–12 660.

57. Brown, M. (1995) The singular value decomposition method applied to the deduction of the emissions and isotopic composition of atmospheric methane. *J. Geophys. Res.* **100D**, 11 425–11 446.

58. Bruhwiler, L., Tans, P. and Ramonet, M. (2000) A time-dependent assimilation and source retrieval technique for atmospheric tracers, pp. 265–277 of *Inverse Methods in Global Biogeochemical Cycles*. Eds. P. Kasibhatla, M. Heimann, P. Rayner, N. Mahowald, R. G. Prinn and D. E. Hartley. Geophysical Monograph No. 114. (American Geophysical Union: Washington.)

59. Buchmann, N. and Schulze, E.-D. (1999) Net CO_2 and H_2O fluxes of terrestrial ecosystems. *Global Biogeochemical Cycles*, **13**, 751–760.

60. Burrows, J. P. (1999) Current and future passive remote sensing techniques used to determine atmospheric constituents, pp. 317–347 of *Approaches to Scaling of Trace Gas Fluxes in Ecosystems*. Ed. A. F. Bouwman. (Elsevier: Amsterdam.)

61. Cannadell, J. G., Mooney, H. A., Baldocchi, D. D., Berry, J. A., Ehleringer, J. R., Field, C. B., Gower, S. T., Hollinger, D. Y., Hunt, J. E., Jackson, R. B., Running, S. W., Shaver, G. R., Steffen, W., Trumbore, S. E., Valentini, R. and Bond, B. Y. (2000) Carbon metabolism of the terrestrial biosphere: a multitechnique approach for improved understanding. *Ecosystems*, **3**, 115–130.

62. Cape, J. N., Methven, J. and Hudson, L. E. (2000) The use of trajectory cluster analysis to interpret trace gas measurements at Mace Head, Ireland. *Atmos. Environ.*, **34**, 3651–3663.

63. Cárdenas, L. M., Austin, J. F., Burgess, R. A., Clemitshaw, K. C., Dorling, S., Penkett, S. A. and Harrison, R. M. (1998) Correlations between CO, NO_y, O_3 and non-methane hydrocarbons and their relationships with meteorology during winter 1993 on the north Norfolk coast, U. K. *Atmos. Environ.*, **32**, 3339–3351.

64. Chameides, W. L. (committee chair), Anderson, J. G., Carroll, M. A., Hales, J. M., Hofmann, D. J., Huebert, B. J., Logan, J. A., Ravishankara, A. R., Schimel, D. and Tolbert, M. A. (2000) Atmospheric chemistry research entering the twenty-first century, pp. 107–337 of *The Atmospheric Sciences Entering the Twenty-First Century*. (National Research Council Report.) (National Academy Press: Washington.)

65. Cho, J. Y. N., Zhu, Y., Newell, R. E., Anderson, B. E., Barrick, J. D., Gregory, G. L., Sachse, G. W., Carroll, M. A. and Albercock, G. M. (1999) Horizontal wavenumber spectra of winds, temperature and trace gases during the Pacific Exploratory Missions: 1. Climatology. *J. Geophys. Res.* **104D**, 5697–5716.

66. Ciais, P., Tans, P. P., White, J. W. C., Trolier, M., Francey, R. J., Berry, J. A., Randall, D. R., Sellers, P. J., Collatz, J. G. and Schimel, D. S. (1995) Partitioning of ocean and land uptake of CO_2 as inferred by $\delta^{13}C$ measurements from the NOAA Climate Monitoring and Diagnostics Laboratory global air sampling network. *J. Geophys. Res.*, **100D**, 5051–5070.

67. Ciais, P., Tans, P. P., Trolier, M., White, J. W. C. and Francey, R. J. (1995) A large northern hemisphere terrestrial CO_2 sink indicated by the $^{13}C/^{12}C$ ratio of atmospheric CO_2. *Science*, **269**, 1098–1102.

68. Ciais, P., Denning, A. S., Tans, P. P., Berry, J. A., Randall, D. A., Collatz, G. J., Sellers, P. J., White, J. W. C., Trolier, M., Meijer, H. A. J., Francey, R. J., Monfray, P. and Heimann, M. (1997) A three-dimensional synthesis study of $\delta^{18}O$ in atmospheric CO_2 1. Surface fluxes. *J. Geophys. Res.*, **102D**, 5857–5872.

69. Ciais, P., Tans, P. P., Denning, A. S., Francey, R. J., Trolier, M., Meijer, H. A. J., White, J. W. C., Berry, J. A., Randall, D. A., Collatz, G. J., Sellers, P. J., Monfray, P. and Heimann, M. (1997) A three-dimensional synthesis study of $\delta^{18}O$ in atmospheric CO_2 2. Simulation with the TM2 transport model. *J. Geophys. Res.*, **102D**, 5873 5883.

70. Ciattaglia, L. (1983) Interpretation of atmospheric CO_2 measurements at Mt. Cimone (Italy) related to wind data. *J. Geophys. Res.*, **88C**, 1331–1338.

71. Cicerone, R. J. and Oremland, R. S. (1988) Biogeochemical aspects of atmospheric methane. *Global Biogeochemical Cycles*, **2**, 299–327.

72. Cleugh, H. A. and Grimmond, C. S. B. (2001) Modelling regional scale surface energy exchanges and CBL growth in a heterogeneous urban–rural landscape. *Boundary-Layer Meteorol.*, **98**, 1–31.

73. Cleveland, W. S. and Devlin, S. J. (1988) Locally-weighted regression: an approach to regression analysis by local fitting. *J. Am. Statist. Assoc.*, **83**, 596–610.

74. Cleveland, W. S., Freeny, A. E. and Graedel, T. E. (1983) The seasonal component of atmospheric CO_2: information from new approaches to the decomposition of seasonal time series. *J. Geophys. Res.*, **88C**, 10 934–10 946.

75. Coddington, E. A. and Levinson, N. (1955) *Theory of Differential Equations*. (McGraw-Hill: New York.)

76. Cohen, R. (and Johnson, H. S.) (1999) Johnson receives Revelle medal. *EOS (Trans. AGU)*, **80(3)**, 28.

77. Conny, J. M. (1998) The isotopic characterization of carbon monoxide in the troposphere. *Atmos. Environ.*, **32**, 2669–2683.

78. Conway, T. J., Tans, P. P., Waterman, L. S., Thoning, K. W., Kitzis, D. R., Masarie, K. A. and Zhang, N. (1994) Evidence for interannual variability of the carbon cycle from the National Oceanic and Atmospheric Administration/Climate Monitoring and Diagnostics Laboratory Global Air Sampling Network. *J. Geophys. Res.*, **99D**, 22 831–22 855.

79. Cornfield, J. (1967) Bayes' theorem. *Rev. Inst. Int. Statist.*, **35**, 34–49.

80. Courant, R. (1962) *Partial Differential Equations*. Volume 2 of *Methods of Mathematical Physics* by R. Courant and D. Hilbert. (Interscience: New York.)

81. Craig, I. J. D. and Brown, J. C. (1986) *Inverse Problems in Astronomy: A Guide to Inversion Strategies for Remotely Sensed Data*. (Adam Hilger: Bristol.)

82. Cramer, W., Kicklighter, D. W., Bondeau, A., Moore, B., Churkina, G., Nemry, B., Ruimy, A. Schloss, A. L. and the participants of the Potsdam

NPP model intercomparison (1999) Comparing global models of terrestrial net primary productivity (NPP): overview and key results. *Global Change Biol.*, **5**, 1–15.

83. Craven, P. and Wahba, G. (1979) Smoothing noisy data with spline functions: estimating the correct degree of smoothing by the method of generalized cross-validation. *Numer. Math.*, **31**, 377–403.

84. Cressie, N. A. C. (1993) *Statistics for Spatial Data.* (John Wiley and Sons: New York.)

85. Criss, R. E. (1999) *Principles of Stable Isotope Distribution.* (OUP: New York.)

86. Crutzen, P. J. (1973) A discussion of the chemistry of some minor constituents in the stratosphere and the troposphere. *PAGEOPH*, **106–108**, 1385–1399.

87. Crutzen, P. J. and Zimmermann, P. H. (1991) The changing photochemistry of the troposphere. *Tellus*, **43AB** *(Bolin-65 symposium issue)*, 136–151.

88. Cunnold, D. M. and Prinn, R. G. (1991) Comment on "Tropospheric OH in a three-dimensional chemical tracer model: an assessment based on observations of CH_3CCl_3" by C. M. Spivakovsky *et al. J. Geophys. Res.*, **96D**, 17 391–17 393.

89. Cunnold, D., Alyea, F. and Prinn, R. (1978) A methodology for determining the atmospheric lifetime of fluorocarbons. *J. Geophys. Res.*, **83C**, 5493–5500.

90. Cunnold, D. M., Prinn, R. G., Rasmussen, R. A., Simmonds, P. G., Alyea, F. N., Cardelino, C. A., Crawford, A. J., Fraser, P. J. and Rosen, R. D. (1983) The Atmospheric Lifetime Experiment. 3. Lifetime methodology and application to three years of $CFCl_3$ data. *J. Geophys. Res.*, **88C**, 8379–8400.

91. Cunnold, D. M., Weiss, R. F., Prinn, R. G., Hartley, D., Simmonds, P. G., Fraser, P. J., Miller, B., Alyea, F. N. and Porter, L. (1997) GAGE/AGAGE measurements indicating reductions in global emissions of CCl_3F and CCl_2F_2 in 1992–1994. *J. Geophys. Res.*, **102D**, 1259–1269.

92. Da Costa, G. and Steele, P. (1997) A low-flow analyser system for making measurements of atmospheric CO_2, pp. 16–20 of *Report of the Ninth WMO Meeting of Experts on Carbon Dioxide Concentration and Related Tracer Measurement Techniques.* Ed. R. Francey. Global Atmospheric Watch Report Series no. 132. (World Meteorological Organization: Geneva.)

93. Dai, A. and Fung, I. Y. (1993) Can climate variability contribute to the "missing" CO_2 sink? *Global Biogeochemical Cycles*, **7**, 599–609.

94. Daley, R. (1991) *Atmospheric Data Analysis.* (CUP: Cambridge.)

95. Daley, R. (1995) Estimating the wind field from chemical constituent observations: experiments with a one-dimensional extended Kalman filter. *Monthly Weather Rev.*, **123**, 181–198.

96. Dargaville, R. (1999) *The Variability of Atmospheric Carbon Dioxide and its Surface Sources.* Ph. D. Thesis, University of Melbourne.

97. Dargaville, R. J. and Simmonds, I. (2000) Calculating CO_2 fluxes by data assimilation coupled to a three dimensional mass balance inversion, pp. 255–264 of *Inverse Methods in Global Biogeochemical Cycles.* Eds.

P. Kasibhatla, M. Heimann, P. Rayner, N. Mahowald, R. G. Prinn and D. E. Hartley. Geophysical Monograph No. 114. (American Geophysical Union: Washington.)

98. Dargaville, R. J., Law, R. M. and Pribać, F. (2000) Implications of interannual variability in atmospheric circulation on modeled CO_2 concentrations and source estimates. *Global Biogeochemical Cycles*, **14**, 931–943.

99. DeFries, R. S., Field, C. B., Fung, I., Justice, C. O., Los, S., Matson, P. A., Matthews, E., Mooney, H. A., Potter, C. S., Prentice, K., Sellers, P. J., Townshend, J. R. G., Tucker, C. J., Ustin, S. L. and Vitousek, P. M. (1995) Mapping the land surface for global atmosphere-biosphere models: towards continuous distributions of vegetation's functional properties. *J. Geophys. Res.*, **100D**, 20 867–20 882.

100. DeMore, W. B., Sander, S. P., Golden, D. M., Hampson, R. F., Kurylo, M. J., Howard, C. J., Ravishankara, A. R., Kolb, C. E. and Molina, M. J. (1997) *Chemical Kinetics and Photochemical Data for Use in Stratospheric Modeling, Eval 12*. (NASA: Pasedena, California.)

101. Denman, K., Hofmann, E. and Marchant, H. (1996) Marine biotic responses to environmental change and feedbacks to climate, pp. 483–516 of *Climate Change 1995: The Science of Climate Change*. Eds. J. T. Houghton, L. G. Meira Filho, B. A. Callander, N. Harris, A. Kattenberg and K. Maskell. (Published for the IPCC by CUP: Cambridge.)

102. Denmead, O. T., Raupach, M. R., Dunin, F. X., Cleugh, H. A. and Leuning, R. (1996) Boundary-layer budgets for regional estimates of scalar fluxes. *Global Change Biol.*, **2**, 255–264.

103. Denmead, O. T., Leuning, R., Griffith, D. W. T. and Meyer, C. P. (1999) Some recent developments in trace gas flux measurement techniques, pp. 69–84 of *Approaches to Scaling of Trace Gas Fluxes in Ecosystems*. Ed. A.F. Bouwman. (Elsevier: Amsterdam.)

104. Denning, A. S. (1994) *Investigations of the Transport, Sources and Sinks of Atmospheric CO_2, Using a General Circulation Model*. (Ph. D. Thesis). Department of Atmospheric Science, paper No. 564. (Colorado State University: Fort Collins, Colorado.)

105. Denning, A. S., Fung, I. Y. and Randall, D. (1995) Latitudinal gradient of atmospheric CO_2 due to seasonal exchange with the land biota. *Nature*, **376**, 240–243.

106. Denning, A. S., Holzer, M., Gurney, K. R., Heimann, M., Law, R. M., Rayner, P. J., Fung, I. Y., Fan, S.-M., Taguchi, S., Friedlingstein, P., Balkanski, Y., Taylor, J., Maiss, M. and Levin, I. (1999) Three-dimensional transport and concentration of SF_6. A model intercomparison study (TransCom 2). *Tellus*, **51B**, 266–297.

107. Derwent, R. G., Simmonds, P. G., O'Doherty, S., Ciais, P. and Ryall, D. B. (1998) European source strengths and northern hemisphere baseline concentrations of radiatively active trace gases at Mace Head, Ireland. *Atmos. Environ.*, **32**, 3703–3715.

108. Derwent, R. G., Carslaw, N., Simmonds, P. G., Bassford, M., O'Doherty, S., Ryall, D. B., Pilling, M. J., Lewis, A. C. and McQuaid, J. B. (1999) Hydroxyl radical

concentrations estimated from measurements of trichloroethylene during the EASE/ACSOE campaign at Mace Head, Ireland during July 1996. *J. Atmos. Chem.*, **34**, 185–205.

109. Deutsch, R. (1965) *Estimation Theory.* (Prentice Hall: Englewood Cliffs, New Jersey.)

110. Dlugokencky, E. J., Masarie, K. A., Lang, P. M., Tans, P. P., Steele, L. P. and Nisbet, E. G. (1994) A dramatic decrease in the growth rate of atmospheric methane in the northern hemisphere during 1992. *Geophys. Res. Lett.*, **21**, 45–48.

111. Dlugokencky, E. J., Steele, L. P., Lang, P. M. and Masarie, K. A. (1994) The growth rate and distribution of atmospheric methane. *J. Geophys. Res.*, **99D**, 17 021–17 043.

112. Dlugokencky, E. J., Masarie, K. A., Lang, P. M. and Tans, P. P. (1998) Continuing decline in the growth rate of the atmospheric methane burden. *Nature*, **393**, 447–450.

113. Dobson, G. M. B. (1951) Recent work on the stratosphere. *Q. J. Roy. Meteorol. Soc.*, **77**, 448–492.

114. Draper, N. R. and Smith, H. (1981) *Applied Regression Analysis.* (2nd edn.) (John Wiley and Sons: New York.)

115. Dunse, B. L., Steele, L. P., Fraser, P. J. and Wilson, S. R. (2001) An analysis of Melbourne pollution episodes observed at Cape Grim from 1995–1998, in *Baseline Atmospheric Program (Australia) 1997–98.* Eds. N. W. Tindale, N. Derek and R. J. Francey. (Bureau of Meteorology and CSIRO Atmospheric Research: Australia.)

116. Ebel, A., Freidrich, R. and Rodhe, H. (Eds.) (1977) *Tropospheric Modelling and Emission Estimation.* Volume 7 of *Transport and Chemical Transformation of Pollutants in the Troposphere.* (Springer-Verlag: Berlin.)

117. ECMWF (1999) *ECMWF and WCRP/GEWEX workshop on Modelling and Data Assimilation for Land Surface Processes: 29 June–2 July 1998.* (ECMWF: Reading, UK.)

118. Elbern, H., Schmidt, H. and Ebel, A. (1997) Variational data assimilation for tropospheric chemistry modelling. *J. Geophys. Res.*, **102D**, 15 967–15 985.

119. Elliott, W. P. and Angell, J. K. (1987) On the relation between atmospheric CO_2 and equatorial sea-surface temperature. *Tellus*, **39B**, 171–183.

120. Elliott, W. P., Angell, J. K. and Thoning, K. W. (1991) Relation of atmospheric CO_2 to tropical sea-surface temperatures and precipitation. *Tellus*, **43B**, 144–155.

121. England, M. H. and Maier-Reimer, E. (2001) Using tracers to assess ocean models. *Rev. Geophys.*, **39**, 29–70.

122. Enting, I. G. (1985) A classification of some inverse problems in geochemical modelling. *Tellus*, **37B**, 216–229.

123. Enting, I. G. (1985) *A Strategy for Calibrating Atmospheric Transport Models.* Division of Atmospheric Research Technical Paper no. 9. (CSIRO: Australia.)

124. Enting, I. G. (1985) Green's functions and response functions in geochemical modelling. *PAGEOPH*, **123**, 328–343.

125. Enting, I. G. (1985) A lattice statistics model for the age distribution of air bubbles in polar ice. *Nature*, **351**, 654–655.

126. Enting, I. G. (1987) On the use of smoothing splines to filter CO_2 data. *J. Geophys. Res.*, **92D**, 10977–10984.

127. Enting, I. G. (1987) The interannual variation in the seasonal cycle of carbon dioxide concentration at Mauna Loa. *J. Geophys. Res.*, **92D**, 5497–5504.

128. Enting, I. G. (1989) Studies of baseline selection criteria for Cape Grim, Tasmania, pp. 51–60 of *The Statistical Treatment of CO_2 Data Records.* Ed. W. P. Elliott. NOAA Technical Memorandum ERL ARL-173 (US Department of Commerce: Washington.)

129. Enting, I. G. (1992) The incompatibility of ice-core CO_2 data with reconstructions of biotic CO_2 sources (II). The influence of CO_2-fertilised growth. *Tellus*, **44B**, 23–32.

130. Enting, I. G. (1993) Inverse problems in atmospheric constituent studies: III. Estimating errors in surface sources. *Inverse Problems*, **9**, 649–665.

131. Enting, I. G. (1999) *Characterising the Temporal Variability of the Global Carbon Cycle.* CSIRO Atmospheric Research Technical Paper no. 40. (CSIRO, Australia.) Electronic edition at http://www.dar.csiro.au/publications/Enting_2000a.pdf.

132. Enting, I. G. (2000) Constraints on the atmospheric carbon budget from spatial distributions of CO_2. Chapter 8 of *The Carbon Cycle*. Ed. T. Wigley and D. Schimel. (CUP: Cambridge.)

133. Enting, I. G. (2000) Green's function methods of tracer inversion, pp. 19–31 of *Inverse Methods in Global Biogeochemical Cycles*. Eds. P. Kasibhatla, M. Heimann, P. Rayner, N. Mahowald, R. G. Prinn and D. E. Hartley. Geophysical Monograph No. 114. (American Geophysical Union: Washington.)

134. Enting, I. G. and Mansbridge, J. V. (1987) The incompatibility of ice-core CO_2 data with reconstructions of biotic CO_2 sources. *Tellus*, **39B**, 318–325.

135. Enting, I. G. and Mansbridge, J. V. (1987) Inversion relations for the deconvolution of CO_2 data from ice cores. *Inverse Problems*, **3**, L63–69.

136. Enting, I. G. and Mansbridge, J. V. (1989) Seasonal sources and sinks of atmospheric CO_2: direct inversion of filtered data. *Tellus*, **41B**, 111–126.

137. Enting, I. G. and Mansbridge, J. V. Latitudinal distribution of sources and sinks of CO_2: results of an inversion study. *Tellus*, **43B**, 156–170.

138. Enting, I. G. and Newsam, G. N. (1990) Inverse problems in atmospheric constituent studies: II. Sources in the free atmosphere. *Inverse Problems*, **6**, 349–362.

139. Enting, I. G. and Newsam, G. N. (1991) An improved approach to one-dimensional modelling of meridional transport in the atmosphere. *Tellus*, **43B**, 76–79.

140. Enting, I. G. and Pearman, G. I. (1987) Description of a one-dimensional carbon cycle model calibrated using techniques of constrained inversion. *Tellus*, **39B**, 459–476.

141. Enting, I. G. and Pearman, G. I. (1993) Average global distributions of CO_2, pp. 31–64 of *The Global Carbon Cycle*. Ed. M. Heimann. (Springer-Verlag: Heidelberg.)

142. Enting, I. G. and Trudinger, C. M. (1992) Modelling studies of the small-scale variability in CO_2, pp. 10–16 of *Baseline Atmospheric Program: Australia 1990*. Ed. S. R. Wilson and J. L. Gras. (Bureau of

Meteorology (DEST) and CSIRO: Australia.)

143. Enting, I. G., Francey, R. J., Trudinger, C. M. and Granek, H. (1993) *Synthesis Inversion of Atmospheric CO$_2$ Using the GISS Tracer Transport Model*. CSIRO Division of Atmospheric Research, Technical Paper no. 29. (CSIRO: Australia.)

144. Enting, I. G., Wigley, T. M. L. and Heimann, M. (1994) *Future Emissions and Concentrations of Carbon Dioxide: Key Ocean/Atmosphere/Land Analyses*. CSIRO Division of Atmospheric Research Technical Paper no. 31. (CSIRO: Australia.)

145. Enting, I. G., Trudinger, C. M. and Francey, R. J. (1995) A synthesis inversion of the concentration and δ^{13}C of atmospheric CO$_2$. *Tellus*, **47B**, 35–52.

146. ETEX participants (1998) ETEX special issue of *Atmos. Environ.*, **32**, Issue 24.

147. Etheridge, D. M., Steele, L. P., Francey, R. J. and Langenfelds, R. L. (1998) Atmospheric methane between 1000 A. D. and present: evidence of anthropogenic emissions and climatic variability. *J. Geophys. Res.*, **103D**, 15 979–15 993.

148. Fabian, P., Borchers, R., Leifer, R., Subbaraya, B. H., Lal, S. and Boy, M. (1996) Global stratospheric distribution of halocarbons. *Atmos. Environ.*, **30**, 1787–1796.

149. Fan, S.-M., Gloor, M., Mahlman, J., Pacala, S., Sarmiento, J., Takahashi, T. and Tans, P. (1998) A large terrestrial carbon sink in North America implied by atmospheric and oceanic carbon dioxide data and models. *Science*, **282**, 442–446.

150. Fan, S.-M., Blaine, T. L. and Sarmiento, J. L. (1999) Terrestrial carbon sink in the northern hemisphere estimated from the atmospheric CO$_2$ difference between Mauna Loa and South Pole since 1959. *Tellus*, **51B**, 863–870.

151. Fan, S.-M., Sarmiento, J. L., Gloor, M. and Pacala, S. W. (1999) On the use of regularization techniques in the inverse modeling of atmospheric carbon dioxide. *J. Geophys. Res.*, **104D**, 21 503–21 512.

152. Farman, J. C., Gardiner, B. G. and Shanklin, J. D. (1985) Large losses of total ozone in Antarctica reveal seasonal ClO$_x$/NO$_x$ interaction. *Nature*, **315**, 207–210.

153. Feichter, J. (2000) Atmospheric chemistry and aerosol dynamics, pp. 353–374 of *Numerical Modeling of the Global Atmosphere*. Eds. P. Mote and A. O'Neill. (Kluwer: Dordrecht.)

154. Fisher, R. A. (1921) On the mathematical foundations of theoretical statistics. *Philos. Trans. Royal Soc. London*, **222**, 309–368.

155. Fisher, M. and Lary, D. J. (1995) Lagrangian four-dimensional variational data assimilation of chemical species. *Q. J. Roy. Meteorol. Soc.*, **121**, 1681–1704.

156. Foulger, B. and Simmonds, P. (1979) Drier for field use in the determination of trace atmospheric gases. *Anal. Chem.*, **51**, 1089–1090.

157. Francey, R. (Ed.) (1997) *Report of the Ninth WMO Meeting of Experts on Carbon Dioxide Concentration and Related Tracer Measurement Techniques*. Global Atmospheric Watch Report Series no. 132. (World Meteorological Organization: Geneva.)

158. Francey, R. J. and Tans, P. P. (1987) Latitudinal variation in oxygen-18 in CO$_2$. *Nature*, **327**, 495–497.

159. Francey, R. J., Tans, P. P., Allison, C. E., Enting, I. G., White, J. W. C. and Trolier, M. (1995) Changes in oceanic and terrestrial carbon uptake since 1982. *Nature*, **373**, 326–330.

160. Francey, R. J., Steele, L. P., Langenfelds, R. L., Lucarelli, M. P., Allison, C. E., Beardsmore, D. J., Coram, S. A., Derek, N., de Silva, F. R., Etheridge, D. M., Fraser, P. J., Henry, R. J., Turner, B., Welch, E. D., Spencer, D. A. and Cooper, L. N. (1996) Global atmospheric sampling laboratory (GASLAB): supporting and extending the Cape Grim trace gas programs, pp. 8–29 of *Baseline Atmospheric Program Australia 1993*. Eds. R. J. Francey, A. L. Dick and N. Derek. (Bureau of Meteorology and CSIRO: Australia.)

161. Franssens, G., Fonteyn, D., De Mazière, M. and Fussen, D. (2000) A comparison between interpolation and assimilation as cartography methods for the SAGE II aerosol product, pp. 155–169 of *Inverse Methods in Global Biogeochemical Cycles*. Eds. P. Kasibhatla, M. Heimann, P. Rayner, N. Mahowald, R. G. Prinn and D. E. Hartley. Geophysical Monograph No. 114. (American Geophysical Union: Washington.)

162. Friend, J. P. and Charlson, R. J. (1969) Double tracer techniques for studying air pollution. *Environmental Sci. Technol.*, **3**, 1181–1182.

163. Fung, I., Prentice, K., Matthews, E., Lerner, J. and Russell, G. (1983) Three-dimensional tracer model study of atmospheric CO_2: response to seasonal exchanges with the terrestrial biosphere. *J. Geophys. Res.*, **88C**, 1281–1294.

164. Fung, I. Y., Tucker, C. J. and Prentice, K. C. (1987) Application of advanced very high resolution radiometer vegetation index to study atmosphere–biosphere exchange of CO_2. *J. Geophys. Res.*, **92D**, 2999–3015.

165. Fung, I. Y., John, J., Lerner, J., Matthews, E., Prather, M., Steele, L. P. and Fraser, P. J. (1991) Three-dimensional model synthesis of the global methane cycle. *J. Geophys. Res.*, **96D**, 13 033–13 065.

166. Fung, I. Y., Field, C. B., Berry, J. A., Thompson, M. V., Randerson, J. T., Malmström, C. M., Vitousek, P. M., Collatz, G. J., Sellers, P. J., Randall, D. A., Denning, A. S., Badeck, F. and John, J. (1997) Carbon 13 exchanges between atmosphere and biosphere. *Global Biogeochemical Cycles*, **11**, 507–533.

167. Ganachaud, A. and Wunsch, C. (2000) Improved estimates of global ocean circulation, heat transport and mixing from hydrographic data. *Nature*, **408**, 453–457.

168. Garcia, R., Hess, P. and Smith, A. (1999) Atmospheric dynamics and transport. Chapter 2 of *Atmospheric Chemistry and Global Change*. Eds. G. P. Brasseur, J. J. Orlando and G. S. Tyndall. (OUP: Oxford.)

169. Garratt, J. R. (1992) *The Atmospheric Boundary Layer.* (CUP: Cambridge.)

170. Gates, W. L. (1992) AMIP: the Atmospheric Model Intercomparison Project. *Bull. Amer. Meteorol. Soc.*, **73**, 1962–1970.

171. Gates, W. L., Boyle, J. S., Covey, C., Dease, C. G., Doutriaux, C. M., Drach, R. S., Fiorino, M., Glecker, P. J., Hnilo, J. J., Marlais, S. M., Phillips, T. J., Potter, G. L., Santer, B. J., Sperber, K. R., Taylor, K. E. and Williams, D. N. (1999) An overview of the

results of the Atmospheric Model Intercomparison Project (AMIP). *Bull. Amer. Meteorol. Soc.*, **80**, 29–55.

172. Gaudry, A., Monfray, P., Polian, G., Bonsang, G., Ardouin, B., Jegou, A. and Lambert, G. (1991) Non-seasonal variations of atmospheric CO_2 concentrations at Amsterdam Island. *Tellus*, **43B**, 136–143.

173. Gelb, A. (Ed.) (1974) *Applied Optimal Estimation*. (MIT Press: Cambridge, Massachusetts.)

174. Genthon, C. and Armengaud, A. (1995) Radon 222 as a comparative tracer of transport and mixing in two general circulation models of the atmosphere. *J. Geophys. Res.*, **100D**, 2849–2866.

175. Gery, M. W., Whitten, G. Z., Killus, J. P. and Dodge, M. C. (1989) A photochemical kinetics mechanism for urban and regional scale computer modeling. *J. Geophys. Res.*, **94D**, 12 925–12 956.

176. Giering, R. (2000) Tangent linear and adjoint biogeochemical models, pp. 33–48 of *Inverse Methods in Global Biogeochemical Cycles*. Eds. P. Kasibhatla, M. Heimann, P. Rayner, N. Mahowald, R. G. Prinn and D. E. Hartley. Geophysical Monograph No. 114. (American Geophysical Union: Washington.)

177. Gill, A. E. (1982) *Atmosphere–Ocean Dynamics*. (Academic Press: New York.)

178. Gleik, J. (1987) *Chaos: Making a New Science*. (Viking: New York.)

179. GLOBAL-VIEW CO_2 (2000) *Cooperative Atmospheric Data Integration Project – Carbon Dioxide*. CD-ROM (NOAA CMDL: Boulder, Colorado.) (Also available by anonymous ftp from ftp.cmdl.noaa.gov on path ccg/co2/GLOBALVIEW.)

180. Gloor, M., Fan, S.-M., Pacala, S. and Sarmiento, J. (1999) A model-based evaluation of inversions of atmospheric transport, using annual mean mixing ratios, as a tool to monitor fluxes of nonreactive trace substances like CO_2 on a continental scale. *J. Geophys. Res.*, **104D**, 14 245–14 260.

181. Gloor, M., Fan, S.-M., Pacala, S. and Sarmiento, J. (2000) Optimal sampling of the atmosphere for purpose of inverse modelling. A model study. *Global Biogeochemical Cycles*, **14**, 407–428.

182. Gordon, R. and Herman, G. T. (1974) Three-dimensional reconstruction from projections: a review of algorithms. *Int. Rev. Cytol.*, **38**, 111–151.

183. Graedel, T. E., Bates, T. S., Bouwman, A. F., Cunnold, D., Dignon, J., Jacob, D. J., Lamb, B. K., Logan, J. A., Marland, G., Middleton, J. M., Placet, M. and Veldt, G. (1993) A compilation of inventories of emissions to the atmosphere. *Global Biogeochemical Cycles*, **7**, 1–26.

184. Green, J. (1999) *Atmospheric Dynamics*. (CUP: Cambridge.)

185. Grubb, M. (1999) *The Kyoto Protocol: A Guide and Assessment*. (Royal Institute for International Affairs: London.)

186. Guthrie, P. D. (1989) The CH_4–CO–OH conundrum: a simple analytic approach. *Global Biogeochemical Cycles*, **3**, 287–298.

187. Haas-Laursen, D. E., Hartley, D. E. and Prinn, R. G. (1996) Optimizing an inverse method to deduce time-varying emissions of trace gases. *J. Geophys. Res.*, **101D**, 22 823–22 831.

188. Haas-Laursen, D. E., Hartley, D. E. and Conway T. J. (1997) Consistent sampling methods for comparing models to CO_2 flask data. *J. Geophys. Res.*, **102D**, 19059–19071.

189. Hall, F. G. (1999) Introduction to the special section: BOREAS in 1999: experiment and science overview. *J. Geophys. Res.*, **104D**, 27627–27639.

190. Hall, T. M. and Plumb, R. A. (1994) Age as a diagnostic of stratospheric transport. *J. Geophys. Res.*, **99D**, 1059–1070.

191. Hall, T. M. and Waugh, D. W. (1997) Timescales for the stratospheric circulation derived from tracers. *J. Geophys. Res.*, **102D**, 8991–9001.

192. Hall, T. M. and Waugh, D. W. (1998) Influence of nonlocal chemistry on tracer distributions: inferring the mean age of air from SF_6. *J. Geophys. Res.*, **103D**, 13327–13336.

193. Hall, B. D., Elkins, J. W., Butler, J. H., Montzka, S. A., Thompson, T. M., del Negro, L., Dutton, G. S., Hurst, D. F., King, D. B., Kline, E. S., Lock, L., MacTaggart, D., Mondeel, D., Moore, F. L., Nance, J. D., Ray, E. A. and Romashkin, P. A. (2001) Halocarbons and other atmospheric trace species. *Climate Monitoring and Diagnostics Laboratory: Summary Report no. 25, 1998–1999*. Ed. R. C. Schnell, D. B. King and R. M. Rosson. (US Department of Commerce: Washington.)

194. Hardt, M. and Scherbaum, F. (1994) The design of optimum networks for aftershock recordings. *Geophys. J. Int.*, **117**, 716–726.

195. Harnisch, J., Borchers, R., Fabian, P. and Maiss, M. (1996) Tropospheric trends for CF_4 and C_2F_6 since 1982

derived from SF_6 dated stratospheric air. *Geophys. Res. Lett.*, **23**, 1099–1102.

196. Hartley, D. and Prinn, R. (1991) Comment on "Tropospheric OH in a three-dimensional chemical tracer model: an assessment based on observations of CH_3CCl_3" by C. M. Spivakovsky *et al. J. Geophys. Res.*, **96D**, 17383–17387.

197. Hartley, D. and Prinn, R. (1993) Feasibility of determining surface emissions of trace gases using an inverse method in a three-dimensional chemical transport model. *J. Geophys. Res.*, **98D**, 5183–5197.

198. Harvey, L. D. D. (2000) *Global Warming: The Hard Science*. (Prentice Hall (Pearson Education Ltd): Harlow, UK.)

199. Heard, D. E. (1998) New directions: measuring the elusive tropospheric hydroxyl radical. *Atmos. Environ.*, **32**, 801–802.

200. Heimann, M. (Ed.) (1993) *The Global Carbon Cycle*. (Springer-Verlag: Heidelberg.)

201. Heimann, M. (1998) A review of the contemporary global carbon cycle and as seen a century ago by Arrhenius and Högbom, pp. 43–57 of *The Legacy of Svante Arrhenius: Understanding the Greenhouse Effect*. Eds. H. Rodhe and R. Charlson (Royal Swedish Academy of Sciences: Stockholm.)

202. Heimann, M. and Maier-Reimer, E. (1996) On the relations between the oceanic uptake of CO_2 and its isotopes. *Global Biogeochemical Cycles*, **10**, 89–110.

203. Heimann, M. and Kaminski, T. (1999) Inverse modelling approaches to infer surface trace gas fluxes from observed atmospheric mixing ratios,

pp. 277–295 of *Approaches to Scaling of Trace Gas Fluxes in Ecosystems*. Ed. A.F. Bouwman. (Elsevier: Amsterdam.)

204. Heimann, M. and Keeling, C. D. (1986) Meridional eddy diffusion model of the transport of atmospheric carbon dioxide 1. Seasonal carbon cycle over the tropical Pacific ocean. *J. Geophys. Res.*, **91D**, 7765–7781.

205. Heimann, M., Keeling, C. D. and Tucker, C. J. (1989) A three-dimensional model of atmospheric CO_2 transport based on observed winds. 3. Seasonal cycle and synoptic time scale variations. In *Aspects of Climate Variability in the Pacific and the Western Americas. Geophysical Monograph 55*. Ed. D. H. Peterson. (AGU: Washington.)

206. Hein, R. and Heimann, M. (1994) Determination of global scale emissions of atmospheric methane using an inverse modelling method, pp. 271–281 of *Non-CO_2 Greenhouse Gases*. Ed. J. van Ham, L. H. J. M. Janssen and R. J. Swart. (Kluwer: Dordrecht.)

207. Hein, R., Crutzen, P. J. and Heimann, M. (1997) An inverse modeling approach to investigate the global methane cycle. *Global Biogeochemical Cycles*, **11**, 43–76.

208. Henderson-Sellers, A., Pitman, A. J., Love, P. K., Irannejad, P. and Chen, T. H. (1995) The project for intercomparison of land-surface parameterization schemes (PILPS): phases 2 and 3. *Bull. Amer. Meteorol. Soc.*, **76**, 489–503.

209. Henry, R. C., Lewis, C. W., Hopke, P. K. and Williamson, H. J. (1984) Review of receptor model fundamentals. *Atmos. Environ.*, **18**, 1507–1515.

210. Hering, W. S. (1966) Ozone and atmospheric transport processes. *Tellus*, **18**, 329–336.

211. Hesshaimer, V., Heimann, M. and Levin, I. (1994) Radiocarbon evidence for a smaller oceanic carbon dioxide sink than previously believed. *Nature*, **370**, 210–230.

212. Holton, J. R. (1992) *An Introduction to Dynamical Meteorology (3rd edn)*. (Academic Press: San Diego, California.)

213. Holton, J. R., Haynes, P. H., McIntyre, M. E., Douglass, A. R., Rood, R. B. and Pfister, L. (1995) Stratosphere–troposphere exchange. *Rev. Geophys.*, **33**, 403–439.

214. Houweling, S. (1999) *Global Modeling of Atmospheric Methane Sources and Sinks*. Doctoral Dissertation, University of Utrecht.

215. Houweling, S., Kaminski, T., Dentener, F., Lelieveld, J., and Heimann, M. (1999) Inverse modeling of methane sources and sinks using the adjoint of a global transport model. *J. Geophys. Res.*, **104D**, 26 137–26 160.

216. Houweling, S., Dentener, F., Lilieveld, J., Walter, B. and Dlugokencky, E. (2000) The modeling of tropospheric methane: how well can point measurements be reproduced by a global model? *J. Geophys. Res.*, **105D**, 8981–9002.

217. Houweling, S., Dentener, F. and Lelieveld, J. (2000) Simulation of preindustrial atmospheric methane to constrain the global source strength of natural wetlands. *J. Geophys. Res.*, **105**, 17 243–17 255.

218. Huang, J. (2000) *Optimal Determination of Global Tropospheric OH Concentrations Using Multiple Trace Gases*. (Ph. D. Thesis). Report

no. 65 of the Center for Global Change Science. (MIT: Cambridge, Massachusetts.)

219. Huber, P. J. (1981) *Robust Statistics.* (Wiley: New York.)

220. Hurrell, J. W., van Loon, H. and Shea, D. J. (1998) The mean state of the troposphere, pp. 1–46 of *Meteorology of the Southern Hemisphere.* Eds. D. J. Karoly and D. G. Vincent. Meteorological Monographs. Vol. 27. No. 49. (American Meteorological Society: Boston, Massachusetts.)

221. Hurst, D. F., Bakwin, P. S. and Elkins, J. W. (1998) Recent trends in the variability of halogenated trace gases over the United States. *J. Geophys. Res.*, **103D**, 25 299–25 306.

222. Ide, K., Courtier, P., Ghil, M. and Lorenc, A. (1997) Unified notation for data assimilation: operational, sequential and variational. *J. Meteorol. Soc. Jap.*, **57**, 181–189.

223. IPCC NGGIP (2000) *Good Practice Guidance and Uncertainty Management in National Greenhouse Gas Inventories.* (IPCC National Greenhouse Gas Inventories Program.) Electronic edition from http://www.ipcc.iges. or.jp/public/gp/gpgaum.htm (links to sections as pdf files).

224. Jackson, D. D. (1972) Interpretation of inaccurate, insufficient and inconsistent data. *Geophys. J. Roy. Astr. Soc.*, **28**, 97–109.

225. Jackson, M. A. (1975) *Principles of Program Design.* (Academic: London.)

226. Jacob, D. J., Prather, M. J., Wofsy, S. C. and McElroy, B. (1987) Atmospheric distribution of ^{85}Kr simulated with a general circulation model. *J. Geophys. Res.*, **92D**, 6614–6626.

227. Jacob, D. J., Prather, M. J., Rasch, P. J., Shia, R.-L., Balkanski, Y. J., Beagley, S. R., Bergmann, D. J., Blackshear, W. T., Brown, M., Chiba, M., Chipperfield, M. P., de Grandpré, J., Dignon, J. E., Feichter, J., Genthon, C., Grose, W. L., Kasibhatla, P. S., Köhler, I., Kritz, M. A., Law, K., Penner, J. E., Ramonet, M., Reeves, C. E., Rotman, D. A., Stockwell, D. Z., van Velthoven, P. F. J., Verver, G., Wild, O., Yang, H. and Zimmermann, P. (1997) Evaluation and intercomparisons of global atmospheric transport models using ^{222}Rn and other short-lived tracers. *J. Geophys. Res.*, **102D**, 5953–5970.

228. Jacobson, M. Z., and Turco, R. P. (1994) SMVGEAR: a sparse-matrix vectorized Gear code for atmospheric models. *Atmos. Environ.*, **28**, 273–284.

229. Jagovkina, S. V., Karol, I. L., Zubov, V. A., Lagun, V. E., Reshetnikov, A. I. and Rozanov, E. V. (2000) Reconstruction of the methane fluxes from the west Siberia gas fields by the 3D regional chemical transport model. *Atmos. Environ.*, **34**, 5319–5328.

230. James, I. N. (1994) *Introduction to Circulating Atmospheres.* (CUP: Cambridge.)

231. Jeffreys, H. (1948) *Theory of Probability* (2nd edn). (Clarendon: Oxford.)

232. Jobson, B. T., McKeen, S. A., Parrish, D. D., Fehsenfeld, F. C., Blake, D. R., Goldstein, A. H., Schauffler, S. M. and Elkins, J. W. (1999) Trace gas mixing ratio variability versus lifetime in the troposphere and stratosphere: observations. *J. Geophys. Res.*, **104D**, 16 091–16 113.

233. Joos, F., Bruno, M., Fink, R., Siegenthaler, U., Stocker, T., Le Quéré, C. and Sarmiento, J. L. (1996) An efficient and accurate representation of complex oceanic and biospheric models of anthropogenic carbon uptake. *Tellus*, **48B**, 397–417.

234. Junge, C. E. (1962) Note on the exchange rate between the northern and southern atmosphere. *Tellus*, **14**, 242–246.

235. Junge, C. E. (1962) Global ozone budget and exchange between stratosphere and troposphere. *Tellus*, **14**, 363–377.

236. Junge, C. E. (1974) Residence time and variability of tropospheric gases. *Tellus*, **26**, 477–487.

237. Junge, C. E. and Czeplak, G. (1968) Some aspects of the seasonal variation of carbon dioxide and ozone. *Tellus*, **20**, 422–434.

238. Kagan, R. L. (1997) *Averaging of Meteorological Fields*. (Trans. from 1979 Russian edition.) Eds. L. S. Gandin and T. M. Smith. (Kluwer: Dordrecht.)

239. Kaminski, T. (1998) *On the Benefit of the Adjoint Technique for Inversion of the Atmospheric Transport Employing Carbon Dioxide as an Example of a Passive Tracer.* Ph. D. Thesis, Universität Hamburg.

240. Kaminski, T., Heimann, M. and Giering, R. (1997) A global scale inversion of the transport of CO_2 based on a matrix representation of an atmospheric transport model derived by its adjoint. Presented at 5th International CO_2 Conference. Cairns, September 1997. pp147–148 of extended abstracts.

241. Kaminski, T., Rayner, P. J., Heimann, M. and Enting, I. G. (2001) On aggregation errors in atmospheric transport inversions. *J. Geophys. Res.*, **106**, 4703–4715.

242. Kandlikar, M. (1997) Bayesian inversion for reconciling uncertainties in global mass balances. *Tellus*, **49B**, 123–135.

243. Karoly, D. J., Vincent, D. G. and Schrage, J. M. (1998) General circulation, pp. 47–85 of *Meteorology of the Southern Hemisphere*. Eds. D. J. Karoly and D. G. Vincent. Meteorological Monographs. Vol. 27. No. 49. (American Meteorological Society: Boston, Massachusetts.)

244. Kasibhatla, P., Heimann, M., Rayner, P., Mahowald, N., Prinn, R. J. and Hartley, D. E. (Eds.) (2000) *Inverse Methods in Global Biogeochemical Cycles.* Geophysical Monograph no. 114. (American Geophysical Union: Washington.)

245. Keeling, C. D. and Heimann, M. (1986) Meridional eddy diffusion model of the transport of atmospheric carbon dioxide 2. Mean annual carbon cycle. *J. Geophys. Res.*, **91D**, 7782–7796.

246. Keeling, C. D., Guenther, P. R. and Moss, D. J. (1986) *Scripps Reference Gas Calibration System For Carbon Dioxide-In-Air Standards: Revision 1985.* Environmental Pollution Monitoring and Research Programme, Publication no. 42 (WMO/TD-125). (WMO: Geneva.)

247. Keeling, C. D., Bacastow, R. B., Bainbridge, A. E., Ekdahl, C. A., Guenther, P. R., Waterman, L. S. and Chin, J. (1976) Atmospheric carbon dioxide variations at Mauna Loa observatory, Hawaii. *Tellus*, **28**, 538–551.

248. Keeling, C. D., Bacastow, R. B., Carter, A. F., Piper, S. C., Whorf, T. P., Heimann, M., Mook, W. G. and

Roeloffzen, H. (1989) A three-dimensional model of atmospheric CO_2 transport based on observed winds. 1: Analysis of observational data. In *Aspects of Climate Variability in the Pacific and the Western Americas. Geophysical Monograph 55.* Ed. D. H. Peterson. (AGU: Washington.)

249. Keeling, C. D., Piper, S. C. and Heimann, M. (1989) A three-dimensional model of atmospheric CO_2 transport based on observed winds. 4: Mean annual gradients and interannual variations. In *Aspects of Climate Variability in the Pacific and the Western Americas. Geophysical Monograph 55.* Ed. D. H. Peterson. (AGU: Washington.)

250. Keeling, C. D., Whorf, T. P., Wahlen, M. and van der Plicht, J. (1995) Interannual extremes in the rate of rise of atmospheric carbon dioxide since 1980. *Nature*, **375**, 666–670.

251. Keeling, R. F. (1988) *Development of an Interferometric Oxygen Analyzer for Precise Measurement of the Atmospheric O_2 Mole Fraction.* Ph. D. Thesis, Harvard.

252. Keeling, R. F., Najjar, R. P., Bender, M. L. and Tans, P. P. (1993) What atmospheric oxygen measurements can tell us about the global carbon cycle. *Global Biogeochemical Cycles*, **7**, 37–67.

253. Keilis-Borok, V. I. and Yanovskaya, T. B. (1967) Inverse problems in seismology. *Geophys. J.*, **13**, 223–234.

254. Kendall, M. G. and Stuart, A. (1967) *The Advanced Theory of Statistics. Volume 2: Inference and Relationship.* (Charles Griffin and Co.: London.)

255. Kernighan, B. W. and Plauger, P. J. (1978) *The Elements of Programming Style* (2nd edn). (McGraw-Hill: New York.)

256. Khalil, M. A. K. and Rasmussen, R. A. (1999) Atmospheric methyl chloride. *Atmos. Environ.*, **33**, 1305–1321.

257. Khalil, M. A. K. and Shearer, M. J. (Eds.) (1993) *Atmospheric Methane: Sources, Sinks and Role in Global Change.* Proceedings of the NATO Advanced Research Workshop, Mt. Hood, Oregon, October 1991. *Chemosphere*, **26**, nos. 1–4. (Pergamon: Oxford.)

258. Khalili, N. R., Scheff, P. A. and Holsen, T. M. (1995) PAH source fingerprints for coke ovens, diesel and gasoline engines, highway tunnels, and wood combustion emissions. *Atmos. Environ.*, **29**, 533–542.

259. Khattatov, B. V., Lamarque, J.-F., Lyjak, L. V., Menard, R., Levelt, P., Tie, X., Brasseur, G. P. and Gille, J. C. (2000) Assimilation of satellite observations of long-lived chemical species in global chemistry transport models. *J. Geophys. Res.*, **105D**, 29 135–29 144.

260. Kheshgi, H. S., Jain, A. K., Kotamarthi, V. R. and Wuebbles, D. J. (1999) Future atmospheric methane concentrations in the context of the stabilization of greenhouse gas concentrations. *J. Geophys. Res.*, **104D**, 19 183–19 190.

261. Knorr, W. and Heimann, M. (1995) Impact of drought stress and other factors on seasonal land biosphere CO_2 exchange studied through an atmospheric tracer transport model. *Tellus*, **47B**, 471–489.

262. Kowalczyk, E. A. and McGregor, J. L. (2000) Modeling trace gas concentrations at Cape Grim using the CSIRO Division of Atmospheric

Research Limited Area Model (DARLAM). *J. Geophys. Res.*, **105D**, 22 167–22 183.

263. Kraus, E. B. and Businger, J. A. (1994) *Atmosphere–Ocean Interaction* (2nd edn). (OUP: New York.)

264. Kroeze, C., Mosier, A. and Bouwman, L. (1999) Closing the global N_2O budget: A retrospective analysis 1500–1994. *Global Biogeochemical Cycles*, **13**, 1–8.

265. Krol, M. S. and van der Woerd, H. J. (1994) Atmospheric composition calculations for evaluation of climate scenarios. *Water, Air and Soil Pollution*, **76**, 259–281. (Reprinted in *IMAGE 2.0: Integrated Modeling of Global Climate Change*. Ed. J. Alcamo. (Kluwer: Dordrecht.))

266. Krol, M., van Leeuwen, P. and Lelieveld, J. (1998) Global OH trend inferred from methylchloroform measurements. *J. Geophys. Res.*, **103D**, 10 697–10 711.

267. Kruijt, B., Dolman, A. J., Lloyd, J., Ehleringer, J., Raupach, M. and Finnigan, J. (2001) Assessing the regional carbon balance: towards an integrated, multiple constraints approach. *Change*, **56** (March–April 2001), 9–12.

268. Langenfelds, R. L., Fraser, P. J., Francey, R. J., Steele, L. P., Porter, L. W. and Allison, C. E. (1996) The Cape Grim Air Archive: the first seventeen years, pp. 53–70 of *Baseline Atmospheric Program (Australia) 1994–95*. Eds. R. J. Francey, A. L. Dick and N. Derek. (Bureau of Meteorology and CSIRO: Melbourne.)

269. Langenfelds, R. L., Francey, R. J., Steele, L. P., Battle, M., Keeling, R. F. and Budd, W. F. (1999) Partitioning of the global fossil CO_2 sink using a 19-year trend in atmospheric O_2. *Geophys. Res. Lett.*, **26**, 1897–1900.

270. Langenfelds, R. L., Francey, R. J., Steele, L. P., Keeling, R. F., Bender, M. L., Battle, M., and Budd, W. F. (1999) Measurements of O_2/N_2 ratio from the Cape Grim air archive and three independent flask sampling programs, pp. 57–70 of *Baseline Atmospheric Program Australia: 1996*. Eds. J. L. Gras, N. Derek, N. W. Tindale and A. L. Dick. (Bureau of Meteorology and CSIRO Atmospheric Research: Australia).

271. Lary, D. J., Hall, S. and Fisher, M. (1999) Atmospheric chemical data assimilation, pp. 99–105 of *SODA Workshop on Chemical Data Assimilation: Proceedings*. (KNMI (Royal Netherlands Meteorological Institute): de Bilt, The Netherlands.)

272. Lassey, K. R., Lowe, D. C. and Manning, M. R. (2000) The trend in atmospheric methane $\delta^{13}C$ and implications for isotopic constraints on the global methane budget. *Global Biogeochemical Cycles*, **14**, 41–49.

273. Law, R. M. (1996) The selection of model-generated CO_2 data: a case study with seasonal biospheric sources. *Tellus*, **48B**, 474–486.

274. Law, R. M. (1999) CO_2 sources from a mass-balance inversion: sensitivity to the surface constraint. *Tellus*, **51B**, 254–265.

275. Law, R. M. and Rayner, P. J. (1999) Impacts of seasonal covariance on CO_2 inversions. *Global Biogeochemical Cycles*, **13**, 845–856.

276. Law, R. M. and Vohralik, P. (2001) *Methane Sources from Mass-Balance Inversions: Sensitivity to Transport*. CSIRO Atmospheric Research

Technical Paper no. 50. Electronic edition at http://www.dar.csiro.au/publications/Law_2000a.pdf.

277. Law, R. M., Rayner, P. J., Denning, A. S., Erickson, D., Fung, I. Y., Heimann, M., Piper, S. C., Ramonet, M., Taguchi, S., Taylor, J. A., Trudinger, C. M. and Watterson, I. G. (1996) Variations in modeled atmospheric transport of carbon dioxide and the consequences for CO_2 inversions. *Global Biogeochemical Cycles*, **10**, 783–796.

278. Lee, K., Wanninkhof, R., Takahashi, T., Doney, S. C., and Feely, R. A. (1998) Low interannual variability in recent oceanic uptake of atmospheric carbon dioxide. *Nature*, **396**, 155–159.

279. Le Quéré, C., Orr, J. C., Monfray, P., Aumont, O. and Madec, G. (2000) Interannual variability of the oceanic sink of CO_2 from 1979 through 1997. *Global Biogeochemical Cycles*, **14**, 1247–1265.

280. Lerner, J., Matthews, E. and Fung, I. (1988) Methane emissions from animals: a global high-resolution data base. *Global Biogeochemical Cycles*, **2**, 139–156.

281. Levin, I. (1987) Atmospheric CO_2 in continental Europe – an alternative approach to clean air CO_2 data. *Tellus*, **39B**, 21–28.

282. Levin, I. and Hesshaimer, V. (1996) Refining of atmospheric transport model entries by the globally observed passive tracer distributions of ^{85}krypton and sulfur hexafluoride (SF_6). *J. Geophys. Res.*, **101D**, 16 745–16 755.

283. Levin, I., Graul, R. and Trivett, N. B. A. (1992) The CO_2 observation network in the Federal Republic of Germany, pp. 47–49 of *Report of the WMO Meeting of Experts on Carbon Dioxide Concentration and Isotopic Measurement Techniques*. Lake Arrowhead, California, 14–19 October 1990. (Coord. P. P. Tans). (WMO: Geneva.)

284. Levin, I., Glatzel-Mattheier, H., Marik, T., Cuntz, M. and Schmidt, M. (1999) Verification of German methane emission inventories and their recent changes based on atmospheric observations. *J. Geophys. Res.*, **104D**, 3447–3456.

285. Lide, D. R. (Ed.) (1997) *Handbook of Chemistry and Physics* (78th edn). (CRC Press: Boca Raton, Florida.)

286. Livingstone, G. P. and Hutchinson, G. L. (1995) Enclosure-based measurement of trace gas exchange: applications and sources of error, pp. 14–51 of *Biogenic Trace Gases: Measuring Emissions from Soil and Water.* Ed. P. A. Matson and R. C. Harriss. (Blackwell Science: Oxford.)

287. Lodge, J. P. Jr (Ed.) (1989) *Methods of Air Sampling and Analysis.* Intersociety Committee: APCA, ACS, AIChE, APWA, ASME, AOAC, HPS, ISA. (Lewis Publishers: Chelsea, Michigan.)

288. Lorenc, A. (1986) Analysis methods for numerical weather prediction. *Q. J. Roy. Meteorol. Soc.*, **112**, 1177–1194.

289. Lorenz, E. N. (1963) Deterministic non-periodic flow. *J. Atmos. Sci.*, **20**, 130–141.

290. Lorenz, E. N. (1995) *The Essence of Chaos.* (UCL Press: London.)

291. Lovelock, J. E. (1974) The electron capture detector: theory and practice. *J. Chromatogr.*, **99**, 3–12.

292. Lovelock, J. E. (1979) *Gaia: A New Look at Life on Earth.* (OUP: Oxford.)

293. Lurmann, F. W., Lloyd, A. C., and Atkinson, R. (1986) A chemical mechanism for use in long-range transport /acid deposition computer modeling. *J. Geophys. Res.*, **91D**, 10905–10936.

294. McCarthy, J. J. (2000) The evolution of the Joint Global Ocean Flux Study project, pp. 3–15 of *The Changing Ocean Carbon Cycle: A Mid-Term Synthesis of the Joint Global Ocean Flux Study*. Eds. R. B. Hanson, H. W. Ducklow and J. G. Field. (CUP: Cambridge.)

295. McIntyre, M. E. (1980) An introduction to the generalized Lagrangian-mean description of wave mean-flow interaction. *PAGEOPH*, **118**, 152–176.

296. Mahlman, J. D. (1997) Dynamics of transport processes in the upper troposphere. *Science*, **276**, 1079–1083.

297. Mahlman, J. D. and Moxim, W. J. (1978) Tracer simulation using a global general circulation model: results from a mid-latitude instantaneous source experiment. *J. Atmos. Sci.*, **35**, 1340–1378.

298. Mahowald, N. M., Prinn, R. G. and Rasch, P. J. (1997) Deducing CCl_3F emissions using an inverse method and chemical transport models with assimilated winds. *J. Geophys. Res.*, **102D**, 28 153–28 168.

299. Maiss, M., Steele, L. P., Francey, R. J., Fraser, P. J., Langenfelds, R. L., Trivett, N. B. A. and Levin, I. (1996) Sulfur hexafluoride – a powerful new atmospheric tracer. *Atmos. Environ.*, **30**, 1621–1629.

300. Mandelbrot, B. B. (1977) *Fractals: Form, Chance and Dimension* (W. H. Freeman: San Francisco, California.)

301. Mankin, W., Atlas, E., Cantrell, C., Eisle, F. and Fried, A. (1999) Observational methods: instruments and platforms. Chapter 11 of *Atmospheric Chemistry and Global Change*. Eds. G. P. Brasseur, J. J. Orlando and G. S. Tyndall. (OUP: Oxford.)

302. Manning, A. C., Keeling, R. F. and Severinghaus, J. P. (1999) Precise atmospheric oxygen measurements with a paramagnetic oxygen analyzer. *Global Biogeochemical Cycles*, **13**, 1107–1115.

303. Manning, M. R. (1993) Seasonal cycles in atmospheric CO_2 concentrations, pp. 65–94 of *The Global Carbon Cycle*. Ed. M. Heimann. (Springer-Verlag: Heidelberg.)

304. Mansbridge, J. V. and Enting, I. G. (1989) *Sensitivity Studies in a Two-Dimensional Atmospheric Transport Model*. Division of Atmospheric Research Technical Paper no. 18. (CSIRO: Australia.)

305. Marland, G. and Rotty, R. M. (1984) Carbon dioxide emissions from fossil fuels: a procedure for estimation and results for 1950–82. *Tellus*, **36B**, 232–261.

306. Martens, P. and Rotmans, J. (1999) *Climate Change: An Integrated Perspective*. (Kluwer: Dordrecht.)

307. Martín, F. and Díaz, A. (1991) Different methods of modeling the variability in the monthly mean concentrations of atmospheric CO_2 at Mauna Loa. *J. Geophys. Res.*, **96D**, 18 689–18 704.

308. Maryon, R. H., and Best, M. J. (1995) Estimating the emissions from a nuclear accident using observations of radioactivity with dispersion model

products. *Atmos. Environ.*, **29**, 1853–1869.

309. Masarie, K. A. and Tans, P. P. (1995) Extension and integration of atmospheric carbon dioxide data into a globally consistent measurement record. *J. Geophys. Res.*, **100D**, 11 593–11 610.

310. Masarie, K., Conway, T., Dlugokencky, E., Novelli, P., Tans, P. P., Vaughn, B., White, J., Trolier, M., Francey, R. J., Langenfelds, R. L., Steele, L. P., and Allison, C. (1997) An update on the on-going flask-air intercomparison program between NOAA CMDL Carbon Cycle Group and the CSIRO DAR Global Atmospheric Sampling Laboratory, pp. 40–44 of *Report of the Ninth WMO Meeting of Experts on Carbon Dioxide Concentration and Related Tracer Measurement Techniques*. Ed. R. Francey. Global Atmospheric Watch Report Series no. 132. (World Meteorological Organization: Geneva.)

311. Matthews, E. and Fung, I. (1987) Methane emission from natural wetlands: global distribution, area, and environmental characteristics of sources. *Global Biogeochemical Cycles*, **1**, 61–86.

312. Matthews, E., Fung, I. and Lerner, J. (1991) Methane emissions from rice cultivation: geographical and seasonal distribution of cultivated area and emissions. *Global Biogeochemical Cycles*, **5**, 3–24.

313. Mazzega, P. (2000) On the assimilation and inversion of small data sets under chaotic regimes, pp. 223–237 of *Inverse Methods in Global Biogeochemical Cycles*. Eds. P. Kasibhatla, M. Heimann, P. Rayner, N. Mahowald, R. G. Prinn and D. E. Hartley. Geophysical Monograph No. 114. (American Geophysical Union: Washington.)

314. Medawar, P. B. (1963) Is the scientific paper a fraud? *Listener*, **70** (12 September). Reprinted in *The Threat and the Glory: Reflections on Science and Scientists*. Ed. D. Pyke. (OUP: Oxford.)

315. Melillo, J. M., Prentice, I. C., Farquhar, G. D., Schultze, E.-D. and Sala, O. E. (1996) Terrestrial biotic responses to environmental change and feedbacks to climate, pp. 445–481 of *Climate Change 1995: The Science of Climate Change*. Eds. J. T. Houghton, L. G. Meira Filho, B. A. Callander, N. Harris, A. Kattenberg and K. Maskell. (Published for the IPCC by CUP: Cambridge.)

316. Ménard, R. (2000) Tracer assimilation, pp. 67–79 of *Inverse Methods in Global Biogeochemical Cycles*. Eds. P. Kasibhatla, M. Heimann, P. Rayner, N. Mahowald, R. G. Prinn and D. E. Hartley. Geophysical Monograph No. 114. (American Geophysical Union: Washington.)

317. Ménard, R. and Chang, L.-P. (2000) Stratospheric assimilation of chemical tracer observations using a Kalman filter. Part II. χ^2 validated results and analysis of variance and correlation dynamics. *Monthly Weather Rev.*, **128**, 2671–2686.

318. Ménard, R., Cohn, S. E., Chang, L.-P. and Lyster, P. M. (2000) Stratospheric assimilation of chemical tracer observations using a Kalman filter. Part I. Formulation. *Monthly Weather Rev.*, **128**, 2654–2671.

319. Midgley, P. (1997) New directions: HCFCs and HFCs: halocarbon replacements for CFCs. *Atmos. Environ.*, **31**, 1095–1096.

320. Miller, B. R., Huang, J., Weiss, R. F., Prinn, R. G. and Fraser, P. J. (1998) Atmospheric trend and lifetime of chlorodifluoromethane (HCFC-22) and the global tropospheric OH concentration. *J. Geophys. Res.*, **103D**, 13 237–13 248.

321. Molina, M. J. and Rowland, F. S. (1974) Stratospheric sink for chlorofluoromethanes: chlorine atoms catalyzed destruction of ozone. *Nature*, **249**, 810–814.

322. Monfray, P., Ramonet, M. and Beardsmore, D. (1996) Longitudinal and vertical CO_2 gradients over the subtropical/subantarctic oceanic sink. *Tellus*, **48B**, 445–456.

323. Montzka, S. A., Spivakovsky, C. M., Butler, J. H., Elkins, J. W., Lock, L. T. and Mondeel, D. J. (2000) New observational constraints for atmospheric hydroxyl on global and hemispheric scales. *Science*, **288**, 500–503.

324. Morgan, M. G., and Henrion, M. (with M. Small) (1990) *Uncertainty: A Guide to Dealing with Uncertainty in Quantitative Risk and Policy Analysis.* (CUP: Cambridge.)

325. Morimoto, S., Nakazawa, T., Higuchi, K. and Aoki, S. (2000) Latitudinal distribution of atmospheric CO_2 sources and sinks inferred by $\delta^{13}C$ measurements from 1985 to 1991. *J. Geophys. Res.*, **105D**, 24 315–24 326.

326. Mosca, S., Graziani, G., Klug, W., Bellasio, R. and Bianconi, R. (1998) A statistical methodology for the evaluation of long-range dispersion models: an application to the ETEX exercise. *Atmos. Environ.*, **32**, 4307–4324.

327. Mote, P. and O'Neill, A. (Eds.) (2000) *Numerical Modeling of the Global Atmosphere.* (Kluwer: Dordrecht.)

328. Mroz, E. J., Alei, M., Cappis, J. H., Guthals, P. R., Mason, A. S. and Rokop, D. J. (1989) Antarctic atmospheric tracer experiments. *J. Geophys. Res.*, **94D**, 8577–8583.

329. Mulholland, M. and Seinfeld, J. H. (1995) Inverse air pollution modelling of urban-scale carbon monoxide emissions. *Atmos. Environ.*, **29**, 497–516.

330. Mulquiney, J. E. and Norton, J. P. (1998) A new inverse method for trace gas flux estimation 1. State-space model identification and constraints. *J. Geophys. Res.*, **103D**, 1417–1427.

331. Mulquiney, J. E., Taylor, J. A. and Norton, J. P. (1993) Identifying global sources and sinks of greenhouse gases, pp. 431–451 of *Modelling Change in Environmental Systems.* Eds. A. J. Jakeman, M. B. Beck and M. J. McAleer. (Wiley: Chichester.)

332. Mulquiney, J. E., Taylor, J. A., Jakeman, A. J., Norton, J. P. and Prinn, R. G. (1998) A new inverse method for trace gas flux estimation 2. Application to tropospheric $CFCl_3$ fluxes. *J. Geophys. Res.*, **103D**, 1429–1442.

333. Nakićenović, N. and Swart, R. (Eds.) (2000) *Special Report on Emission Scenarios.* (IPCC WG III Special Report.) (CUP: Cambridge.)

334. NASA (1986) *Earth System Science: A Program for Global Change.* (NASA: Washington.)

335. Nemry, B., Francois, L., Gérard, J.-C., Bondeau, A., Heimann, M. and the participants of the Potsdam NPP model intercomparison (1999) Comparing global models of terrestrial net primary productivity (NPP): analysis of the

seasonal atmospheric CO_2 signal. *Global Change Biol.*, **5** (suppl.), 65–76.

336. Neu, J. L. and Plumb, R. A. (1999) Age of air in a "leaky pipe" model of stratospheric transport. *J. Geophys. Res.*, **104D**, 19 243–19 255.

337. Newell, R. E. (1963) The general circulation of the atmosphere and its effects on the movement of trace substances. *J. Geophys. Res.*, **68**, 3949–3962.

338. Newell, R. E., Vincent, D. G. and Kidson, J. W. (1969) Interhemispheric mass exchange from meteorological and trace substance observations. *Tellus*, **21**, 641–647.

339. Newsam, G. N. and Enting, I. G. (1988) Inverse problems in atmospheric constituent studies: I. Determination of surface sources under a diffusive transport approximation. *Inverse Problems*, **4**, 1037–1054.

340. NOAA/NASA (1976) *U. S. Standard Atmosphere, 1976.* (US Government Printing Office: Washington.)

341. Noble, B. and Daniel, J. W. (1988) *Applied Linear Algebra.* (Prentice-Hall: Englewood Cliffs, New Jersey.)

342. Noilhan J., Bazile, E., Champeaux, J.-L. and Giard, D. (2000) Examples of the use of satellite data in numerical weather prediction models, pp. 61–69 of *Observing Land from Space: Science, Customers and Technology.* Eds. M. M. Verstraete, M. Menenti and J. Peltoniemi. (Kluwer: Dordrecht.)

343. Oeschger, H. and Heimann, M. (1983) Uncertainties of predictions of future atmospheric CO_2 concentrations. *J. Geophys. Res.*, **88C**, 1258–1262.

344. Oeschger, H., Siegenthaler, U., Schotterer, U. and Gugelmann, A. (1975) A box diffusion model to study the carbon dioxide exchange in nature. *Tellus*, **27**, 168–192.

345. Olivier, J. G. J., Bouwman, A. F., van der Maas, C. M. W., Berdowski, J. J. M., Veldt, C., Bloos, J. P. J., Visschedijk, A. J. H., Zandveld, P. Y. J. and Haverlag, J. L. (1996) *Description of EDGAR Version 2.0: A Set of Global Emission Inventories of Greenhouse Gases and Ozone Depleting Substances for All Anthropogenic and Most Natural Sources on a Per Country Basis and on 1° × 1° Grid.* Report 771 060-002. (National Institute for Public Health and the Environment: Bilthoven, The Netherlands.)

346. Olson, J., Prather, M., Berntsen, T., Carmichael, G., Chatfield, R., Connell, P., Derwent, R., Horowitz, L., Jin, S., Kanakidou, M., Kasibhatla, P., Kotamarthi, R., Kuhn, M., Law, K., Penner, J., Perliski, L., Sillman, S., Stordal, F., Thompson, A. and Wild, O. (1997) Results from the Intergovernmental Panel on Climatic Change photochemical model intercomparisons (Photocomp). *J. Geophys. Res.*, **102D**, 5979–5991.

347. Orlanski, J. (1975) A rational subdivision of scales for atmospheric processes. *Bull. Amer. Meteorol. Soc.*, **56**, 527–530.

348. Parker, R. L. (1977) Understanding inverse theory. *Ann. Rev. Earth Plan. Sci.*, **5**, 35–64.

349. Parkinson, S. and Young, P. (1998) Uncertainty and sensitivity in global carbon cycle modelling. *Climate Res.*, **9**, 157–174.

350. Pavlis, G. L. and Booker, J. R. (1980) The mixed discrete–continuous inverse problem: application to the simultaneous determination of

earthquake hypocenters and velocity structure. *J. Geophys. Res.*, **85B**, 4801–4810.

351. Peacock, S., Visbeck, M. and Broecker, W. (2000) Deep water formation rates inferred from global tracer distributions: an inverse approach, pp. 185–195 of *Inverse Methods in Global Biogeochemical Cycles*. Eds. P. Kasibhatla, M. Heimann, P. Rayner, N. Mahowald, R. G. Prinn and D. E. Hartley. Geophysical Monograph No. 114. (American Geophysical Union: Washington.)

352. Pearman, G. I. (1977) Further studies of the comparability of baseline atmospheric carbon dioxide measurements. *Tellus*, **29**, 171–181.

353. Pearman, G. I. (1980) *Atmospheric CO_2 Concentration Measurements. A Review of Methodologies, Existing Programmes and Available Data.* WMO Project on Research and Monitoring of Atmospheric CO_2. Report no. 3. (WMO: Geneva.)

354. Perry, M. V., Percival, I. C. and Weiss, N. D. (Eds.) (1987) *Dynamical Chaos.* (Royal Society: London.)

355. Petit, J. R., Jouzel, J., Raynaud, D., Barkov, N. I., Barnola, J.-M., Basile, I., Bender, M., Chappellaz, J., Davis, M., Delaygue, G., Delmotte, M., Kotlyakov, V. M., Legrand, M., Lipenkov, V. Y., Lorius, C., Ritz, C., Saltzman, E. and Stievenard, M. (1999) Climate and atmospheric history of the past 420,000 years from the Vostok ice core, Antarctica. *Nature*, **399**, 429–436.

356. Peylin, P., Bousquet, P., Ciais, P. and Monfray, P. (2000) Differences of CO_2 flux estimates based on a 'time-independent' versus a 'time-dependent' inversion method, pp. 295–309 of *Inverse Methods in Global Biogeochemical Cycles*. Eds. P. Kasibhatla, M. Heimann, P. Rayner, N. Mahowald, R. G. Prinn and D. E. Hartley. Geophysical Monograph No. 114. (American Geophysical Union: Washington.)

357. Pielke, R. A. (1984) *Mesoscale Meteorological Modeling.* (Academic: Orlando, Florida.)

358. Plumb, R. A. and Ko, M. K. W. (1992) Interrelationships between mixing ratios of long-lived stratospheric constituents. *J. Geophys. Res.*, **97D**, 10 145–10 156.

359. Plumb, R. A. and MacConalogue, D. D. (1988) On the meridional structure of long-lived tropospheric constituents. *J. Geophys. Res.*, **93D**, 15 897–15 913.

360. Plumb, R. A. and Mahlman, J. D. (1987) The zonally averaged transport characteristics of the GFDL general circulation/transport model. *J. Atmos. Sci.*, **44**, 298–327.

361. Plumb, R. A. and Zheng, X. (1996) Source determination from trace gas observations: an orthogonal function approach and results for long-lived gases with surface sources. *J. Geophys. Res.*, **101D**, 18 569–18 585.

362. Poisson, N., Kanakidou, M. and Crutzen, P. J. (2000) Impact of non-methane hydrocarbons on tropospheric chemistry and the oxidising power of the troposphere: 3-dimensional modelling results. *J. Atmos. Chem.*, **36**, 157–230.

363. Polian, G., Lambert, G., Ardouin, B. and Jegou, A. (1986) Long-range transport of continental radon in subantarctic and antarctic areas. *Tellus*, **38B**, 178–189.

364. Prahm, L. P., Conradsen, K. and Nielsen, L. B. (1980) Regional source

quantification model for sulphur oxides in Europe. *Atmos. Environ.*, **14**, 1027–1054.

365. Prather, M. J. (1985) Continental sources of halocarbons and nitrous oxide. *Nature*, **317**, 221–225.

366. Prather, M. J. (1994) Lifetimes and eigenstates in atmospheric chemistry. *Geophys. Res. Lett.*, **21**, 801–804.

367. Prather, M. J. (1996) Time-scales in atmospheric chemistry: theory for GWPs for CH_4 and CO, and runaway growth. *Geophys. Res. Lett.*, **23**, 2597–2600.

368. Prather, M., McElroy, M., Wofsy, S., Russell, G. and Rind, D. (1987) Chemistry of the global troposphere: fluorocarbons as tracers of air motion. *J. Geophys. Res.*, **92D**, 6579–6613.

369. Prather, M., Derwent, R., Ehhalt, D., Fraser, P., Sanhueza, E. and Zhou, X. (1995) Other trace gases and atmospheric chemistry, pp. 73–126 of *Climate Change 1994: Radiative Forcing of Climate Change and An Evaluation of the IPCC IS92 Emission Scenarios*. Eds. J. T. Houghton, L. G. Meira Filho, J. Bruce, Hoesung Lee, B. A. Callander, E. Haites, N. Harris and K. Maskell. (Published for the IPCC by CUP: Cambridge.)

370. Prather, M., Ehhalt, D., Dentener, F., Derwent, R., Dlugokencky, E., Holland, E., Isaksen, I., Katima, J., Kirchoff, V., Matson, P., Midgely, P. and Wang, M. (2001) Atmospheric chemistry and greenhouse gases. Chapter 4 of *Climate Change 2001: The Scientific Basis*. Eds. J. T. Houghton, Y. Ding, D. J. Griggs, N. Noguer, P. J. van der Linden, X. Dai, K. Maskell and C. A. Johnson. (Published for the IPCC by CUP: Cambridge.)

371. Prentice, I. C., Farquhar, G. D., Fasham, M. J. R., Goulden, M. L., Heimann, M., Jaramillo, V. J., Kheshgi, H. S., Le Quéré, C., Scholes, R. J. and Wallace, D. W. R. (2001) The carbon cycle and atmospheric carbon dioxide. Chapter 3 of *Climate Change 2001: The Scientific Basis*. Eds. J. T. Houghton, Y. Ding, D. J. Griggs, N. Noguer, P. J. van der Linden, X. Dai, K. Maskell and C. A. Johnson. (Published for the IPCC by CUP: Cambridge.)

372. Press, F. (1968) Earth models obtained by Monte Carlo inversion. *J. Geophys. Res.*, **73**, 5223–5234.

373. Press, W. H., Flannery, B. P., Teukolsky, S. A. and Vetterling, W. T. (1986) *Numerical Recipes: The Art of Scientific Computing*. (CUP: Cambridge.)

374. Priestley, M. B. (1981) *Spectral Analysis and Time Series*. (Academic: London.)

375. Prinn, R. G. (2000) Measurement equation for trace chemicals in fluids and solution of its inverse, pp. 3–18 of *Inverse Methods in Global Biogeochemical Cycles*. Eds. P. Kasibhatla, M. Heimann, P. Rayner, N. Mahowald, R. G. Prinn and D. E. Hartley. Geophysical Monograph No. 114. (American Geophysical Union: Washington.)

376. Prinn, R. and Hartley, D. (1995) Inverse methods in atmospheric chemistry, pp. 172–197 of *Progress and Problems in Atmospheric Chemistry*. Ed. J. R. Barker. (World Scientific: Singapore.)

377. Prinn, R., Cunnold, D., Rasmussen, R., Simmonds, P., Alyea, F., Crawford, A., Fraser, P. and Rosen, R. (1987) Atmospheric trends in methylchloroform and the global

average for the hydroxyl radical. *Science*, **238**, 945–950.

378. Prinn, R., Cunnold, D., Rasmussen, R., Simmonds, P., Alyea, F., Crawford, A., Fraser, P. and Rosen, R. (1990) Atmospheric emissions and trends of nitrous oxide deduced from 10 years of ALE-GAGE data. *J. Geophys. Res.*, **95D**, 18 369–18 385.

379. Prinn, R., Cunnold, D., Simmonds, P., Alyea, F., Boldi, R., Crawford, A., Fraser, P., Gutzler, D., Hartley, D., Rosen, R. and Rasmussen, R. (1992) Global average concentration and trend for hydroxyl radicals deduced from ALE/GAGE tricholoroethane (methyl chloroform) data for 1978–1990. *J. Geophys. Res.*, **97D**, 2445–2461.

380. Prinn, R. G., Weiss, R. F., Miller, B. R., Huang, J., Alyea, F. N., Cunnold, D. M., Fraser, P. J., Hartley, D. E. and Simmonds, P. G. (1995) Atmospheric trends and lifetime of CH_3CCl_3 and global OH concentrations. *Science*, **269**, 187–192.

381. Prinn, R., Jacoby, H., Sokolov, A., Wang, C., Xiao, X., Yang, Z., Eckhaus, R., Stone, P., Ellerman, D., Melillo, J., Fitzmaurice, J., Kicklighter, D., Holian, G. and Liu, Y. (1999) Integrated global system model for climate policy assessment: Feedbacks and sensitivity studies. *Climatic Change*, **41**, 469–546.

382. Prinn, R. G., Weiss, R. F., Fraser, P. J., Simmonds, P. G., Cunnold, D. M., Alyea, F. N., O'Doherty, S., Salameh, P., Miller, B. R., Huang, J., Wang, R. H. J., Hartley, D. E., Harth, C., Steele, L. P., Sturrock, G., Midgley, P. M. and McCulloch, A. (2000) A history of chemically and radiatively important gases in air deduced from ALE/GAGE/AGAGE. *J. Geophys. Res.*, **105D**, 17 751–17 792.

383. Prinn, R., Cunnold, D., Fraser, P., Weiss, R., Simmonds, P., Alyea, F., Steele, L. P., Hartley, D. and Wang, R. H. J. (2000) Chlorofluorocarbons. In *WMO WDCGG Data Report*. WDCGG no. 21. (Japan Meteorological Agency for WMO: Tokyo.) (Data available on CD-ROM.)

384. Pryor, S. C., Davies, T. D., Hoffer, T. E. and Richman, M. B. (1995) The influence of synoptic scale meteorology on transport of urban air to remote locations in the southwestern United States of America. *Atmos. Environ.*, **29**, 1609–1618.

385. Pudykiewicz, J. A. (1998) Application of adjoint tracer transport equations for evaluating source parameters. *Atmos. Environ.*, **32**, 3039–3050.

386. Quay, P. D., King, S. L., Stutsman, J., Wilbur, D. O., Steele, L. P., Fung, I., Gammon, R. H., Brown, T. A., Farwell, G. W., Grootes, P. M. and Schmidt, F. H. (1991) Carbon isotopic composition of atmospheric CH_4: Fossil and biomass burning source strengths. *Global Biogeochemical Cycles*, **5**, 25–47.

387. Quay, P. D., Tilbrook, B. and Wong, C. S. (1992) Oceanic uptake of fossil fuel CO_2: carbon-13 evidence. *Science*, **256**, 74–79.

388. Quay, P. D., Stutsman, J., Wilbur, D., Snover, A., Dlugokencky, E. and Brown, T. (1999) The isotopic composition of atmospheric methane. *Global Biogeochemical Cycles*, **13**, 445–461.

389. Quay, P., King, S., White, D., Brockington, M., Plotkin, B., Gammon, R., Gerst, S. and Stutsman, J. (2000) Atmospheric ^{14}CO: a tracer of OH concentration and mixing rates. *J. Geophys. Res.*, **105D**, 15 147–15 166.

390. Radon, J. (1917) Über die Bestimmung von Funktionen durch ihre integralwerte längs gewisser Mannigfaltigkeiten. [On the determination of functions from their integrals along certain manifolds.] *Ber. Sächs. Akad. Wiss. Leipzig, Math. Phys. Kl.*, **69**, 262–277.

391. Rahn, T. and Wahlen, M. (2000) A reassessment of the global isotopic budget of atmospheric nitrous oxide. *Global Biogeochemical Cycles*, **14**, 537–543.

392. Raich, J. W. and Potter, C. S. (1995) Global patterns of carbon dioxide emissions from soils. *Global Biogeochemical Cycles*, **9**, 23–36.

393. Ramonet, M. and Monfray, P. (1996) CO_2 baseline concept in 3-D atmospheric transport models. *Tellus*, **48B**, 502–520.

394. Ramonet, M., Tans, P. P. and Masarie, K. (1997) CO_2 data assimilation. Presented at 5th International CO_2 Conference. Cairns, September 1997, pp. 191 of extended abstracts.

395. Rasch, P. J., Mahowald, N. M. and Eaton, B. E. (1997) Representations of transport, convection, and the hydrologic cycle in chemical transport models: implications for the modeling of short-lived and soluble species. *J. Geophys. Res.*, **102D**, 28 127–28 138.

396. Raupach, M. R., Denmead, O. T., and Dunin, F. X. (1992) Challenges in linking atmospheric CO_2 concentrations to fluxes at local and regional scales. *Aust. J. Botany*, **40**, 697–716.

397. Rayner, P. J. (2001) Atmospheric perspectives on the ocean carbon cycle. *Global Biogeochemical Cycles in the Climate System*. Eds. E.-D. Schulze, M. Heimann, S. Harrison, E. Holland, J. Lloyd, I. C. Prentice and D. Schimel. (Academic: San Diego, California.)

398. Rayner, P. J. and Law, R. M. (1995) *A Comparison of Modelled Responses to Prescribed CO_2 Sources*. CRCSHM Technical Paper no. 1 (and CSIRO Division of Atmospheric Research Technical Paper no. 36). (CSIRO: Australia.)

399. Rayner, P. J. and O'Brien, D. (2001) The utility of remotely sensed CO_2 concentration data in surface source inversions. *Geophys. Res. Lett.*, **28**, 175–178.

400. Rayner, P. J., Enting, I. G. and Trudinger, C. M. (1996) Optimizing the CO_2 observing network for constraining sources and sinks. *Tellus*, **48B**, 433–444.

401. Rayner, P. J., Law, R. M. and Dargaville, R. (1999) The relationship between tropical CO_2 fluxes and the El Niño–Southern Oscillation. *Geophys. Res. Lett.*, **26**, 493–496.

402. Rayner, P. J., Enting, I. G., Francey, R. J. and Langenfelds, R. (1999) Reconstructing the recent carbon cycle from atmospheric CO_2, $\delta^{13}C$ and O_2/N_2 observations. *Tellus*, **51B**, 213–232.

403. Rayner, P. J., Giering, R., Kaminski, T., Ménard, R., Todling, R. and Trudinger, C.M. (2000) Exercises, pp. 81–106 (plus CD) of *Inverse Methods in Global Biogeochemical Cycles*. Eds. P. Kasibhatla, M. Heimann, P. Rayner, N. Mahowald, R. G. Prinn and D. E. Hartley. Geophysical Monograph No. 114. (American Geophysical Union: Washington.)

404. Reichle, H. G., Connors, V. S., Holland, J. A., Hypes, W. D., Wallio, H. A., Casas, J. C., Gormsen, B. B.,

Saylor, M. S. and Hesketh, W. D. (1986) Middle and upper tropospheric carbon monoxide mixing ratios as measured by a satellite-borne remote sensor during November 1981. *J. Geophys. Res.*, **91**, 10 865–10 887.

405. Reiter, E. R. (1978) *Atmospheric Transport Processes. Part 4: Radioactive Tracers.* (US AEC and DOE: Washington.)

406. Revelle, R. and Suess, H. E. (1957) Carbon dioxide exchange between atmosphere and ocean and the question of an increase of atmospheric CO_2 during the past decades. *Tellus*, **9**, 18–27.

407. Rind, D., Lerner, J., Shah, K. and Suozzo, R. (1999) Use of on-line tracers as a diagnostic tool in general circulation model development: 2. Transport between the troposphere and the stratosphere. *J. Geophys. Res.*, **104D**, 9151–9167.

408. Robertson, L. and Langner, J. (1998) Source function estimates by means of variational data assimilation applied to the ETEX-I tracer experiment. *Atmos. Environ.*, **32**, 4219–4225.

409. Robertson, J. E. and Watson, A. J. (1992) Thermal skin effect on the surface ocean and its implications for CO_2 uptake. *Nature*, **358**, 738–740.

410. Robertson, L., Langner, J. and Engardt, M. (1996) *MATCH – Mesoscale Atmospheric Transport and Chemistry Model. Basic Transport Model Description and Control Experiment with* ^{222}Rn. RMK no. 70. (Swedish Meteorological and Hydrological Institute (RMK).)

411. Rodgers, C. D. (1976) Retrieval of atmospheric temperature and composition from remote measurements of thermal radiation.

Rev. Geophys. Space Phys., **14**, 609–624.

412. Roemer, M., van Loon, M. and Boersen, G. (2000) *Methane Emission Verification on a National Scale.* TNO report TNO-MEP – R 2000/308. (Netherlands Organization for Applied Scientific Research (TNO): Apeldoom, The Netherlands.)

413. Rood, R. (1987) Numerical advection algorithms and their role in atmospheric transport and chemistry models. *Rev. Geophys.*, **25**, 71–100.

414. Roscoe, H. K. and Clemitshaw, K. C. (1997) Measurement techniques in gas-phase tropospheric chemistry: a selective view of the past, present and future. *Science*, **276**, 1065–1072.

415. Ross, B. (1974) A brief history and exposition of the fundamental theory of fractional calculus, pp. 1–36 of *Fractional Calculus and Its Applications*. Lecture Notes in Mathematics, no. 457.

416. Running, S. W., Baldocchi, D. D., Turner, D. P., Gower, S. T., Bakwin, P. S. and Hibbard, K. A. (1999) A global terrestrial monitoring network integrating tower fluxes, flask sampling, ecosystem modeling and EOS satellite data. *Remote Sens. Environ.*, **70**, 108–127. (Springer-Verlag: Berlin.)

417. Ryall, D. B., Derwent, R. G., Manning, A. J., Simmonds, P. G. and O'Doherty, S. (2001) Estimating source regions of European emissions of trace gases from observations at Mace Head. *Atmos. Environ.*, **35**, 2507–2523.

418. Saeki, T., Nakazawa, T., Tanaka, M. and Higuchi, K. (1998) Methane emissions deduced from a two-dimensional atmospheric transport model and surface measurements.

J. Meteorol. Soc. Jap., **76**, 307–324.

419. Sanak, J., Lambert, G., and Ardouin, B. (1985) Measurement of stratosphere-to-troposphere exchange in Antarctica by using short-lived cosmonuclides. *Tellus*, **37B**, 109–115.

420. Sarmiento, J. L. and Sundquist, E. T. (1992) Revised budget for the oceanic uptake of anthropogenic carbon dioxide. *Nature*, **356**, 589–593.

421. Sarmiento, J. L., Monfray, P., Maier-Reimer, E., Aumont, O., Murnane, R. J. and Orr, J. C. (2000) Sea–air CO_2 fluxes and carbon transport: a comparison of three ocean general circulation models. *Global Biogeochemical Cycles*, **14**, 1267–1281.

422. Schimel, D., Enting, I. G., Heimann, M., Wigley, T. M. L., Raynaud, D., Alves, D. and Siegenthaler, U. (1995) CO_2 and the carbon cycle, pp. 35–71 of *Climate Change 1994: Radiative Forcing of Climate Change and An Evaluation of the IPCC IS92 Emission Scenarios*. Eds. J. T. Houghton, L. G. Meira Filho, J. Bruce, Hoesung Lee, B. A. Callander, E. Haites, N. Harris and K. Maskell. (Published for the IPCC by CUP: Cambridge.)

423. Schlesinger, S. (1979) Terminology of model credibility. *Simulation*, **32**, 103–104.

424. Schlesinger, W. H. (1991) *Biogeochemistry: An Analysis of Global Change*. (Academic: San Diego, California.)

425. Schlitzer, R. (2000) Applying the adjoint method for biogeochemical modeling: export of particulate organic matter in the world ocean, pp. 107–124 of *Inverse Methods in Global Biogeochemical Cycles*. Eds. P. Kasibhatla, M. Heimann, P. Rayner, N. Mahowald, R. G. Prinn and D. E. Hartley. Geophysical Monograph No. 114. (American Geophysical Union: Washington.)

426. Schmitt, R., Schreiber, B. and Levin, I. (1998) Effects of long-range transport on atmospheric trace constituents at the baseline station Tenerife (Canary Islands). *J. Atmos. Chem.*, **7**, 335–351.

427. Schnell, R. C., King, D. B. and Rosson, R. M. (2001) *Climate Monitoring and Diagnostics Laboratory: Summary Report no. 25, 1998–1999.* (US Department of Commerce: Washington.)

428. Schulze, E.-D., Heimann, M., Harrison, S., Holland, E., Lloyd, J., Prentice, I. C. and Schimel, D. (Eds.) (2001) *Global Biogeochemical Cycles in the Climate System.* (Academic: San Diego, California.)

429. Segers, A., Heemink, A., Verlaan, M. and van Loon, M. (2000) Nonlinear Kalman filters for atmospheric chemistry models, pp. 139–146 of *Inverse Methods in Global Biogeochemical Cycles*. Eds. P. Kasibhatla, M. Heimann, P. Rayner, N. Mahowald, R. G. Prinn and D. E. Hartley. Geophysical Monograph No. 114. (American Geophysical Union: Washington.)

430. Seibert, P. (2000) Inverse modelling of sulfur emissions in Europe based on trajectories, pp. 147–154 of *Inverse Methods in Global Biogeochemical Cycles*. Eds. P. Kasibhatla, M. Heimann, P. Rayner, N. Mahowald, R. G. Prinn and D. E. Hartley. Geophysical Monograph No. 114. (American Geophysical Union: Washington.)

431. Seibert, P., Kromp-Kolb, H.,
Baltensperger, U., Jost, D. T.,
Schikowski, M., Kasper, A. and
Puxbaum, H. (1994) Trajectory
analysis of aerosol measurements at
high alpine sites, pp. 689–693 of
*Transport and Transformation of
Pollutants in the Troposphere.* Eds.
P. M. Borrell, P. Borrell, T. Cvitas and
W. Seiler. (Academic: The Hague.)

432. Sharada, M. K. and Yajnik, K. S.
(2000) Comparison of simulations of a
marine ecosystem model with CZCS
data in the north Indian Ocean,
pp. 197–204 of *Inverse Methods in
Global Biogeochemical Cycles.* Eds.
P. Kasibhatla, M. Heimann, P. Rayner,
N. Mahowald, R. G. Prinn and D. E.
Hartley. Geophysical Monograph
No. 114. (American Geophysical
Union: Washington.)

433. Shine, K. P., Derwent, R. G.,
Wuebbles, D. J. and Morcrette, J.-J.
(1990) Radiative forcing of climate,
pp. 41–68 of *Climate Change: The
IPCC Scientific Assessment.* Eds. J. T.
Houghton, G. J. Jenkins and J. J.
Ephraums. (Published for the IPCC
by CUP: Cambridge.)

434. Siegenthaler, U. and Oeschger, H.
(1987) Biospheric CO_2 emissions
during the past 200 years reconstructed
by deconvolution of ice core data.
Tellus, **39B**, 140–154.

435. Silverman, B. W. (1984) Spline
smoothing: the equivalent variable
kernel method. *Annals Statist.*, **12**,
896–916.

436. Silverman, B. W. (1985) Some aspects
of the spline smoothing approach to
non-parametric regression curve fitting
(with discussion). *J. Roy. Statist. Soc.*,
B47, 1–52.

437. Singh, H. B. and Jacob, D. J. (2000)
Future directions: satellite
observations of tropospheric
chemistry. *Atmos. Environ.*, **34**,
4399–4401.

438. Smith, S. J. and Wigley, T. M. L.
(2000) Global warming potentials:
2. Accuracy. *Climatic Change*, **44**,
459–469.

439. SODA (1999) *SODA Workshop on
Chemical Data Assimilation:
Proceedings.* (KNMI (Royal
Netherlands Meteorological Institute):
de Bilt, The Netherlands.)

440. Solomon, S. (1990) Progress towards a
quantitative understanding of Antarctic
ozone depletion. *Nature*, **347**,
347–354.

441. Solomon, S. and Albritton D. L. (1992)
Time-dependent ozone depletion
potentials for short- and long-term
forecasts. *Nature*, **357**, 33–37.

442. Solomon, S., Garcia, R. R., Rowland,
F. S. and Wuebbles, D. J. (1986) On
the depletion of Antarctic ozone.
Nature, **321**, 755–758.

443. Solomon, S., Kiehl, J. T., Garcia, R. R.
and Grose, W. (1986) Tracer transport
by the diabatic circulation deduced
from satellite observations. *J. Atmos.
Sci.*, **43**, 1603–1617.

444. Spivakovsky, C. M. (1991) Reply.
J. Geophys. Res., **96D**, 17 395–17 398.

445. Spivakovsky, C. M., Yevich, R.,
Logan, J. A., Wofsy, S. C. and
McElroy, M. B. (1990) Tropospheric
OH in a three-dimensional chemical
tracer model: an assessment based on
observations of CH_3CCl_3. *J. Geophys.
Res.*, **95D**, 18 441–18 471.

446. Spivakovsky, C. M., Yevich, R.,
Logan, J. A., Wofsy, S. C. and
McElroy, M. B. (1991) Reply.
J. Geophys. Res., **96D**, 17 389–17 390.

447. Spivakovsky, C. M., Logan, J. A.,
Montzka, S. A., Balkanski, Y. J.,

Foreman-Fowler, M., Jones, D. B. A., Horowitz, L. W., Fusco, A. C., Brenninkmeijer, C. A. M., Prather, M. J., Wofsy, S. C. and McElroy, M. B. (2000) Three-dimensional climatological distribution of tropospheric OH: update and evaluation. *J. Geophys. Res.*, **105D**, 9831–9880.

448. Steele, L. P., Dlugokencky, E. J., Lang, P. M., Tans, P. P., Martin, R. C. and Masarie, K. A. (1992) Slowing down of the global accumulation of atmospheric methane during the 1980s. *Nature*, **358**, 313–316.

449. Stephens, B. B., Wofsy, S. C., Keeling, R. F., Tans, P. P. and Potosnak, M. J. (2000) The CO_2 budget and rectification airborne study: strategies for measuring rectifiers and regional fluxes, pp. 311–324 of *Inverse Methods in Global Biogeochemical Cycles*. Eds. P. Kasibhatla, M. Heimann, P. Rayner, N. Mahowald, R. G. Prinn and D. E. Hartley. Geophysical Monograph No. 114. (American Geophysical Union: Washington.)

450. Stijnen, J., Heemink, A. W., Janssen, L. H. J. M. and van der Wal, J. T. (1997) Estimation of methane emissions in Europe using Kalman smoothing, pp. 133–142 of *Proceedings of the CKO/CCB Workshop on Bottom-up and Top-down Estimates of Greenhouse Gases in Bilthoven on 27-6-1997*. Report no. 728 001 006. (Rijksinstituut voor Volkgezendheid en Milieu: The Netherlands.)

451. Stockwell, W. R. (1986) A homogeneous gas phase mechanism for use in a regional acid deposition model. *Atmos. Environ.*, **20**, 1615–1632.

452. Stockwell, W. R., Middleton, P.,

Chang, J. S. and Tang, X. (1990) The second generation Regional Acid Deposition Model chemistry mechanism for regional air quality modeling. *J. Geophys. Res.*, **95D**, 16 343–16 367.

453. Stockwell, W. R., Kirchner, F., Kuhn, M. and Seefeld, S. (1997) A new mechanism for regional atmospheric chemistry modeling. *J. Geophys. Res.*, **102D**, 25 847–25 879.

454. Stohl, A. (1996) Trajectory statistics – a new method to establish source–receptor relationships of air pollutants and its application to the transport of particulate sulfate in Europe. *Atmos. Environ.*, **30**, 579–587.

455. von Storch, H. and Zwiers, F. W. (1999) *Statistical Analysis in Climate Research*. (CUP: Cambridge.)

456. Sundquist, E. T. (1985) Geological perspectives on carbon dioxide and the carbon cycle, pp. 5–59 of *The Carbon Cycle and Atmospheric CO_2: Natural Variations Archean to Present*. Eds. E. T. Sundquist and W. S. Broecker. Geophysical Monograph 32. (AGU: Washington.)

457. Surendran, S. and Mulholland, R. J. (1986) Estimation of atmospheric CO_2 concentration using Kalman filtering. *Int. J. System. Sci.*, **17**, 897–909.

458. Surendran, S. and Mulholland, R. J. (1987) Modeling the variability in measured atmospheric CO_2 data. *J. Geophys. Res.*, **92D**, 9733–9739.

459. Taguchi, S. (1996) A three-dimensional model of atmospheric CO_2 transport based on analyzed winds: model description and simulation results for TRANSCOM. *J. Geophys. Res.*, **101D**, 15 099–15 109.

460. Taguchi, S. (2000) Synthesis inversion of atmospheric CO_2 using the NIRE chemical transport model, pp. 239–253 of *Inverse Methods in Global Biogeochemical Cycles*. Eds. P. Kasibhatla, M. Heimann, P. Rayner, N. Mahowald, R. G. Prinn and D. E. Hartley. Geophysical Monograph No. 114. (American Geophysical Union: Washington.)

461. Takahashi, T., Wanninkhof, R. H., Feely, R. A., Weiss, R. F., Chipman, D. W., Bates, N., Olafsson, J., Sabine, C. and Sutherland, S. C. (1999) Net sea–air CO_2 flux over the global oceans: an improved estimate based on the sea–air pCO_2 difference, pp. 9–15 of *Proceedings of the 2nd International Symposium on CO_2 in the Oceans*. Ed. Y. Nojiri. (Centre for Global Environmental Research: Tsukuba.)

462. Tans, P. P. (1980) On calculating the transfer of carbon-13 in reservoir models of the carbon cycle. *Tellus*, **32**, 464–469.

463. Tans, P. P. (1997) A note on isotopic ratios and the global atmospheric methane budget. *Global Biogeochemical Cycles*, **11**, 77–81.

464. Tans, P. P. and Wallace, D. W. R. (1999) Carbon cycle research after Kyoto. *Tellus*, **51B**, 562–571.

465. Tans, P. P., Conway, T. J. and Nakazawa, T. (1989) Latitudinal distribution of the sources and sinks of atmospheric carbon dioxide derived from surface observations and an atmospheric transport model. *J. Geophys. Res.*, **94D**, 5151–5172.

466. Tans, P. P., Thoning, K. W., Elliott, W. P. and Conway, T. J. (1989) Background atmospheric CO_2 patterns from weekly flask samples at Barrow, Alaska. Optimal signal recovery and error estimates, pp. 112–123 of *The Statistical Treatment of CO_2 Data Records*. Ed. W. P. Elliott. NOAA Technical Memorandum ERL ARL-173. (US Department of Commerce: Washington.)

467. Tans, P. P., Fung, I. Y. and Takahashi, T. (1990) Observational constraints on the global atmospheric CO_2 budget. *Science*, **247**, 1431–1438.

468. Tans, P. P., Thoning, K. W., Elliott, W. P. and Conway, T. J. (1990) Error estimates of background atmospheric CO_2 patterns from weekly flask samples. *J. Geophys. Res.*, **95D**, 14 063–14 070.

469. Tans, P. P., Berry, J. A. and Keeling, R. F. (1993) Oceanic $^{13}C/^{12}C$ observations: a new window on oceanic CO_2 uptake. *Global Biogeochemical Cycles*, **7**, 353–368.

470. Tans, P. P., Bakwin, P. S. and Guenther, D. W. (1996) A feasible global carbon cycle observing system: a plan to decipher today's carbon cycle based on observations. *Global Change Biol.*, **2**, 309–318.

471. Tarantola, A. (1987) *Inverse Problem Theory: Methods for Data Fitting and Model Parameter Estimation*. (Elsevier: Amsterdam.)

472. Taylor, J. A. (1989) A stochastic Lagrangian atmospheric transport model to determine global CO_2 sources and sinks – a preliminary discussion. *Tellus*, **41B**, 272–285.

473. Thompson, A. M. and Cicerone, R. J. (1982) Clouds and wet removal as causes of variability in the trace-gas composition of the marine troposphere. *J. Geophys. Res.*, **87C**, 8811–8826.

474. Thompson, M. L., Enting, I. G., Pearman, G. I. and Hyson, P. (1986) Interannual variation of atmospheric

CO_2 concentrations. *J. Atmos. Chem.*, **4**, 125–155.

475. Thoning, K. W. (1989) Selection of NOAA/GMCC CO_2 data from Mauna Loa Observatory, pp. 1–26 of *The Statistical Treatment of CO_2 Data Records*. Ed. W. P. Elliott. NOAA Technical Memorandum ERL ARL-173. (US Department of Commerce: Washington.)

476. Thoning, K. W., Tans, P. P. and Komhyr, W. D. (1989) Atmospheric carbon dioxide at Mauna Loa observatory 2. Analysis of the NOAA GMCC data 1974–1985. *J. Geophys. Res.*, **94D**, 8549–8565.

477. Thouret, V., Cho, J. Y. N., Newell, R. E., Marenco, A. and Smit, H. G. J. (2000) General characteristics of tropospheric trace constituent layers observed in the MOZAIC program. *J. Geophys. Res.*, **105D**, 17 379–17 392.

478. Tikhonov, A. N. (1963) On the solution of incorrectly posed problems. *Soviet Math. Dokl.*, **4**, 1035–1042.

479. Tindale, N. W., Derek, N. and Francey, R. J. (Eds.) (2001) *Baseline Atmospheric Program Australia 1997–1998*. (Bureau of Meteorology and CSIRO Atmospheric Research: Australia.)

480. Todling, R. (2000) Estimation theory and atmospheric data assimilation, pp. 49–65 of *Inverse Methods in Global Biogeochemical Cycles*. Eds. P. Kasibhatla, M. Heimann, P. Rayner, N. Mahowald, R. G. Prinn and D. E. Hartley. Geophysical Monograph No. 114. (American Geophysical Union: Washington.)

481. Townshend, J. R. G. and Justice, C. O. (1988) Selecting the spatial resolution of satellite sensors for global monitoring of land transformations. *Int. J. Remote Sensing*, **9**, 187–236.

482. Townshend, J. R. G. and Justice, C. O. (1990) The spatial variation of vegetation at very coarse scales. *Int. J. Remote Sensing*, **11**, 149–157.

483. Trabalka, J. A. (Ed.) (1985) *Atmospheric Carbon Dioxide and the Global Carbon Cycle*. Oak Ridge National Laboratory report TN 37 831. (US Department of Energy: Washington.)

484. Trampert, J. and Snieder, R. (1996) Model estimation biased by truncated expansions: possible artifacts in seismic tomography. *Science*, **271**, 1257–1260.

485. Trenberth, K. E. (Ed.) (1992) *Climate System Modeling*. (CUP: Cambridge.)

486. Trenberth, K. E. and Guillemot, C. J. (1994) The total mass of the atmosphere. *J. Geophys. Res.*, **99D**, 23 079–23 088.

487. Trudinger, C. M. (2000) *The Carbon Cycle Over the Last 1000 Years Inferred from Ice Core Data*. (Ph. D. Thesis, Monash University). Electronic edition at http://www.dar.csiro.au/publications/ Trudinger_2001a0.htm.

488. Twomey, S. (1977) *Introduction to the Mathematics of Inversion of Remote Sensing and Indirect Measurements*. (Elsevier: Amsterdam.)

489. UNEP (1999) *Synthesis of the Scientific, Environmental Effects and Economic Assessment Panels of the Montreal Protocol*. Eds. D. L. Albritton and L. Kuijpers. (United Nations Environment Programme Ozone Secretariat: Nairobi.)

490. UNEP/OECD/IEA/IPCC (1995) *IPCC Guidelines for National Greenhouse Gas Emission Inventories* (three volumes). (IPCC: Paris.)

491. Vermeulen, A. T., Hensen, A.,
Erisman, J. W. and Slanina, J. (2000)
Sensitivity study of inverse modelling
of non-CO_2 greenhouse gas emissions
in Europe, pp. 515–521 of *Non-CO_2
Greenhouse Gases: Scientific
Understanding, Control and
Implementation*. Eds. J. van Ham, A. P.
M. Baede, L. A. Meyer and R. Ybema.
(Kluwer: Dordrecht.)

492. Verstraete, M. and Pinty, B. (2000)
Environmental information extraction
from satellite remote sensing data,
pp. 125–137 of *Inverse Methods in
Global Biogeochemical Cycles*. Eds.
P. Kasibhatla, M. Heimann, P. Rayner,
N. Mahowald, R. G. Prinn and D. E.
Hartley. Geophysical Monograph
No. 114. (American Geophysical
Union: Washington.)

493. Volk, C. M., Elkins, J. W., Fahey,
D. W., Salawitch, R. J., Dutton, G. S.,
Gilligan, J. M., Proffitt, M. H.,
Loewenstein, M., Podolske, J. R.,
Minschwaner, K., Margitan, J. J. and
Chan K. R. (1996) Quantifying
transport between the tropical and
mid-latitude lower stratosphere.
Science, **272**, 1763–1768.

494. Vukićević, T. and Hess, P. (2000)
Analysis of tropospheric transport in
the Pacific Basin using the adjoint
technique. *J. Geophys. Res.*, **105D**,
7213–7230.

495. Wahba, G. (1990) *Spline Models for
Observational Data*. (SIAM:
Philadelphia.)

496. Walton, J. J., MacCracken, M. C. and
Ghan, S. J. (1988) A global-scale
Lagrangian trace species model of
transport, transformation and removal
processes. *J. Geophys. Res.*, **93D**,
8339–8354.

497. Wang, C., Prinn, R. G. and Sokolov, A.
(1998) A global interactive chemistry
and climate model: formulation and
testing. *J. Geophys. Res.*, **103D**,
3399–3417.

498. Warneck, P. (1999) *Chemistry of the
Natural Atmosphere* (2nd edn).
(Academic: San Diego, California.)

499. Watson, R. T., Rodhe, H., Oeschger, H.
and Siegenthaler, U. (1990)
Greenhouse gases and aerosols, pp.
1–40 of *Climate Change: The IPCC
Scientific Assessment*. Eds. J. T.
Houghton, G. J. Jenkins and J. J.
Ephraums. (Published for the IPCC
by CUP: Cambridge.)

500. Watson, R. T., Noble, I. R., Bolin, B.,
Ravindranath, N. H., Verardo, D. J.
and Dokken, D. J. (Eds.) (2000) *Land
Use, Land-Use Change, and Forestry*
(IPCC Special Report). (Published for
the IPCC by CUP: Cambridge.)

501. Wayne R. P. (2000) *Chemistry of
Atmospheres. An Introduction to the
Chemistry of the Atmospheres of the
Earth, the Planets and Their Satellites*
(3rd edn). (OUP: Oxford.)

502. WDCGG (1999) *WMO WDCGG Data
Catalogue: GAW Data*. WDCGG
no. 19. (Japan Meteorological Agency
for WMO: Tokyo.)

503. Weizenbaum, J. (1976) *Computer
Power and Human Reason: From
Judgment to Calculation*. (W. H.
Freeman: San Francisco, California.)

504. Weyant, J., Davidson, O., Dowlatabadi,
H., Edmonds, J., Grubb, M., Parson, E.
A., Richels, R., Rotmans, J., Shulka, P.
R., Tol, R. S. J., Cline, W. and
Fankhauser, S. (1996) Integrated
assessment of climate change: an
overview and comparison of results,
pp. 367–396 of *Climate Change 1995:
Economic and Social Dimensions of
Climate Change*. Eds. J. P. Bruce, H.
Lee and E. F. Haites. (Published for the
IPCC by CUP: Cambridge.)

505. Wiens, D., Florence, L. Z. and Hiltz, M. (2000) Robust estimation of chemical profiles of air-borne particulate matter. *Environmetrics*, **12**, 25–40.

506. Wigley, T. M. L. (1991) A simple inverse carbon cycle model. *Global Biogeochemical Cycles*, **5**, 373–382.

507. Wigley, T. M. L. and Schimel, D. S. (Eds.) (2000) *The Carbon Cycle*. (CUP: Cambridge.)

508. Williamson, D. L. and Laprise, R. (2000) Numerical approximations for global atmospheric general circulation models, pp. 127–219 of *Numerical Modeling of the Global Atmosphere*. Eds. P. Mote and A. O'Neill. (Kluwer: Dordrecht.)

509. Wilson, S. R., Dick, A. L., Fraser, P. J. and Whittlestone, S. (1997) Nitrous oxide flux estimates for south eastern Australia. *J. Atmos. Chem.*, **26**, 169–188.

510. Winguth, A. M. E., Archer, D., Maier-Reimer, E. and Mikolajewicz, U. (2000) Paleonutrient data analysis of the glacial Atlantic using an adjoint ocean general circulation model, pp. 171–183 of *Inverse Methods in Global Biogeochemical Cycles*. Eds. P. Kasibhatla, M. Heimann, P. Rayner, N. Mahowald, R. G. Prinn and D. E. Hartley. Geophysical Monograph No. 114. (American Geophysical Union: Washington.)

511. WMO (1987) *Report of the NBS/WMO Expert Meeting on Atmospheric Carbon Dioxide Measurement Techniques: Gaithersburg, Maryland, 15–17 June 1987*. Environmental Pollution Monitoring and Research Programme, Publication no. 51. (WMO: Geneva.)

512. WMO (1990) *Provisional Daily Atmospheric Carbon Dioxide Concentrations as Measured at Global Atmosphere Watch (GAW)– BAPMoN sites for the year 1989*. Environmental Pollution Monitoring and Research Programme: Publication no. 69. (WMO: Geneva.)

513. WMO, UNEP, NASA, NOAA, UKDoE (1990) *Scientific Assessment of Ozone Depletion: 1989* (two volumes). Assessment co-chairs: D. L. Albritton and R. T. Watson. World Meteorological Organization, Global Ozone Research and Monitoring Project, Report no. 20. (WMO: Geneva.)

514. WMO, UNEP, NASA, NOAA, UKDoE (1992) *Scientific Assessment of Ozone Depletion: 1991*. Assessment co-chairs: D. L. Albritton and R. T. Watson. World Meteorological Organization, Global Ozone Research and Monitoring Project, Report no. 25. (WMO: Geneva.)

515. WMO, UNEP, NOAA, NASA (1995) *Scientific Assessment of Ozone Depletion: 1994*. Assessment co-chairs: D. L. Albritton, R. T. Watson and P. J. Auchamp. World Meteorological Organization, Global Ozone Research and Monitoring Project, Report no. 37. (WMO: Geneva.)

516. WMO, UNEP, NOAA, NASA, EC (1999) *Scientific Assessment of Ozone Depletion: 1998*. Assessment co-chairs: D. L. Albritton, P. J. Auchamp, G. Mégie and R. Watson. World Meteorological Organization, Global Ozone Research and Monitoring Project, Report no. 44. (Also 1998 edition in two volumes.) (WMO: Geneva.)

517. Woodwell, G. M. and MacKenzie, F. T. (Eds.) (1955) *Biotic Feedbacks in the Global Climatic System: Will*

the Warming Feed the Warming?
(OUP: New York.)

518. Wratt, D. S., Gimson, N. R.,
Brailsford, G. W., Lassey, K. R.,
Bromley, A. M. and Bell, M. J. (2001)
Estimating regional methane emissions
from agriculture using aircraft
measurements of concentration
profiles. *Atmos. Environ.*, **35**, 497–508.

519. Wuebbles, D. J. (1983) Chlorocarbon
emission scenarios: potential impact
on stratospheric ozone. *J. Geophys.
Res.*, **88C**, 1433–1443.

520. Wuebbles, D. J. and Hayhoe, K. (2000)
Atmospheric methane: trends and
impacts, pp. 1–44 of *Non-CO$_2$
Greenhouse Gases: Scientific
Understanding, Control and
Implementation.* Eds. J. van Ham,
A. P. M. Baede, L. A. Meyer and R.
Ybema. (Kluwer: Dordrecht.)

521. Wunsch, C. (1996) *The Ocean
Circulation Inverse Problem.*
(CUP: Cambridge.)

522. Wunsch, C. and Minster, J.-F. (1982)
Methods for box models and ocean
circulation tracers: mathematical
programming and non-linear inverse
theory. *J. Geophys. Res.*, **87C**,
5647–5662.

523. Yamamoto, S., Kondo, H., Gamo, M.,
Murayama, S., Kaneyasu, N. and
Hayashi, M. (1996) Airplane
measurements of carbon dioxide
distribution on Iriomote Island in
Japan. *Atmos. Environ.*, **30**,
1091–1097.

524. Yoshida, N. and Toyoda, S. (2000)
Constraining the atmospheric N$_2$O
budget from intramolecular site
preference in N$_2$O isoptomers.
Nature, **405**, 330–334.

525. Young, P. (1984) *Recursive Estimation
and Time Series: An Introduction.*
(Springer-Verlag: Berlin.)

526. Young, P. C. (Ed.) (1993) *Concise
Encyclopedia of Environmental
Systems.* (Pergamon: Oxford.)

527. Young, P., Parkinson, S. and Lees, M.
(1996) Simplicity out of complexity
in environmental modelling: Occam's
razor revisited. *J. Appl. Statist.*,
23, 165–210.

528. Zaveri, R. A. and Peters, L. K. (1999)
A new lumped structure
photochemical mechanism for
large-scale applications. *J. Geophys.
Res.*, **104D**, 30 387–30 415.

529. Zhang, X. F., Heemink, A. W. and van
Eijkeren, J. C. H. (1997) Data
assimilation in transport models. *Appl.
Math. Modelling*, **21**, 2–14.

530. Zhang, X. F., Heemink, A. W.,
Janssen, L. H. J. M., Janssen, P. H. M.
and Sauter, F. J. (1999) A
computationally efficient Kalman
smoother for the evaluation of the CH$_4$
budget in Europe. *Appl. Math.
Modelling*, **23**, 109–129.

531. Zhao, X., Turco, R. P. and Shen, M.
(1999) A new family Jacobian solver
for global three-dimensional modeling
of atmospheric chemistry. *J. Geophys.
Res.*, **104D**, 1767–1799.

Index

adjoint methods, 122, 186
adjustment time (CH_4), 257
age spectrum, 302
aggregation error, 144
atmosphere
  composition, 23
  thermal structure, 23
atmospheric budget
  carbon, 248
  CH_4–$^{13}CH_4$, 184
  CH_4 (from IPCC), 256
  CO_2–O_2, 240
atmospheric lifetime
  CH_3CCl_3, 276
  estimation, 274
  Junge relation, 23
atmospheric transport model
  estimating transport, 299
  inter-comparison, 155
  off-line, 36
  on-line, 36
  tuning, 303
autocovariance, 62
autoregressive model, 63
  of noise in CO_2 data, 94
  spectral density, 63

baseline data
  concept, 86
  matching to models, 157
  model statistics, 284
  operational definitions, 86
basis functions, 171

choice, 176
  truncation error, 144
Bayes-theorem, 43
Bayesian estimation, 42
  mode of posterior distribution,
    52
boundary-integral method
  application to CO_2, 295
  definition, 289
boundary-layer, 27
boundary-layer accumulation, 297
butterfly effect, 126

Cape Grim, 288, 289
  air archive, 240
  baseline conditions, 288
  location map, 289
carbon budget
  flux form, 234
  storage form, 234
carbon dioxide (CO_2)
  classification of fluxes, 237
  interannual variation, 252
carbon monoxide (CO)
  ^{14}CO, 308
  as reference source strength, 293
  inversions, 282
  role in CO_2 inversions, 239
CFC-11 (CCl_3F)
  model tuning, 303
chaos, 126
Chernobyl incident
  estimates of release, 294

chlorofluorocarbons (CFCs), 271
    estimating lifetimes, 274
    estimating sources, 272
    nomenclature, 268
    tracer-ratio data, 289
climate model, 36
convolution of response, 100, 112
    inversion for source, 113

data assimilation, 122
    assimilation cycle, 123
    chemical, 127
    objective analysis, 124
    ozone, 126, 129
    spin-up problem, 126
data selection
    model analogue, 157
    operational, 86
deconvolution, 112
delta notation for isotopes, 182
digital filtering, 65

earth-system modelling, 104
El Niño/Southern Oscillation, 27
    effect on transport, 307
    influence on CO_2, 246, 253
emission inventories
    electronic data access, 321
    for FCCC, 101
estimators
    bias, 47
    consistent, 47
    efficient, 47
    linear, 48
    maximum-likelihood, 48
    MPD, 52
experimental design
    network design, 227
    schematic, 122

FCCC, 101
fingerprint method, 217
flux measurements, 107
fractional differentiation, 139

general circulation model (GCM)
    atmosphere–ocean (AOGCM), 36
    atmospheric, 33
global tracer inversions
    carbon monoxide, 282
    CFC-11, 273
    krypton-85, 282
    methyl chloride, 282
    nitrous oxide, 282
global-warming potential (GWP), 102
    absolute GWP, 102
    in Kyoto Protocol, 101
    relation to natural response, 105

GlobalView (data extension), 90
Green's function, 164
    for long-lived tracers, 168
    inhomogeneous b.c., 164
    synthesis inversion, 171
gross flux
    carbon, 238
    definition of isoflux, 184

halocarbons, 267
    atmospheric lifetimes, 274
    inverse problems, 5
    nomenclature, 268
    regional inversions, 293
hydroxyl radical (OH)
    regional inversions, 294
    sink strength for CH_3CCl_3, 275

ill-conditioning, 11, 131
    classification, 136
    origin (schematic), 11
integrated assessment modelling,
    103
integrating factor, 13
interannual variability
    carbon cycle, 246
inter-comparisons, 155
    TransCom, 155
inventories, 106
    electronic data access, 321
inverse problems
    atmospheric transport, 129
    classification of difficulty, 136
    ocean biogeochemistry, 129
    ocean circulation, 129
    origin of ill-conditioning, 11
    remote sensing, 129
inversion techniques
    for model parameters, 219
    hybrid, 204
    mass-balance, 121, 192
    Monte Carlo, 121
    regularisation, 133
    synthesis, 119, 171
    time-stepping, 192, 204
isoflux, 185
    vector representation, 185
isotopes
    CO_2, 240
    delta notation, 182
isotopic budget
    CH_4 and $^{13}CH_4$, 183
    principles, 181

Kalman filter
    estimation equations, 74
    extended, 77
    recursive regression, 75, 198
Kalman smoothing, 76

krypton-85
    estimating emission, 282
    model tuning, 304
Kyoto Protocol, 101

linear programming, 214
    in robust estimation, 212

Mace Head
    location map, 289
mass-balance inversions, 192
    advantages, 195
    application to CH_4, 264
    application to CO_2, 252
    disadvantages, 195
    schematic, 120
    state-space representation, 202
matrix inversion
    block-matrix form, 225
matrix-inversion lemma, 60, 338
MESA, 72
methane (CH_4)
    adjustment time, 257
    atmospheric budget, 256
    CH_4 and $^{13}CH_4$ budget, 183
    pre-industrial, 255
    regional estimates, 293, 294
methyl chloroform (CH_3CCl_3)
    destruction by OH, 276
mode of posterior distribution (MPD), 52
model error, 150
    expressed as data error, 152
    expressed as parameter uncertainty, 153
    in statistical model, 160
    TRANSCOM assessment, 155
Monte Carlo, 46
    algorithms, 118
    inversion, 121
Monte Carlo inversion
    hydroxyl, 278
    schematic, 120
Montreal Protocol, 270
moving-average model, 65
    of noise in CO_2 data, 95

network design, 227
nitrous oxide (N_2O), 282
    global inversions, 282
    regional inversions, 293
norms, 212
null space, 49
numerical differentiation, 136

objective analysis, 124
observing network
    design, 227
    map, 81
    programmes, 82

oxygen (O_2)
    CO_2–O_2 budget, 181
    measurement, 83, 84
ozone (O_3)
    as tracer of transport, 308
    assimilation, 129
    destruction, 269
    hole, 271
ozone-depletion potential, 275

perfect-model experiments,
    227
periodogram, 71
posterior distribution, 44
process models, 108
    inter-comparisons, 155
    inversion, 219
pseudo-inverse, 134

radiocarbon (^{14}C)
    for selection of data, 87
    in carbon cycle, 83
    in CO, 308
    units, 182
radon
    as reference source strength,
        293
    for baseline selection, 87
    for selection in models, 158
    selection using ^{220}Rn, 293
random walk, 65
rectifier effect, 156
regional inversion, 284
    carbon dioxide, 295
    halocarbons, 293
    hydroxyl, 294
    methane, 294
    nitrous oxide, 293
    sulfur, 294
    techniques, 286
regularisation, 133
    comparative studies, 178
remote sensing
    as inverse problem, 129
    of trace-gas concentrations, 80
resolution, 142
response functions
    global budget, 100
    in synthesis inversion, 173

sampling distribution, 42, 47
signal attenuation
    in atmospheric transport, 35
singular-value decomposition,
    50
    use for regularisation, 134
smoothing splines, 68, 69
source–receptor methods, 291

spatial scales
  c.f. time-scales, 26
spectral density, 62
  normalisation, 64
spectral estimation, 69
  maximum entropy (MESA), 72
  periodogram, 71
  smoothing, 70
  tapering, 71
state-space model
  definition, 73
  Kalman filter, 74
  Kalman smoothing, 76
  recursive regression, 75
statistical model
  state-space model, 73
stratosphere–troposphere exchange, 307
stratospheric transport
  age spectrum, 302
sulfur
  regional inversion, 294
sulfur hexafluoride (SF_6)
  input to stratosphere, 303
  observed gradient, 306
  TRANSCOM, 155
surface fluxes
  classification, 98
  process modelling, 108
synthesis inversion, 171
  adjoint methods, 186
  application to CFCs, 273
  cyclo-stationary, 175
  discretisation error, 144
  schematic, 119
  time-dependent of CO_2, 253

tangent linear model, 187
time horizon, 102
time series, 61
  models, 63

models of CO_2 data, 94
  stationarity, 62
time-scales
  c.f. spatial scales, 26
  interhemispheric transport, 302
tomography (axial), 290
  analogy to back-trajectories,
    290
toy models, 12
  advective, 15
  CH_4–CO–OH, 17
  diffusive, 14, 35
  four-zone, 127
  two-box, 34
trace-gas fluxes
  functional role, 99
trace-gas measurement
  calibration, 84
  sampling, 79
  techniques, 81
tracer-ratio method, 217
  applications, 293
  illustrative data, 288
TRANSCOM, 155
transport times
  applications, 277, 305
  statosphere, 302
  theory, 168
turnover time
  hemispheric box, 34
two-box atmosphere model,
    34
  transport times, 170

uncertainty
  analysis, 8
  causes, 91
  communicating, 56

Vienna Convention, 270